Sustainable Increase of Marine Harvesting: Fundamental Mechanisms and New Concepts

Developments in Hydrobiology 167

Series editor
H. J. Dumont

Sustainable Increase of Marine Harvesting: Fundamental Mechanisms and New Concepts

Proceedings of the 1st Maricult Conference held in Trondheim, Norway, 25–28 June 2000

Edited by

Olav Vadstein & Yngvar Olsen

Trondhjem Biological Station, Norwegian University of Science and Technology, Trondheim, Norway

Reprinted from Hydrobiologia, volume 484 (2002)

Kluwer Academic Publishers
Dordrecht / Boston / London

Library of Congress Cataloging-in-Publication Data

A C.I.P. Catalogue record for this book is available from the Library of Congress.

ISBN 978-1-4020-1118-4

Published by Kluwer Academic Publishers,
P.O. Box 17, 3300 AA Dordrecht, The Netherlands

Sold and distributed in North, Central and South America
by Kluwer Academic Publishers,
101 Philip Drive, Norwell, MA 02061, U.S.A.

In all other countries, sold and distributed
by Kluwer Academic Publishers Group,
P.O. Box 322, 3300 AH Dordrecht, The Netherlands

Printed on acid-free paper

TABLE OF CONTENTS

Hydrobiologia **484:** vii–ix, 2002.
O. Vadstein & Y. Olsen (eds), Sustainable Increase of Marine Harvesting: Fundamental Mechanisms and New Concepts.

Preface

How can we increase the production and harvest of living marine resources in the new millennium? Bridging potentials and environmental constraints

The 1st Maricult Conference was arranged in Trondheim, Norway in June 2000 in connection with termination of the Norwegian research programme Maricult (1996–2000). Ninety-seven participants from 12 countries and 4 continents attended the conference. The goal of the research programme was "To provide a basis for evaluation of potentials and environmental constraints for increased sustainable provision of food, raw material and energy from the ocean". In addition to its own generated projects, Maricult served as an umbrella for activities funded by other sources – including EU funded projects. The final scientific and management report (Olsen et al., 2001) is available on Maricult's home page (www.maricult.org).

The conference theme was "How can we increase the production and harvest of living marine resources in the new millennium? Bridging potentials and environmental constraints." The intention for the conference was not only to bring up a status of the research conducted as part of the programme, but also to bring up others experiences and to discuss the ideas and the scientific status for the programme with the international scientific community. The presentations at the conference therefore reflected a blend of results from the programme and presentations of key topics by invited international experts. A total of 50 oral communications and 8 posters were presented at the conference during the opening session, five thematic sessions and a poster session. Sixteen of these communications are included in the present volume. The Conference aimed at a multidisciplinary approach, were both short and long-term perspectives are of relevance and with relevance to all sectors and levels of society.

The opening session presented background, status and main conclusions from the programme (Olsen, 2002), and was followed by a stimulating and thought provoking presentation by Michael Crawford on the importance of seafood for human brain evolution and human health. Parts of this talk is published elsewhere (Crawford, 1992; Crawford et al., 1993, 1999, and references within).

Session 1

"Marine productivity and the underlying fundamental mechanisms" had as its main objective: To contribute to better understanding of the fundamental mechanisms involving nutrient supply, marine productivity, and unintended coastal eutrophication. The keynote lecture given by John G. Field on physical and biological characteristics of the Benguela upwelling system has already been published in parts (Graham et al., 1992; Hutchings & Field, 1997; Richardson et al., 1998; Plaganyi et al., 1999). However, his overall analysis of this natural high productivity ecosystem revealed that if we attempt to increase productivity artificially we can obviously learn a lot from natures "existing solutions". Six of the 14 presentations in this session are presented in the present volume. They cover a wide range of fundamental mechanisms related to marine productivity and ecosystem functioning (Sommer et al., 2002; Kristiansen & Hoell, 2002; Escarvage & Prins, 2002; Fiksen et al., 2002; Nerheim et al., 2002; Gismervik et al., 2002).

Session 2

"New concepts of mariculture and harvesting" had as its main objective: To contribute to approaches and methods for cultivation and harvesting of marine biological resources at low trophic level. The keynote presentation of this session "Aquaculture research and development priorities with perspectives for low food web methods" was given by Patrick Sorgeloos. He focused on constrains and challenges that the aquaculture industry will meet if they try to bridge the gap between market demand for aquatic products and the supply from traditional fisheries. He claimed that this challenge cannot be met by traditional "business- and feed-based" aquaculture, but require a fundamentally new perspective where ideas from Maricult may be essential. Four of the 17 presentations given during this session are presented in the present volume. Two of the presentations are related to the mussel cultivation in Europe, with a presentation of status and future perspectives (Smaal, 2002) and an analysis of relationships between nutrient supply, primary production and mussel growth in the upwelling system outside Galicia, Spain (Figueiras et al., 2002). Two papers presents the effect of experimental fertilization of the land-locked bay Hopavågen on growth of blue mussel (Øie et al., 2002) and scallop (Reitan et al., 2002).

Session 3

"Nutrient supplies, production and environmental impact of open marine systems" had as its main objective: to critically examine the experiences gained from controlled and uncontrolled nutrient enrichment to open marine systems. Eutrophication as a problem has been the target topic in environmental concern and research related to man-made effects on nutrient cycling. However, recently it has been shown that man-made oligotrophication may also be a severe problem (e.g. Stockner & Macisaac, 1996; Stockner et al., 2000). In his keynote lecture John G. Stockner elaborated on this mostly overlooked problem in his presentation "The perils of oligotrophication: Implications for global fisheries resources". This presentation clearly focused on nutrients as a resource, and documented that both too little and too much is a problem that may require counteraction. Five of the nine presentations given during this session are presented in the present volume. Colijn et al. (2002) presents an analysis of data from the North Sea during the last decades, which may be regarded as an uncontrolled fertilization "experiment". Kraufvelin et al. (2002) and Bokn et al. (2002) present two studies of the impact of nutrient supply to rocky shore communities. Finally, Berntsen et al. (2002) and McClimans et al. (2002) discuss the possibility of creating artificial upwelling in fjords by submeged fresh water discharge.

Session 4

"Management and sustainable exploitation of marine biological resources" had as its main objective: To address management related questions related to sustainable exploitation of marine biological resources and to identify constraints and main challenges for authorities, science, industry and the public. Representatives from industry, management and the scientific community gave four presentations covering various aspects of this topic. Non of the presentations are printed in the proceedings.

Session 5

"The role of national and regional policies in sustainable exploitation of biological marine resources" had as its main objective: To address political questions and constraints related to an increased sustainable exploitation of marine biological resources. This final session had four presentations by representatives from the scientific community and various national and international agencies. None of these are presented in this volume, but the concluding remarks of Patrick Sorgeloos, former president World Aquaculture Society, is available on p. 18 in the Maricult final report (Olsen et al., 2001).

Of course, the 1st Maricult Conference did not fully answer the conference theme "How can we increase the production and harvest of living marine resources in the new millennium? Bridging potentials and environmental constraints", but it provided pieces to the puzzle that finally will reveal the answer.

References

Berntsen, J., D. L. Aksnes & A. Foldvik, 2002. Production enhancement by artificial upwelling – a simulation study. Hydrobiologia 484/Dev. Hydrobiol. 167: 177–190.

Bokn, T. L., F. E. Moy, H. Christie, S. Engelbert, R. Karez, K. Kersting, P. Kraufvelin, C. Lindblad, N. Marba, M. F. Pedersen & K. Sørensen, 2002. Are rocky shore ecosystems affected by nutrient enriched seawater? Some preliminary results from a mesocosm experiment. Hydrobiologia 484/Dev. Hydrobiol. 167: 167–175.

Colijn, F., K.-J. Hesse, N. Ladwig & U. Tillmann, 2002. Effects of the large-scale uncontrolled fertilisation process along the continental coastal North Sea. Hydrobiologia 484/Dev. Hydrobiol. 167: 133–148.

Crawford, M. A., 1992. The role of dietary fatty-acids in biology – their place in the evolution of the human brain. Nutrition Reviews, 50: 3–11.

Crawford, M. A., M. Bloom, C. L. Broadhurst, W. F. Schmidt, S. C. Cunnane, C. Galli, K. Gehbremeskel, F. Linseisen, J. Lloyd-Smith & J. Parkington, 1999. Evidence for the unique function of docosahexaenoic acid during the evolution of the modern hominid brain. Lipids 34: S39–S47 Suppl. S.

Crawford, M. A., S. C. Cunnane & L. S. Harbige, 1993. A new theory of evolution: quantum theory. In Sinclair, A. J. & R. Gibson (eds), IIIrd International Congress on Essential Fatty Acids and Eicosnoids. Am. Oil Chem. Soc., Adelaide: 87–95.

Figueiras, F. G., U. Labarta & M. J. Fernández Reiriz. 2002. Coastal upwelling, primary production and mussel growth in the Rias Baizxas of Galicia. Hydrobiologia 484/Dev. Hydrobiol. 167: 121–131.

Graham, W. M., J. G. Field & D. C. Potts, 1992. Persistent upwelling shadows and their influence on zooplankton distributions. Mar. Biol. 114: 561–570.

Hutchings, L. & J. G. Field, 1997. Biological oceanography in South Africa, 1896–1996: Observations, mechanisms, monitoring and modelling. Trans. Roy. Soc. S. Afr. 52: 81–120.

Kraufvelin, P., H. Christie & M. Olsen, 2002. Littoral macrofauna (secondary) responses to experimental nutrient addition to rocky shore mesocosms and a coastal lagoon. Hydrobiologia 484/Dev. Hydrobiol. 167: 149–166.

McClimans, T. A., G. Eidnes & J. Aure, 2002. Controlled artificial upwelling in a fjord using a submerged fresh water discharge: computer and laboratory simulations. Hydrobiologia 484/Dev. Hydrobiol. 167: 191–202.

Øie, G., K. I. Reitan, O. Vadstein & H. Reinertsen, 2002. Effect of nutrient supply on growth of blue mussels (*Mytilus edulis*) in a landlocked bay. Hydrobiologia 484/Dev. Hydrobiol. 167: 99–109.

Olsen, Y., 2002. MARICULT Research Programme: Background, status and main conclusions. Hydrobiologia 484/Dev. Hydrobiol. 167: 1–10.

Olsen, Y., T. Bøckmann, T. Bokn, S. Bremdal, E. Hoell, V. Øiestad, H. R. Skjoldal, E. Svendsen & O. Vadstein, 2001. MARICULT Research Programme (1996–2000). Final Scientific and Management Report from the Steering Committee. ISBN: 82-996202: 32 pp. (available at www.maricult.org).

Plaganyi, E. E., L. Hutchings, J. G. Field & H. M. Verheye, 1999. A model of copepod population dynamics in the Southern Benguela upwelling region. J. Plankton Res. 21: 1691–1724.

Reitan, K. I., G. Øie, O. Vadstein & H. Reinertsen, 2002. Response on scallop culture to enhanced nutrient supply by experimental fertilisation of a landlocked bay. Hydrobiologia 484/Dev. Hydrobiol. 167: 111–120.

Richardson, A. J., B. A. Mitchell-Innes, J. L. Fowler, S. F. Bloomer, H. M. Verheye, J. G. Field, L. Hutchings & S. J. Painting, 1998. The effect of sea temperature and food availability on the spawning success of Cape anchovy *Engraulis capensis* in the southern Benguela. South African J. Mar. Sci. 19: 275–290

Smaal, A.C. 2002. European mussel cultivation along the Atlantic coast – production status, problems & perspectives. Hydrobiologia 484/Dev. Hydrobiol. 167: 89–98.

Sommer, U., H. Stibor, A. Katechakis, F. Sommer & T. Hansen, 2002. Pelagic food web configurations at different levels of nutrient richness and their implications for the ratio fish. Hydrobiologia 484/Dev. Hydrobiol. 167: 11–20.

Stockner, J. G. & E. A. Macisaac, 1996. British Columbia lake enrichment programme: Two decades of habitat enhancement for sockeye salmon. Regulated Rivers-Research & Management 12: 547–561.

Stockner, J. G., E. Rydin & P. Hyenstrand, 2000, Cultural oligotrophication: Causes and consequences for fisheries resources. Fisheries 25: 7–14.

OLAV VADSTEIN & YNGVAR OLSEN
Norwegian University of Science and Technology,
Trondhjem Biological Station,
N-7491 Trondheim, Norway

Hydrobiologia **484:** xi–xii, 2002.
O. Vadstein & Y. Olsen (eds), Sustainable Increase of Marine Harvesting: Fundamental Mechanisms and New Concepts.

In memoriam: Arne Jensen (1926–2000)

In August 2000, just after the 1st Maricult Conference, Arne Jensen died after a life devoted to science. Profesor Arne Jensen became internationally recognized for his studies on the chemical composition and practical application of seaweeds. He later became a pioneer in the Norwegian aquaculture and a driver of aquaculture research in Trondheim and Norway.

Arne Jensen finished his education in organic chemistry at the College of Engineering (NTH, later incorporated into the University of Trondheim) in 1950, and became employed as a scientist at the Norwegian Institute of Seaweed Research (NITT) in Trondheim in 1951. NITT became a part of the University of Trondheim in 1973–74, as Department of Marine Biochemistry. The multidisciplinary staff of the department, with Arne Jensen as a main force, focused on the chemical composition of seaweed and made assessments of the standing stocks of the common Norwegian seaweeds. This work became highly recognized internationally.

Arne Jensen told later that "even with the practical perspective we had for our studies, we needed to base them on thorough fundamental research". He obtained his Dr.techn. degree at NTH in 1966 for the dissertation "Carotenoids of Norwegian Brown Seaweeds and of Seaweed Meals", and was appointed Professor of Marine Biochemistry in his department in 1974.

Arne Jensen became strongly involved in the steering board of national research programmes on marine pollution (1976–84) and inland eutrophication (1978–88). His attention, however, gradually turned towards aquaculture. In an invited talk at the annual meeting in the Research Council NTNF in 1984, he succeeded in bringing aquaculture into all political and industrial agendas. With research in technology and biology as the driving force, Arne

Jensen advocated that aquaculture had the potential to become a modern industry in future decades. Time shows that he was right. He became the head of the aquaculture planning group of the NTNF Research Council, and later a member of the board of the research programme on aquaculture (1986–93). From 1991 he was instrumental in the establishment of Maricult Research Programme, in cooperation with Norsk Hydro and Norwegian scientists (1996–2000). He took an active part in the planning of the 1st Maricult Conference, but was not able to attend. His last major contribution was a feasibility study of the future prospects of the Norwegian aquaculture industry in 2000– 2030. It was written in cooperation with the two academies of science in Trondheim. These predictions are his testimony for the future.

Arne Jensen had a strong international commitment throughout his career. He worked for the UN, he published a large number of scientific papers, and he acted for a long time as editor of Aquatic Toxicology. Moreover, he gave numerous talks in scientific institutions worldwide, and he was involved with the board of the International Seaweed Association through many decades. He served as Head of Department for more than 18 years and as elected head of the Faculty of Chemistry during his last years.

Arne Jensen made a strong impression on scientific aquaculture teaching and research in Trondheim. He was always enthusiastic and optimistic, with an impressive working capacity and ability to communicate. He had clear and brave visions for his work, and he will always be remembered as the social and inspiring advisor of science and life that he was.

YNGVAR OLSEN & OLAV VADSTEIN
Norwegian University of Science and Technology

Hydrobiologia **484:** 1–10, 2002.
O. Vadstein & Y. Olsen (eds), Sustainable Increase of Marine Harvesting: Fundamental Mechanisms and New Concepts.

MARICULT Research Programme: background, status and main conclusions

Yngvar Olsen
Trondhjem Biological Station, Norwegian University of Science and Technology (NTNU), 7491 Trondheim, Norway
Tel.: +47-73591592. Fax: +47 73591597. E-mail: yngvar.olsen@vm.ntnu.no

Key words: MARICULT Programme, sustainable exploitation, food web, resources, environment

Abstract

The human population is expected to reach about 10 billion by 2050, and it can be questioned how we can provide the food and the essential nutrients that will be needed. The ultimate goal of MARICULT was to provide a basis for the evaluation of environmental constraints and the potential for increased sustainable provision of food, raw materials and energy from the ocean. The Programme has primarily addressed fundamental questions on the marine ecosystem, because new concepts of harvesting and mariculture must be based on knowledge, and the environmental impact of new concepts should always be elaborated together with their production potential. The work has involved studies of food web structure and its functioning in the marine ecosystem as well as more applied research and testing of concepts for increasing the harvesting potential. There is a relatively tight connection between the nutrient supply rate and the primary production of phytoplankton, but the response of single species of animals is less predictable, especially for single populations on high trophic levels. There is, however, an empirical positive relationship between nutrient supply and secondary production, including fishery yields, on a regional scale. The low contribution of fisheries and aquaculture to human food compared to the contribution by agriculture is due to the fact that we primarily exploit first and higher carnivores. The general strategy to increase the marine harvesting potential is therefore to harvest and cultivate organisms at lower trophic levels. A number of harvesting and production concepts that may allow enhanced availability of marine resourses are proposed. These involve (1) harvesting of herbivore zooplankton resources, (2) cultivation of benthic, herbivore/omnivore animals, (3) enhancement of production potential by the creation of artificial up-welling systems, (4) strategic fertilization or use of available nutrient resources, and (5) further exploitation of biomass from macro-algae. All concepts involve harvesting or cultivation at low trophic levels, and some may readily be combined. It is important to note that some methods are not feasible for all coastal regions. In spite of great similarities, there are some clear differences in how the ecosystems react to enhanced nutrients in European marginal seas (Baltic Sea, Mediterranean Sea) and in Atlantic waters. Negative ecological effects caused by enhanced nutrient supply to pelagic ecosystems may occur if the primary production is not being grazed efficiently by zooplankton or benthic grazers. Dispersion by advection or retention in deepwater are important modifying factors. Loss of integrity in food web function seem to occur above a certain nutrient loading rate, suggesting a critical nutrient-loading rate for the maintenance of food web integrity. The Programme has contributed to a conceptual platform for assessment criteria of the state of the ecosystem and the water quality in pelagic coastal waters that are potential components in nutrient management models. The western principles of removing nutrient wastes from biological cycling is hardly sustainable. The counter-measure strategy that is still not explored and implemented is to reduce the ecological effects of uncontrolled nutrient emission in the coastal zone by using these nutrients for mariculture production. A more holistic ecological view in management and resource exploitation, where we recognize that we are also components in the ecosystem, is well accepted in eastern Asian countries, but not in the West. A holistic perspective may contribute to more concern for the marine environment.

Introduction

The global human population is expected to reach about 10 billion by 2050, and it can be questioned how humans will be able to provide the food and essential nutrients needed to feed that population (Doos & Shaw, 1999). An increase in agriculture will require considerable investments in research, wider implementation of intensive agricultural technology, use of genetically improved plants and animals, more efficient distribution of produced food, adequate availability of sufficient freshwater, and a significant increase in new agricultural land (Berndes & Wirsenius, 1996). It will also involve major environmental challenges (Tilman et al., 2001). The situation calls for a precautionary approach, and we should search for options for food production also outside the agricultural sector.

More seafood and food production based on biotechnological methods is among the most obvious alternative options available. Seafood contains essential lipids and mineral nutrients that are greatly demanded from nutritional and health perspectives (e.g., Sinclair & Gibson, 1992). Prospected bio-protein sources cannot so far replace food of aquatic origin. Marine fats contain $n - 3$ fatty acids that are essential for humans, as well as for cultured marine fish, and a sub-optimal supply may affect health and the quality of life. This is because it may result in a higher frequency of heart and coronary diseases and in some extreme cases mental health problems, such as menthal deficiencies (Nelson, 1991; Crawford, 1992; Hunter& Roberts, 2000).

A main motivation for strengthening of the marine sector is the fact that only a very low proportion of human food originates from the ocean. FAO statistics (www.fao.org) show that only about 1 and 5% of human energy and protein consumption, respectively, originate from the fisheries and aquaculture. This is surprising because marine and terrestrial primary production are of the same magnitude, and it implies that the traditional mode of harvesting the marine ecosystem is less efficient than our cultivation and harvesting of the terrestrial ecosystem. It is well known that the production of the world's oceans is temporarily and spatially very patchy. The classic analysis of Ryther (1969) suggested that some coastal up-welling areas are highly productive, whereas the majority of the open ocean is highly oligotrophic, and may be considered as marine deserts. The coastal marine systems are normally intermediate. Ryther (1969) estimated that 50% of the global fish and squid production take place in the major coastal up-welling systems of the Earth which constitute only 0.1% of the ocean surface. The next 50% of the production potential was located in the 10% which are coastal waters, whereas the 90% surface which was defined as open ocean, was responsible for <1% of the production. These numbers cannot be directly evaluated based on fishery statistics because fish migrate. Even though they most likely overestimates the potentials of up-welling systems (see Pauly & Christensen, 1995), they are still representative. An ultimate question is if such high-productive marine ecosystems can be created artificially.

MARICULT Research Programme has examined the possibilities for increasing the sustainable culture, harvesting and use of marine resources in the future. A further exploitation of marine biological resources should be based on knowledge of how the marine ecosystem functions, and the core activity of MARICULT has therefore been fundamental, strategic research on the marine ecosystem. New concepts of harvesting and mariculture have to be both sustainable and environmentally acceptable. One main question is whether increased activity in mariculture will be feasible and acceptable from an environmental point of view. The production potential of new methods should always be elaborated together with the environmental impact of the method.

The 1st MARICULT Conference was arranged to communicate perspectives and results of the Programme to the international scientific community, to learn about experience in other parts of the world, and to disseminate preliminary ideas and results for the seafood industry, policy makers, and coastal management authorities. The theme of the Conference was: How can we increase the production and harvest of living marine resources in the new millennium? Bridging potentials and environmental constrains.

Background

A very simple comparative analysis of production and harvesting strategies in agriculture *versus* fisheries and aquaculture clearly indicates why a major fraction of the food is produced in agriculture. Figure 1 illustrates the principal relationship between secondary production potential (stocks which are harvested) as a function of the primary production and the number of trophic transfers of energy (based on Ryther, 1969). The upper left corner of the figure is representative for

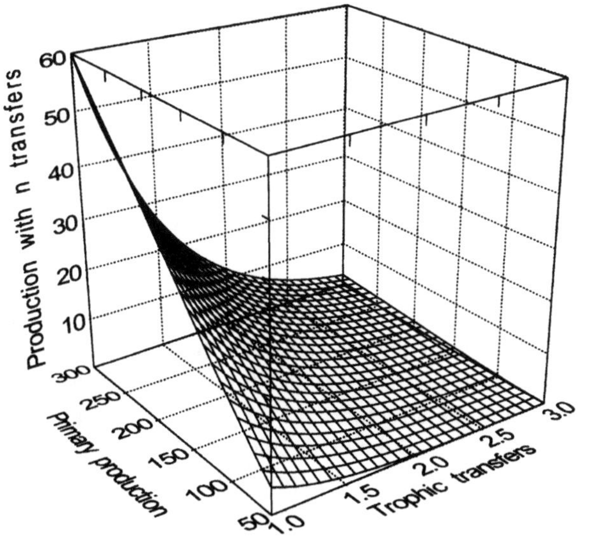

Figure 1. Empirical relationship showing animal production as a function of primary production and the number of trophic transfers that the energy has to pass through before reaching the stock that are harvested. The coefficients used are based on the three cases of systems defined by Ryther (1969). Interpolation is made by assuming coefficients to be continuous functions of primary production and number of trophic transfers through their entire range.

the production conditions of coastal up-welling systems, whereas the lower right corner represents the open ocean. The coastal ocean is intermediate. The figure illustrates how the secondary production is proportional to the primary production when the number of trophic transfers remains constant. Moreover, it illustrates the exponential decrease in the secondary and higher production as the number of trophic transfers increases (i.e., as we move up the food web) for a constant rate of primary production.

The animal production yield is more sensitive to variations in the number of trophic transfers than to variation in the primary production, because as much as 80–90% of the assimilated energy is lost for each trophic transfer of the energy. The total range of variability in secondary production caused by the total variability in primary production is comparable to the variability caused by one single trophic transfer (Fig. 1). The figure therefore reveals that the high harvesting potential of the up-welling systems is primarily an effect of the short food webs. The physical dynamics of up-welling systems are complex, and these systems are characterized by high natural nutrient supply from deep water and short food webs. This implies fewer trophic transfers of the assimilated energy and harvesting on lower trophic levels than in other marine ecosystems (Ryther, 1969; Hutchings & Field, 1997).

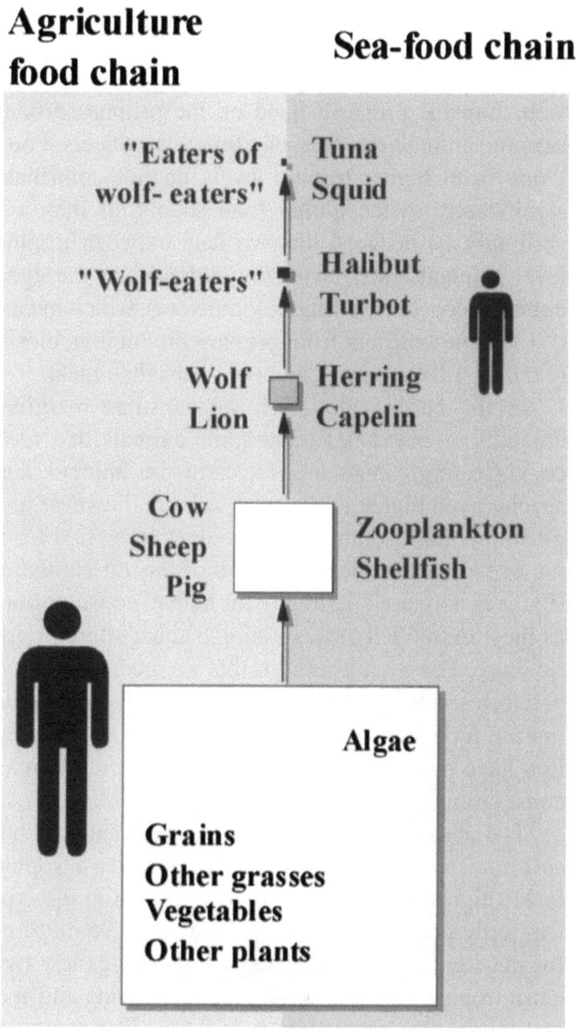

Figure 2. Schematic illustration of the human trophic level in the agriculture food chain and the seafood chain for humans. Our food from the agriculture food chain is harvested approximately two levels lower than in the seafood chain.

This conceptual analysis is simple and therefore also robust to variability in coefficients and empirical input data. Trophic dynamical models are in fact included in most textbooks on ecology. The above conceptual relationships, which are 'laws of nature', are, however, not seriously enough considered in evaluations of harvesting and mariculture potential. It may be questioned if our comprehension fails to fully realize the implications of an exponential loss of energy as we move up the food web.

The above considerations explain the marginal contribution of the marine food web in human food supply (1–2% in 1997, www.fao.org). Figure 2 illustrates human feeding habits for the products from seafood and agriculture, or the human trophic position for these types of food. In the agricultural food web, humans primarily feed on the primary producers and animals that feed on primary producers. Food items from higher trophic levels do not contribute significantly to the global food supply. It may accordingly be deduced that we feed between trophic level 1 (plants) and 2 (grazing animals) in the agricultural food web (herbivore-omnivore), which means 0–1 trophic transfers from primary production, closer to 0 than 1 because we eat more plants than meat.

In the marine food web we consume marginal amounts of macroalgae and some animals that feed on algae (e.g., mussels). 1st carnivore animals and species even higher in the food web are, however, the most important commercial resources harvested from the ocean. A significant fraction of the 1st carnivore stocks is only used indirectly for human consumption, as they are refined through marine aquaculture or agriculture. The overall result is that we most likely feed between levels 3 and 4 (1st and 2nd carnivores) in the marine food web, meaning that the primary production have passed 2–3 trophic transfers before human consumption.

The above considerations are simplifications, but still relatively robust. Humans feed about two trophic levels higher in the marine food web than in the typical agricultural one, meaning that some 96–99% of the marine primary production is lost during these two extra trophic transferes. Because the marine and terrestrial primary production is of the same magnitude, these figures are compatible with FAO statistics showing that only 1–2% of the total energy available to humans originates from marine primary production. The contribution of proteins is higher, because we primarily eat marine animals. The huge potential to enhance the marine harvesting of bio-resources is to catch the steadily declining assimilated carbon before conversion to carbon dioxide during respiratory processes. In other words this implies harvesting and cultivation of organisms at lower trophic levels. It is, however, important that virtually all fish stocks, also those of 1st carnivores, are already overexploited and that the pressure on these stocks should be released rather than increased (Pauly et al., 1998).

Main results and achievements

The work of MARICULT have involved fundamental research related to food web structure and functioning in the marine ecosystem as well as more applied research and testing of concepts for increasing the harvesting potential (see bibliography in Olsen et al. (2001a) and papers of this special issue).

Response of nutrients on lower food web structure and function

Many projects in MARICULT have examined various aspects of the lower pelagic food web, whereas a few projects have treated fundamental aspects of changing the structure and function of benthic communities (e.g., Bokn et al., 2001; Olsen et al., 2001a). The project work has supported earlier findings on many points, but have also provided substantial new information. Some examples of results and conclusions generated in the projects are summarised below (see Sommer et al., 2002, and other papers of this issue).

Many species of phytoplankton, although not all, respond positively and in a predictable way to nutrient enhancement. Balanced N:P:Si perturbations of NE Atlantic waters support normally fast-growing diatoms, but this is not necessarily true in for example Baltic waters. It is notable that pico-cyanobacteria form a ubiquitous dominant group in all types of European coastal waters during mid summer and early autumn. This group does often not respond to nutrient perturbations, most likely because they are controlled by grazing or viral infection. Cell lysis of phytoplankton is an important process in oligotrophic Mediterranean waters, but it is not that important in the NE Atlantic.

Heterotrophic nano-flagellates, ciliates and mesozooplankton all contribute significantly to consumption of primary production, and the grazers harvest approximately 50% of the production in euphotic waters. Smaller grazers feeding on bacteria and picocyanobacteria are, in relative terms, more important in oligotrophic waters than in meso- and eutrophic waters, where larger grazers and ciliates become more important. Little energy and materials are transferred from the microbial food web to the main food chain, because the microbial pathways through heterotrophic nano-plankton and ciliates are rather insignificant. It seems as the energy is transferred mainly from primary producers through three parallel food chains rather than through a food web, meaning that autotrophs were the most important food items for all heterotrophic groups. Increased nutrient supply enhance the autotrophic fluxes more strongly than the heterotrophic ones.

There is a relatively tight coupling between the nutrient supply rate to productive waters and the primary production of phytoplankton, which is not very surprising. This has been repeatedly demonstrated both in closed experimental ecosystems and in open natural coastal ecosystems. The responses of animals feeding on phytoplankton and other food sources are positive as well, but not easily predictable on the species level. Horizontal water transport, vertical mixing processes and interactions with benthic communities make registrations of the slower animal responses difficult during experimental exercises, but this does not mean that there is a lack of response. It is, however, important that the predictability of responses on population level will become gradually less as we move up the food web. There is certainly an positive empirical relationship between nutrient supply and secondary production, including fishery yields, on a regional scale, but the predictability is questionable for single populations feeding high up in the food web. This is among others a result of activity and response of invertebrate predators that interract with components of the lower food web.

Production perspectives

The above background discussion illustrated why the contribution of fisheries and aquaculture to human food was low compared to the contribution by agriculture. Our current harvesting of the marine production potential is inefficient because we primarily exploit first and higher carnivores, and mariculture has restricted potential because most cultured marine species are secondary or higher carnivores that require marine food. A general strategy to increase the marine harvesting potential is therefore to harvest at a lower trophic level. First carnivore fish stocks are already overexploited, and the fishing pressure on these resources should be released rather than strenghtened (Pauly et al., 1998). There are obvious technological and economical constraints in harvesting small particles like phytoplankton and zooplankton. On the other hand, marine macro-algae represent an easily harvestable resource. It is notable that the strategy of harvesting at low trophic levels is widely accepted in agriculture.

Actions to increase the potential of mariculture is to (1) increase the culture efforts of herbivore/omnivore animals (algal eating/omnivore, e.g., mussels) and (2) to use food resources for traditional carnivore aquaculture that are harvested at primary producer or herbivore level. The following concepts of harvesting and cultivation are proposed:

Concept 1 – Harvesting of herbivore zooplankton resources, including copepods and krill

Concept 2 – Cultivation of benthic, herbivore/omnivore animals, for example blue mussel

Concept 3 – Enhancement of production potential by creation of artificial up-welling systems

Concept 4 – Strategic fertilization or use of available nutrient resources

Concept 5 – Further exploitation of biomass from macro-algae

All concepts involve harvesting or cultivation at low trophic levels, and may readily be combined. It is important to note that some methods are not feasible for all coastal regions. The concepts should be further evaluated based on their potential of enhanced harvesting or cultivation and on their potential environmental impact in a wide sense. Other important questions are if there is an economic potential that can become realized through further research and developments, and if there is a market or a potential market for the products. The ultimate requirement is that the activity is sustainable in a wide sense.

Methods of sea ranching have not been included because of the legal problems that are involved. These problems are presently becoming solved for some benthic species in Norway, and it will be important to include options in sea ranching at a later stage.

Concept 1 – Harvesting of herbivore zooplankton resources

Some regions of the world's oceans have large stocks of herbivore copepods and krill that are abundant and potentially exploitable, depending on their distribution in time and space. The fact that these groups of zooplankton are food for important fish stocks brings up management and political dimensions, and the potential interaction with fisheries must therefore have a main focus. It is paramount to clarify the possibilities, constraints, and consequences of large-scale harvesting of marine zooplankton. It is also important to learn from experience in the current krill fisheries (Nicol & Endo, 1999).

Zooplankton is a bioresource for fish feed and an industrial raw material. It is important to note that 10% of the annual herbivore production will be comparable

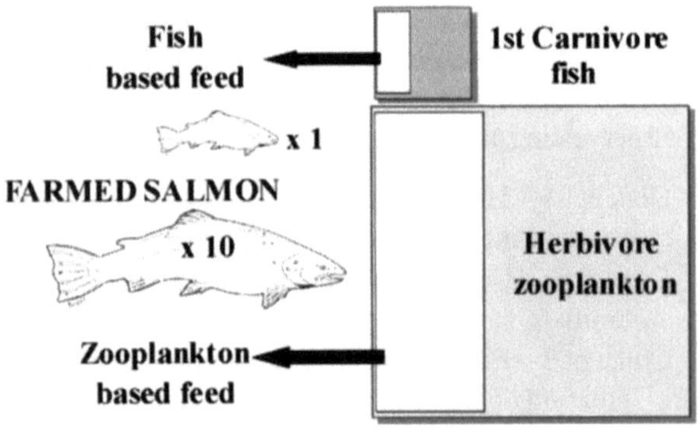

Figure 3. Theoretical scheme illustrating the potential benefits of harvesting resources for salmon feed on herbivore zooplankton level. Feed for farmed salmon is currently based on 1st carnivore fish (arbitrary amount for fish farming is illustrated by white square). If the zooplankton produced on the trophic level below is harvested instead – this zooplankton is the food of 1st carnivore fishes that are used for feed – the salmon production can be increased by a factor of 10 without further needs for fish resources.

to the entire production of planktivore (1st carnivore) fish, and that 1% harvesting of this herbivore production is still a significant resource. If the zooplankton production in the sea tends to be food limited, we may expect that moderate harvesting primarily will result in less mortality of zooplankton and not necessarily in reduced standing stocks and availability for planktivore fish. If, on the other hand, temperature or advective transport is more important than food limitation, harvesting will have greater impact. These questions are among the major challenges to explore.

It is in particular interesting to use herbivore zooplankton instead of 1st carnivore fish as a principal raw material for fish feed. This will theoretically move cultured carnivore fish that depend on marine food at least one step down in the food web (Fig. 3). An important question is then if the release on fishing pressure on 1st carnivore fish then made possible is more important that the reduction in food for these species.

Concept 2 – Cultivation of benthic, herbivore/omnivore animals

The theoretical 'down in food web' potential of culturing benthic herbivore animals in the coastal ocean is high, but there are obstacles that restrict the exploitation of primary production in practical mariculture operations. Cultivation requires space in the coastal zone, and appropriate locations must be biologically rich and physically feasible to support economically feasible production. It is of particular interest that if we employ appropriate technologies, we can utilize some of the nutrient resources that are residue from agriculture and intense aquaculture for mussel production. These resources, which are released into coastal waters, may be brought back into biological cycling through controlled mussel cultivation. Knowledge of the fate of anthropogenic nutrients and hygienic aspects in the coastal zone will most likely allow such applications, as demonstrated in a project in southern Norway.

The blue mussel is a species with growth potential (Smaal, 2002, this issue), high enough to become a potential feed resource for marine aquaculture. This will require production and refinement technologies that are still not invented. It is an attractive option to bring nutrient discharge from intensive aquaculture into blue mussel biomass and then back into fish feed. This will contribute to move the cultured fish species down the food web and secondly to enhance the production potential of cultured fish. The need for knowledge on energy and material cycling in fish farming, its interaction in the coastal ocean, knowledge on threats of disease spreading, and the need to improve production technology suggest that these options are not immediately available.

Concept 3 – Enhance production potential by creation of artificial up-welling systems

The large up-welling areas of the world's oceans are very productive (Ryther, 1969; Hutchings & Field, 1997; Figueiras et al., 2002, this issue). The fundamental physical mechanisms that result in strong fertilization by deep water and short food webs, and accordingly high harvesting potential, have for a long time been a challenge for science. It is important to note that these high productive areas are not recognized as environmental problems. It is a question if small-scale highly productive up-welling systems can be created artificially in suitable locations. An ultimate question is how the physical and nutritional regimes of these systems result in short and highly productive food webs.

The Programme has supported an initiative to explore the option of improving productivity, primarily as a means of enhancing the potential to cultivate herbivore animals such as mussels, and reduce the problems of toxic algae during mussel cultivation (Berntsen et al., 2002, this issue; McClimans et al., 2002, this issue). A submerged freshwater outlet generates the up-welling energy. Preliminary model simulations and background studies have shown that artificial up-welling of nutrient rich deep water in fjords have a potential to increase the phytoplankton production by a factor of 3–4 after the spring bloom. Such up-welling may also reduce the occurrence of toxic flagellates in the phytoplankton. The further development of this method will still require research and development. Many deep Norwegian fjords with available nutrient-rich deep water will most likely be suitable for this treatment. Related methods, adapted to smaller geographic scale, may be developed and used to detoxify mussels locally.

Concept 4 – Strategic fertilization or use of available nutrient resources

There is little doubt that the productivity of marine waters becomes stimulated by nutrient supply from deepwater. Data derived from closed experimental ecosystems and an open coastal lagoon ecosystem exposed to sustained nutrient addition show in detail how the lower trophic levels of the pelagic food web respond to and interact during enhanced nutrient supply. The pattern of responses in food web structure and function are comparable to those found in the southern North Sea, which is affected by anthropogenic nutrient supply (Colijn et al., 2002, this issue).

The ability to control the responses of nutrients at species level is limited for attractive organisms at higher trophic levels. Many alternative food web pathways above herbivore zooplankton may become stimulated, such as 1st carnivore fish, jellyfish, other invertebrate predators, or benthic animals. Production yields are therefore not adequately controlled at the species or population level, and it will generally not be economically feasible to fertilize marine systems to harvest carnivore species. This is in particular the case in oligotrophic systems where such fertilization also will lead to a pronounced change in species diversity (Chisholm et al., 2001).

On the other hand, in semi-closed productive coastal locations, the production yields following nutrient additions may be highly predictable and efficient for macro-algae and for herbivore benthic animals like the blue mussel. Projects have documented growth responses for blue mussels and scallops following the fertilization of an open coastal lagoon (Reitan et al., 2002, this issue; Oie et al., 2002, this issue).

Concept 5 – Further exploitation of macro-algal biomass

Marine macro-algae represent a poorly exploited primary biomass. Only their polysaccharides are systematically used for industrial purposes. Plant breeding has seldomly been applied to these plants and there is a large potential for improvement. Space and access to water are unlikely to limit the cultivation of marine macro-algae. These facts alone make seaweeds particularly interesting as a source of renewable raw materials. It is a main challenge to invent competitive technology for direct utilization of $n - 3$ rich lipids and the conversion of biomass to, for example, fish feed, food ingredients, bulk chemicals, energy carriers and other useful commodities. Another interesting aspect is to culture macro-algae for combined biomass production and bio-remediation measures for coastal waters.

Environmental perspectives

Nutrients are natural elements that are essential for all living organisms. Because they stimulate phytoplankton and macro-algal growth, major ecological changes may take place if nutrients are supplied at too high rates (see review by Cloern, 2001). Some ecological changes following enhanced nutrient emission are undesirable and not esthetically accepted because they are associated with ecosystem degradation or harmful eutrophication. Issues related to coastal eutrophication has been the main focus of the Programme (Bokn et al., 2002, this issue; Kraufvelin et al., 2002, this issue; Olsen et al., 2002b), and some general conclusions are as follows.

Phytoplankton biomass and production respond almost linearly to an enhanced nutrient loading rate during the first weeks of exposure. The response on a seasonal or long-term scale is more moderate and the biomass response is most likely non-linear with increasing loading rate. This suggests that a biological modification of food web structure is taking place. Such pelagic and benthic buffer mechanisms work to reduce the negative ecological effects of enhanced nutrient emission in the coastal zone.

Both the structure and function of the plankton food web are altered by enhanced nutrient supply. There are some clear differences how the ecosystems reacts in European marginal seas (Baltic Sea, Mediterranean Sea) and in Atlantic waters, but there are also great similarities. Negative ecological effects caused by enhanced nutrient supply to pelagic ecosystems may occur if zooplankton or benthic grazers are not grazing the primary production efficiently enough. Other buffering processes are dispersion by advection or retention in deep water.

There is evidence for losses of integrity in food web function above a certain nutrient loading rate to coastal waters, suggesting a critical nutrient-loading rate for the maintenance of food web integrity. Although still at a preliminary stage, the Programme has contributed to a conceptual platform for assessment criteria of ecosystem state and water quality of pelagic coastal waters (Olsen et al., 2001a,b). Many criteria show a common, regionally independent pattern of variation. Primary production, meso-zooplankton grazing and growth, fraction of primary production consumed by grazers, bacterial production relative to primary production, as well as more integrated measures like cycling indices and path lengths, are all well related to the nutrient loading rate. Autotrophic biomass, the ratio of autotrophic to heterotrophic biomass, and the fraction of pico-cyanobacteria of the total autotrophic biomass are all well related to the total nutrient concentration. These relationships, or other integrated variables based on these, are potential candidates of response functions for nutrient management models for coastal waters.

The general response of the phytoplankton and the preliminary framework of a dose–response model for nutrients in the coastal zone are illustrated in Figure 4. The nutrients stimulate phytoplankton growth whereas the water dilution rate, which is inversely related to the water retention time, is a major buffering process for the actual ecological effects of the nutrients.

All kinds of food production require a clean and unpolluted environment, and pollution of coastal waters by persistent toxins is a potential threat against mariculture and fisheries. Heavy metals and organic toxins have been steadily phased down in western countries during the last decades, but organic toxins are still a potential problem. We have a long tradition in using the ocean as deposit for wastes, but we cannot afford to continue that practice. The mechanisms of uptake, transport and cycling of toxins in the lower marine food web are not well known, but lipid-soluble components may become concentrated on higher trophic levels (Barrier et al., 1992). This calls for attention, because harvesting and cultivation involve first and higher carnivores that may accumulate high contents of such toxins. Marine pollution has not been specifically treated in the Programme, but the issue is undoubtedly very important for the fishing and mariculture industries.

Interaction between resources and environmental issues

Harmful eutrophication is a recognised problem, but nutrients that are natural components in the biosphere are not threatening human health. The countermeasures against harmful eutrophication are, however, very expensive for society, and it is therefore important to find a sound balance between the use and protection of coastal ecosystems. The common principles of nutrient management in Western countries are hardly sustainable, because we isolate and remove a considerable part of nutrient wastes from the biological cycle instead of putting them back into cycling. The principle of reduction at source is in many cases inadequate, because a considerable fraction of the nutrient emission is from diffuse sources that cannot be efficiently reduced. A counter-measure strategy that is still not explored is to reduce the ecological effects in the coastal zone by using the nutrient resources for controlled mariculture production where such nutrient sources are available.

Blue mussel cultivation is an efficient means of retaining anthropogenic nutrients in the coastal zone. The species very efficiently clear phytoplankton from the water and the food is rapidly converted to mussel biomass. Nutrient retention in mussel biomass obviously reduces the nutrient concentrations in the water, and this is a vital method in situations where no other water purification technologies are available. An example of the particle retention efficiency of mussels is demonstrated in the shallow Öresund Strait between Denmark and Sweden where the massive banks of blue mussels remove practically all the phytoplankton from the water drained from the Baltic Sea (Haamer & Rodhe, 2000).

The attractive point of combining mussel production and water purification is that nutrient resources that are out of our control are utilized for food production instead of creating environmental problems and costs for society. This also represents a more sustainable way of nutrient utilization as it contributes to the closing of biological nutrient cycles. The use of anthropogenic nutrients and run-off for blue mussel cultivation has been demonstrated experimentally in the Programme. Additionally, the Limfjord in Denmark and the coastal zone of The Netherlands are both examples where high anthropogenic nutrient emmision has created excellent conditions for mussel production.

Another potential is the nutrients that are released from fish in cages, which in Norway represents an

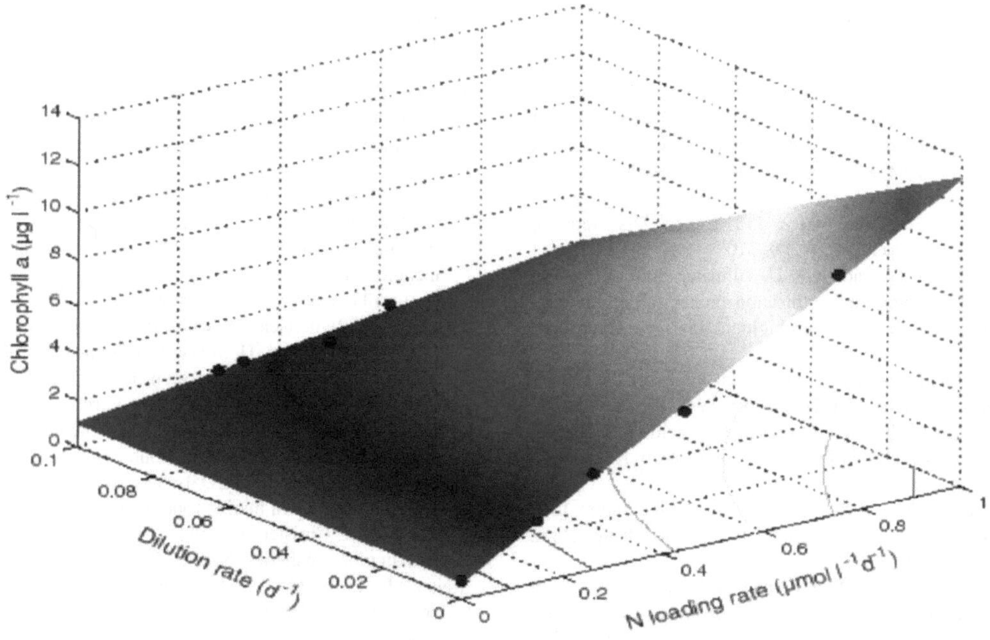

Figure 4. Preliminary conceptual nutrient loading model for management purposes, showing phytoplankton biomass (μg chlorophyll a l^{-1}) as a function of nutrient loading rate (here μmol N l^{-1} day^{-1}, P and Si is added in Redfield proportions) and water dilution rate (day^{-1}).

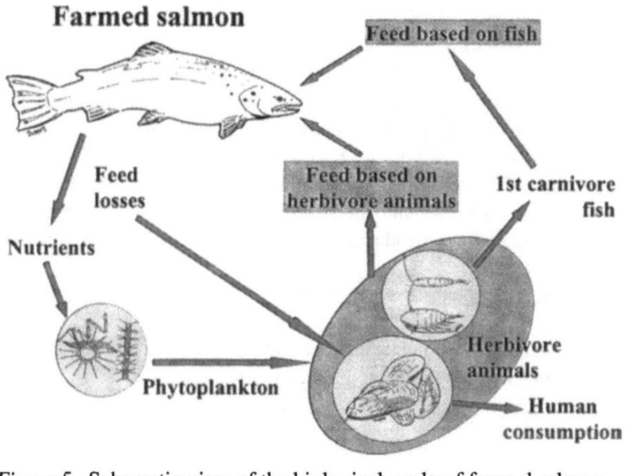

Figure 5. Schematic view of the biological cycle of farmed salmon. Zooplankton and benthic animals like blue mussel are both herbivore/omnivore animals that will compete for and consume small particles of feed and, more important, phytoplankton produced on nutrients released by the salmon. Humans consume blue mussel, and blue mussel–salmon thus forms a true poly-culture system. Zooplankton is normally the food of 1st carnivore fish, the primary resource currently used for fish feed. Both blue mussel and zooplankton are future potential 'low in food web' resources for fish feed, and such application could reduce pressure on stocks of 1st carnivore fish.

annual production of >2 million tonnes of mussels if converted to blue mussel biomass. A significant part of these nutrients can be retained by controlled mussel production. Blue mussel-salmon forms a poly-culture system where the blue mussel may take advantage of the discarges from the salmon farms (Fig. 5). Both blue mussel and zooplankton that will eat a considerable part of the phytoplankton, are potential resources for fish feed instead of 1st carnivore fish.

A more holistic ecological view of management and resource exploitation, where we recognize that humans are also components of the ecosystem, is well accepted in eastern Asian countries, but seem to be forgotten in the West. It is important to have a holistic perspective, and it may in turn contribute to more concern about the marine environment. Most people understand the need to avoid pollution of land areas used for agriculture. The same kind of concern, which is presently not there, is also needed for the marine environment.

References

Barrie, L. A., D. Gregor, B. Hargrave, R. Lake, D. Muir, R. Shearer, B. Tracey & T. Bidleman. 1992. Arctic contaminants – Sources, occurence and pathways. Sci. total Envir. 122: 1–74.

Berndes, G. & S. Wirsenius, 1996. Biomass for energy, food and materials in an industrial society of 10 billion people. Renew. Energy 9: 926–929.

Berntsen, J., D. L. Aksnes & A. Foldvik, 2002. Production enhancement by artificial upwelling – A simulation study. Hydrobiologia 484/Dev. Hydrobiol. 167: 177–190.

Bokn, T. L., E. E. Hoell, K. Kersting, F. E. Moy & K. Sørensen, 2001. Methods applied in the large littorale mesocosm study of nutrient-enrichment in rocky shore ecosystems – EULIT. Cont. Shelf Res. 21: 1925–1936.

Bokn, T. L., F. E. Moy, H. Christie, S. Engelbert, R. Karez, K. Kersting, P. Kraufvelin, C. Lindblad, N. Marba, M. F. Pedersen & K. Sørensen, 2002. Are rocky shore ecosystems affected by nutrient-enriched seawater? Some preliminary results from a mesocosm experiment. Hydrobiologia 484/Dev. Hydrobiol. 167: 167–175.

Chisholm, S. W., P. G. Falkowski & J. J. Cullen. Dis-Crediting ocean fertilisation. Science 294: 309–310.

Colijn, F., K.-J. Hesse, N. Ladwig & U. Tillmann, 2002. Effects of the large-scale uncontrolled fertilisation process along the continental coastal North Sea. Hydrobiologia 484/Dev. Hydrobiol. 167: 133–148.

Cloern, J. E., 2001. Our evolving conceptual model of the coastal eutrophication problem. Mar. Ecol. Progr. Ser. 210: 223–253.

Crawford, M. A., 1992. The role of dietary fatty-acids in biology – their place in the evolution of the human brain. Nutr. Rev. 50 (4): 3–11.

Doos, B. R. & R. Shaw, 1999. Can we predict the future food production? A sensitivity analysis. Glob. Environ. Change Hum. Policy Dimens. 9: 261–283.

Figueiras, F. G., U. Labarta & M. J. Fernández Reiriz, 2002. Coastal upwelling, primary production and mussel growth in the Rías Baixas of Galicia. Hydrobiologia 484/Dev. Hydrobiol. 167: 121–131.

Haamer, J. & J. Rodhe, 2000. Mussel *Mytilus edulis* (L.) filtering of the Baltic Sea outflow through the Oresund – An example of a natural, large-scale ecosystem restoration. J. Shellfish Res. 19: 413–421.

Hunter, B. J. & D. C. K. Roberts, 2000. Potential impact of the fat composition of farmed fish on human health. Nutr. Res. 20: 1047–1058.

Hutchings, L. & J. G. Field, 1997. Biological oceanography in South Africa, 1896–1996: Observations, mechanisms, monitoring and modelling. Trans. r. Soc. S. Afr. 52: 81–120.

Kraufvelin, P., H. Christie & M. Olsen, 2002. Littoral macrofauna (secondary) responses to experimental nutrient addition to rocky shore mesocosms and a coastal lagoon. Hydrobiologia 484/Dev. Hydrobiol. 167: 149–166.

McClimans, T. A., G. Eidnes & J. Aure, 2002. Controlled artificial upwelling in a fjord using a submerged fresh water discharge: computer and laboratory simulations. Hydrobiologia 484/Dev. Hydrobiol. 167: 191–202.

Nelson, G. J. (ed.), 1991. Health effects of dietary fatty acids. American Oil Chemist's Society, ISBN 0-935315-44-6.

Nicol, S. & Y. Endo, 1999. Krill fisheries: Development, management and ecosystem implications. Aquat. Living Resour. 12: 105–120.

Øie, G., K. I. Reitan, O. Vadstein & H. Reinertsen, 2002. Effect of nutrient supply on growth of blue mussels (*Mytilus edulis*) in a landlocked bay. Hydrobiologia 484/Dev. Hydrobiol. 167: 99–109.

Olsen, Y., T. Bøckmann, T. Bokn, S. Bremdal, E. Hoell, V. Øiestad, H. R. Skjoldal, E. Svendsen & O. Vadstein, 2001a. MARICULT Research Programme (1996–2000). Final Scientific and Management Report from the Steering Committee. ISBN: 82-996202, 32 pp. (available at www.maricult.org).

Olsen Y., H. Reinertsen, O. Vadstein, T. Andersen, I. Gismervik, C. Duarte, S. Agusti, H. Stibor, U. Sommer, R. Lignell, T. Tamminen, C. Lancelot, V. Rousseou. E. Hoell & K. A. Sanderud, 2001b. Comparative analysis of food webs based on flow networks: effects of nutrient supply on structure and function of coastal plankton communities. Cont. Shelf Res. 21: 2043–2053.

Pauly, D., V. Christensen, J. Dalsgaard, R. Froese, & F. Torres Jr., 1998. Fishing down marine food webs. Science 279: 860–863.

Pauly, D. & V. Christensen, 1995. Primary production required to sustain global fisheries. Nature 374: 255.

Reitan, K. I., G. Øie, O. Vadstein & H. Reinertsen, 2002. Response on scallop culture to enhanced nutrient supply by experimental fertilisation of a landlocked bay. Hydrobiologia 484/Dev. Hydrobiol. 167: 111–120.

Ryther, J. H., 1969. Photosynthesis and fish production in the sea. Science 166: 72–76.

Sinclair, A. & R. Gibson (eds), 1992. The third international congress on essential fatty acids and eicosanoids. Congress Congress proceedings, Adelaide, Australia, 1–5th March, 1992. ISBN 0-935314-43-8.

Smaal, A. C., 2002. European mussel cultivation along the Atlantic coast – production status, problems and perspectives. Hydrobiologia 484/Dev. Hydrobiol. 167: 89–98.

Sommer, U., H. Stibor, A. Katechakis, F. Sommer & T. Hansen, 2002. Pelagic food web configurations at different levels of nutrient richness and their implications for the ratio fish production:primary production. Hydrobiologia 484/Dev. Hydrobiol. 167: 11–20.

Tilman, D., J. Fargione, B. Wolff, C. d'Antonio, A. Dobson, R. Howarth, D. Schindler, W. H. Schlesinger, D. Simberloff & D. Swackhamer. Forecasting agriculturally driven global environmental change. Science. 292: 281–284.

Hydrobiologia **484:** 11–20, 2002.
O. Vadstein & Y. Olsen (eds), Sustainable Increase of Marine Harvesting: Fundamental Mechanisms and New Concepts.

Pelagic food web configurations at different levels of nutrient richness and their implications for the ratio fish production:primary production

Ulrich Sommer[1], Herwig Stibor[2], Alexis Katechakis[2], Frank Sommer[1] & Thomas Hansen[1]
[1]*Institut für Meereskunde, Düsternbrooker Weg 20, 24105 Kiel, Germany*
[2]*Zoologisches Institut, Ludwig-Maximilians-Universität, Karlstraße 23–25, 80333 München, Germany*
Tel.: +431-600-4400. Fax: +431-600-4402. E-mail: usommer@ifm.uni-kiel.de

Key words: plankton, food web, trophic level, ecological efficiency, copepods, fish production

Abstract

Based on existing knowledge about phytoplankton responses to nutrients and food size spectra of herbivorous zooplankton, three different configurations of pelagic food webs are proposed for three different types of marine nutrient regimes: (1) upwelling systems, (2) oligotrophic oceanic systems, (3) eutrophicated coastal systems. Upwelling systems are characterised by high levels of plant nutrients and high ratios of Si to N and P. Phytoplankton consists mainly of diatoms together with a subdominant contribution of flagellates. Most phytoplankton falls into the food spectrum of herbivorous, crustacean zooplankton. Therefore, herbivorous crustaceans occupy trophic level 2 and zooplanktivorous fish occupy trophic level 3. Phytoplankton in oligotrophic, oceanic systems is dominated by picoplankton, which are too small to be ingested by copepods. Most primary production is channelled through the 'microbial loop' (picoplankton – heterotrophic nanoflagellates – ciliates). Sporadically, pelagic tunicates also consume a substantial proportion of primary production. Herbivorous crustaceans feed on heterotrophic nanoflagellates and ciliates, thus occupying a food chain position between 3 and 4, which leads to a food chain position between 4 and 5 for zooplanktivorous fish. By cultural eutrophication, N and P availability are elevated while Si remains unaffected or even declines. Diatoms decrease in relative importance while summer blooms of inedible algae (*Phaeocystis*, toxic dinoflagellates, toxic prymnesiophyceae, etc.) prevail. The spring bloom may still contain a substantial contribution of diatoms. The production of the inedible algae enters the pelagic energy flow via the detritus food chain: DOC release by cell lysis – bacteria – heterotrophic nanoflagellates – ciliates. Accordingly, crustacean zooplankton occupy food chain position 4 to 5 during the non-diatom seasons. Ecological efficiency considerations lead to the conclusion that fish production:primary production ratios should be highest in upwelling systems and substantially lower in oligotrophic and in culturally eutrophicated systems. Further losses of fish production may occur when carnivorous, gelatinous zooplankton (jellyfish) replace fish.

Introduction

The relationship between pelagic primary productivity and pelagic fish production critically depends on the number of trophic links between phytoplankton and planktivorous fish, such as herring, anchovy or pilchard. Classic ecological common-sense has started from two assumptions: an ecological efficiency of 10% and planktivorous fish at the 3rd trophic level. This would predict that fish production should amount to 1% of primary production. Some early authors have even assumed a considerable amount of herbivory by planktivorous fish of upwelling areas, which would place the trophic position of fish below 3 and the FP:PP quotient >1% (Ryther, 1969). A more recent analysis by Iverson (1990) leads to considerably lower estimates. His regression equation predicts, that nekton production (fish plus squid production) should be non-linearly related to primary production. NP:PP amounts to ca. 1% in eutrophic upwelling areas but <0.5% in the oligotrophic ocean. This dependence of NP:PP on PP could have two reasons: a lower ecological efficiency or longer food chains under more oligotrophic conditions. The first hypothesis seems improbable, because the ecological efficiency at the phytoplankton:zooplankton link has been found to be slightly higher under more oligotrophic conditions (Cushing, 1971).

In this article, we will use the available knowledge about the phytoplankton response to nutrients and about mesozooplankton food selection to develop some hypotheses about differences in the food web structure at different levels of nutrient richness. Instead of just contrasting oligotrophic and eutrophic conditions, we will develop three scenarios. A low nutrient scenario, characteristic for the oligotrophic ocean, a high Si – high N – high P scenario characteristic for upwelling regions and temperate/boreal seas immediately after winter circulation, and a low Si – high N – high P scenario, characteristic for summer stratification under the eutrophicating influence of land-borne nutrient sources.

The pelagic food web

The image of the pelagic food web has undergone drastic change during the last generation (Fig. 1). The old view was of a simple food chain with phytoplankton (mainly diatoms) at the base, herbivorous mesozooplankton (mainly copepods) at the second trophic level, and planktivorous fish at the third trophic level c.f. the central, vertical chain in Fig. 1). Phytoplankton feeding by planktivorous fish has been assumed in several textbook illustrations and by Ryther's (1969) calculation of potential fish production, but a detailed study on the feeding of *Engraulis capensis* and *Engraulis mordax* in the Benguela upwelling system has found only a minor importance of phytoplankton in the diet (James & Chiappa-Carrara, 1990).

The image of the pelagic food web changed fundamentally with the discovery of increasingly smaller phytoplankton species (nanoplankton <20 μm, picoplankton <2 μm), their protozoan feeders, an increasing awareness of the importance of bacteria, and the study of bacterivory by protozoans (Pomeroy, 1974). Initially, this newly discovered compartment of the food web was called a 'microbial loop' while it has now become more customary to call it a 'microbial food web' (lower, left hand portion of Fig. 1). Initially, the debate centred on the question whether it was a 'link' or a 'sink' in the energy transfer to higher trophic levels (Sieburth & Davis, 1982; Azam et al., 1983). The answer is quite straightforward. The microbial loop is a link to the extent to which DOC consumed by bacteria is made available to higher trophic levels by bacterivory. It is a sink to the extent to which primary production by pico- and nanoplankton is consumed by protozoans instead of metazooans, particularly crustaceans, which are the main food item of planktivorous fish.

A further extension of the traditional food web image can be seen on the right-hand side of Figure 1, with pelagic tunicates taking the position of the herbivores/bacterivores and jellyfish (Cnidaria and Ctenophora) the position of the primary carnivores. Those organisms have a much higher water content and lower protein content in the fresh biomass than crustaceans and fish. Therefore, the food chain tunicates-jellyfish has been called the 'jelly food chain' as opposed to the 'muscle food chain' leading from crustaceans to fish. Gelatinous zooplankton are usually considered a dead end in the pelagic food web (Verity & Smetacek, 1996) because their low nutritional value makes them a minor food item for vertebrates, although there are some minor exceptions (e.g. the moon fish *Mola mola* and sea turtles).

Phytoplankton response to nutrients

The response of microalgal community composition to nutrient supply has been the focus of numerous competition experiments in the laboratory, field mesocosm experiments and comparative field surveys. It is one of the major findings of competition experiments, that diatoms are favoured by high Si:N and Si:P ratios, while non-siliceous algae are favoured by low ones (Sommer 1983, 1994, 1996a,b, 1998; Tilman 1982; Tilman et al. 1986). Field mesocosms with Si-enrichment (elevated Si:N or Si:P) or with N- and P-enrichment (lowered Si:N or Si:P) ratios have usually confirmed this finding (Escaravage et al., 1996; Schöllhorn & Granéli, 1996; Egge & Jacobsen 1997), though during summer there have been some cases when diatoms did not respond as expected. Figure 2 shows that in laboratory microcosms the transition from diatom to non-siliceous algae dominance takes place at Si:N ratios from 0.2:1 to 2:1 and at Si:P ratios from 3:1 to 30:1 (all ratios in molar units). This range is important, because Si:N ratios in the winter waters of seasonally stratified temperate and boreal seas and in upwelling regions fall into a range, which would give rise to a phytoplankton community consisting of at least 50% diatoms (Si:N = 0.7 to 1.5:1; Barber & Smith, 1981).

The eutrophication related changes in phytoplankton biomass composition apparent in the coastal waters of the North Sea (Cadée, 1986; Radach & Berg,

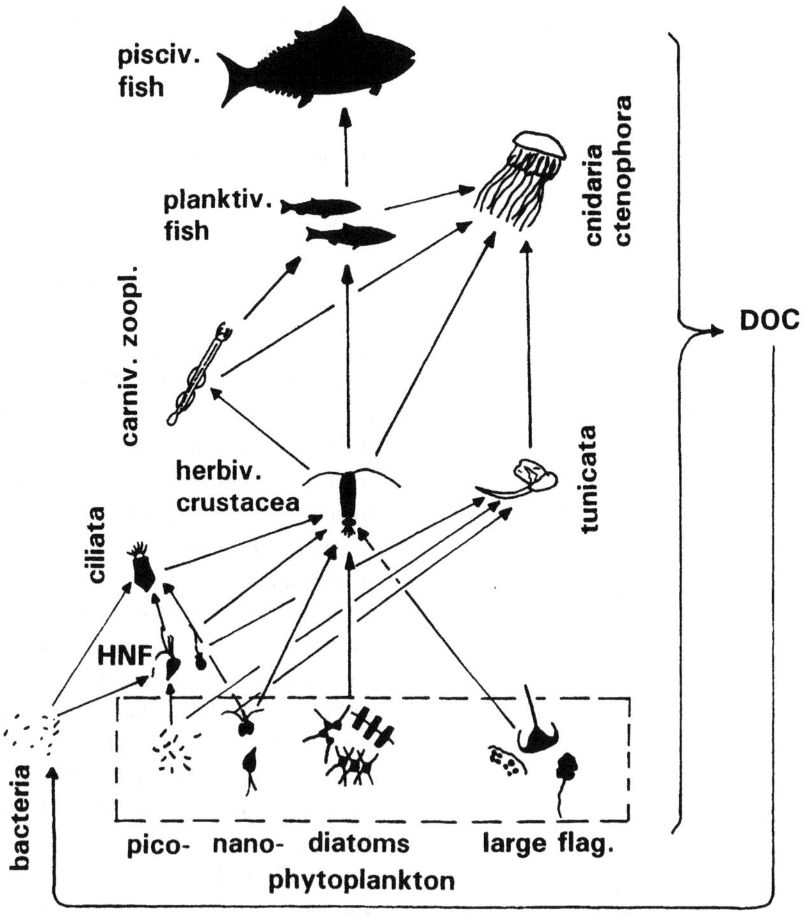

Figure 1. Generalised pelagic food web.

1986; Cadée & Hegeman, 1991) seem to be a good field example for decreasing diatom importance under decreasing Si:N or Si:P ratios. Here, we refer to the Helgoland Straits time series because of its long duration (1962–1984 in Radach & Berg, 1986; but still continuing) and its extraordinary temporal resolution (daily sampling): ratios of the annual mean concentrations of dissolved Si to dissolved inorganic N have declined from 0.66:1 in 1964 to 0.17:1 in 1982; annual mean Si:P declined from 19:1 to 3.4:1 during the same period. At the same time, the relative biomass of diatoms decreased from 71 to 16%. This was partly due to an increase of relative flagellate biomass during the spring bloom. More importantly, more dissolved N and P were left over after the spring bloom had depleted Si and gave rise to summer blooms of the colonial flagellate *Phaeocystis* and other flagellates. More details about the proper interpretation of nutrient ratio – microalgal relationships can be obtained from Sommer (1999).

At the low end of the Si:N gradient, phytoplankton species composition differs considerably, depending on the absolute nutrient levels. In the oligotrophic ocean, phytoplankton is dominated by picoplankton (<2 μm; mainly the prokaryotes *Synechococcus* and *Prochlorococcus*) and the smallest representatives of eukaryotes (usually <5 μm). They frequently make up $>90\%$ of biomass and even higher shares of primary productivity (Stockner & Antia, 1986; Li et al., 1993, Lindell & Post, 1995). Under the appropriate conditions, the large, N_2-fixing cyanobacterium *Trichodesmium* might also be important. Smallness is usually seen as an adaptation to take up extremely low nutrient concentrations because of a favourable surface:volume ratio (Raven, 1986). A recent experimental study with phytoplankton of the ultraoligotrophic northern Red Sea has supported the hypothesis, that low nutrient concentrations exclude larger algae (Sommer, 2000).

Eutrophic waters with low Si:N during summer stratification ratios are characterised by large unicel-

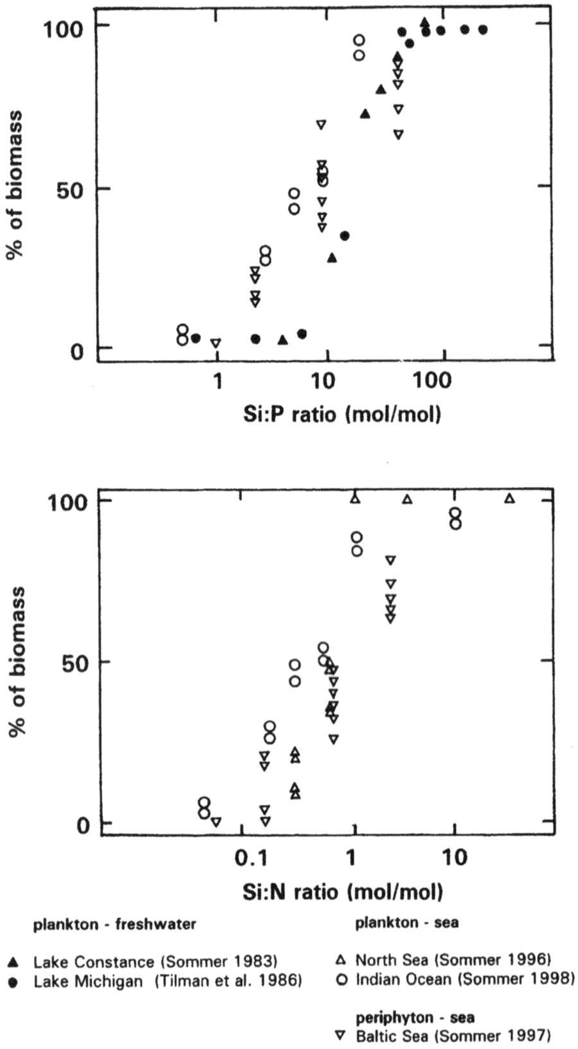

Figure 2. Taxonomic outcome of competition experiments in continuous culture: Relative biomass of diatoms in response to stoichiometric Si:P and Si:N ratios of nutrient supply.

lular flagellates (mainly dinoflagellates), the large (>1 mm) colonial flagellate *Phaeocystis*, or toxic nanoflagellates (e.g. *Chrysochromulina polylepis*). Many of the large flagellates are also potentially toxic (Granéli et al., 1989; Smayda, 1990). Large size is selected for by steep vertical gradients of nutrients and light (migrating flagellates), strong temporal variability in nutrient concentrations (Turpin & Harrison, 1980; Suttle et al., 1987) and size escapes from herbivory. The latter is particularly true for large *Phaeocystis* colonies and for the colonial cyanobacteria in the Baltic Sea (*Aphanizomenon, Nodularia*). Mass developments of such algae can lead to red tides (dinoflagellates), foams (*Phaeocystis*), or surface scums (Cyanobacteria) and are considered one of the most undesirable consequences of eutrophication.

Mesozooplankton feeding selectivity

Feeding experiments with radioactively labelled algae of defined size show that copepods do not feed on the smallest phytoplankton (Fig. 3). The lower size limit is particularly large for *Acartia* from the western Mediterranean Sea (Katechakis, 1999) and for *Calanus* from the Red Sea (H. Stibor; unpubl. data), but also *Paracalanus* and *Pseudocalanus* from the Baltic Sea show strongly reduced clearance rates in the 5–10 μm range (Sommer et al., 2000). The lower nanoplankton size range is covered by the cladoceran *Penilia*, however, this zooplankter only exists in coastal waters south of the British Channel. More important feeders of small phytoplankton are the pelagic tunicates, which are able to feed on particles as small as 0.5 μm (Silver & Bruland, 1981). The larger ones (salps and doliolids) have broad food size ranges including microplankton (20–200 μm) or even larger particles, while the smaller ones (Appendicularia) are mainly pico- and nanoplankton feeders. Seasonally, bivalve veliger larvae might be important as feeders in the smallest phytoplankton size ranges. A calculation for the Baltic Sea shows (Fig. 4) that copepods including nauplii are the dominant mesozooplankton feeders on seston >10 μm, while picoplankton feeding by mesozooplankton is mainly due to the Appendicularia and veligers.

Copepod food selection does not only depend on food particle size. It has been shown that copepods can strongly select between similarly sized food particles because of chemical properties ('taste'; DeMott, 1988), an ability which is lacking in true filter feeders such as tunicates. This means that copepods are able to avoid toxic or otherwise chemically unfavourable algae, even if they have the appropriate size.

Recently, doubts have come up as to whether diatoms are really a suitable food for 'herbivorous' copepods. It has been shown in laboratory experiments that the hatching success of copepod eggs is reduced when copepods are reared on a diatom diet (Ianora et al., 1995). It is not yet clear whether this apparent contradiction to good copepod growth in diatom rich seas results from experimental artefacts, from nutritional inadequacy of diatoms (Jonasdottir et al., 1998) or from adverse chemical properties of

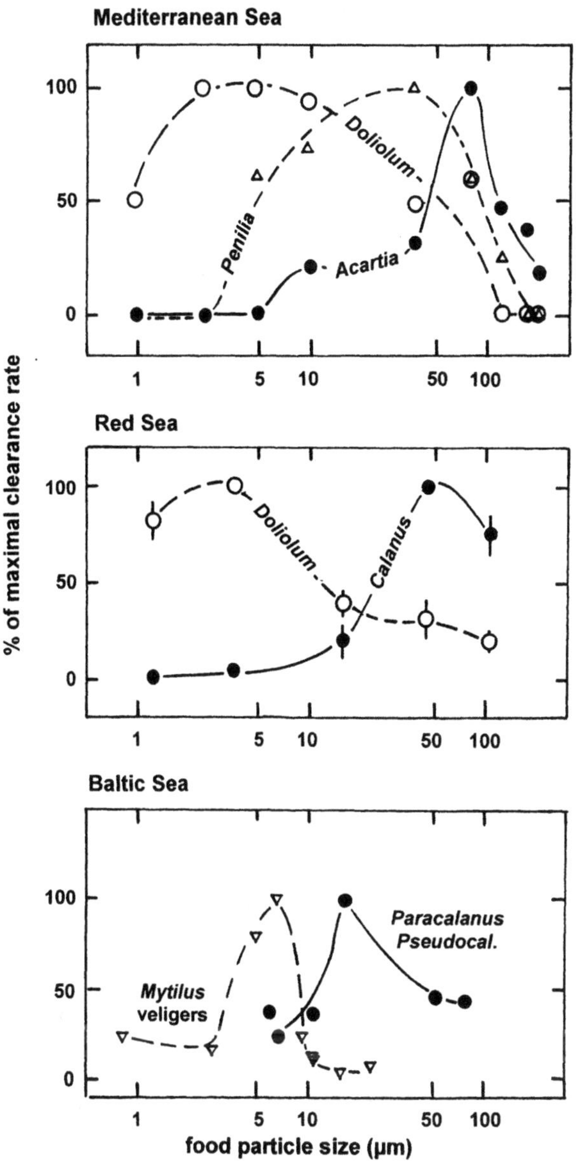

Figure 3. Food size spectra of mesozooplankton: Clearance rate in% of the optimum.

diatoms (Ianora et al., 1999). Nutritional inadequacy might result from a shortage of the essential fatty acid C22:6ω3, which is abundant in flagellate and protozoan biomass. Diatoms, on the other hand, are usually rich in C20:5ω3. The rotifer *Brachionus plicatilis* has shown the best conversion of food production into secondary production at a roughly 50:50 mixture of diatoms and flagellates (Sommer, 1998), which provided a balanced mixture of both fatty acids. The feeding behaviour of presumably herbivorous copepods is flexible enough to ingest an adequate mixture of diatoms and flagellates or protozoa, even if plankton composition is one-sided (Kiørboe et al., 1996; Nejstgaard, 1997). Thus, it seems more adequate to consider them omnivorous (Kleppel, 1993).

Configuration of the lower food web

The nutritional flexibility of copepods and the low abundance of suitable phytoplankton food under certain nutrient conditions makes their trophic level assignment flexible. If we assume that planktivorous fish do not significantly feed on protozoa and tunicates, the trophic level of fish will depend on the trophic level of the copepods. Short-cuts of the food chain by other crustaceans (mainly Cladocera and krill) are possible local phenomena, but will be neglected in the following study. The different scenarios will be described in an exaggerated way, which emphasises the contrasts (Fig. 5).

Scenario 1 is characterised by high supply of all nutrients, including silicon. This is typical for nutrient supply from deep water, i.e. for upwelling on continental margins and for seasonal mixing in temperate and boreal seas. The nutrient regime favours a phytoplankton composition with an important contribution of diatoms. Together with nanoplanktonic phytoflagellates they form a good food basis for copepods, which occupy trophic level 2. As a consequence, planktivorous fish will occupy trophic level 3, and primary predatory fish (e.g. cod, tuna) will occupy level 4. In fact, the trophic levels will be slightly higher, because copepods feed on protozoa as well.

Scenario 2 is characterised by low supply of all nutrients. This results from stable stratification, the absence of land-based nutrient sources and a predominance of recycling by animal excretion as a nutrient source. Such conditions occur under permanent stratification in the subtropical gyres and after prolonged, stable summer stratification in open, temperate oceans. High nutrient-low chlorophyll areas where iron limits production (Coale et al., 1996) might behave similarly. In the oligotrophic ocean, phytoplankton is dominated by algae <5 μm, which is not suitable as a food source for copepods. Copepods will feed on large protozoa (mainly ciliates) and small protozoa (heterotrophic nanoflagellates; HNF). The trophic level of copepods will be between 3 (phytoplankton – protozoa – copepods) or 4 (phytoplankton – HNF – ciliates – copepods). Consequently, planktivorous fish will occupy trophic level 4 to 5. In addition to the energy losses at the trophic links in the microbial

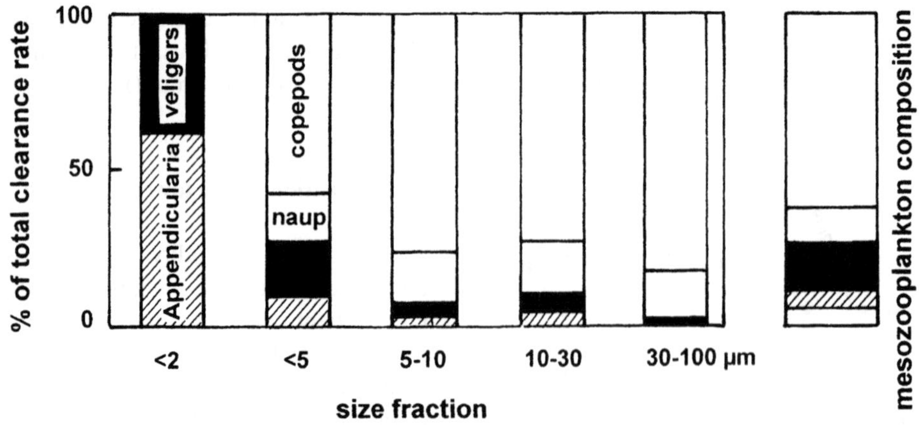

Figure 4. Contribution of different mesozooplankton groups to total mesozooplankton grazing in the Baltic Sea.

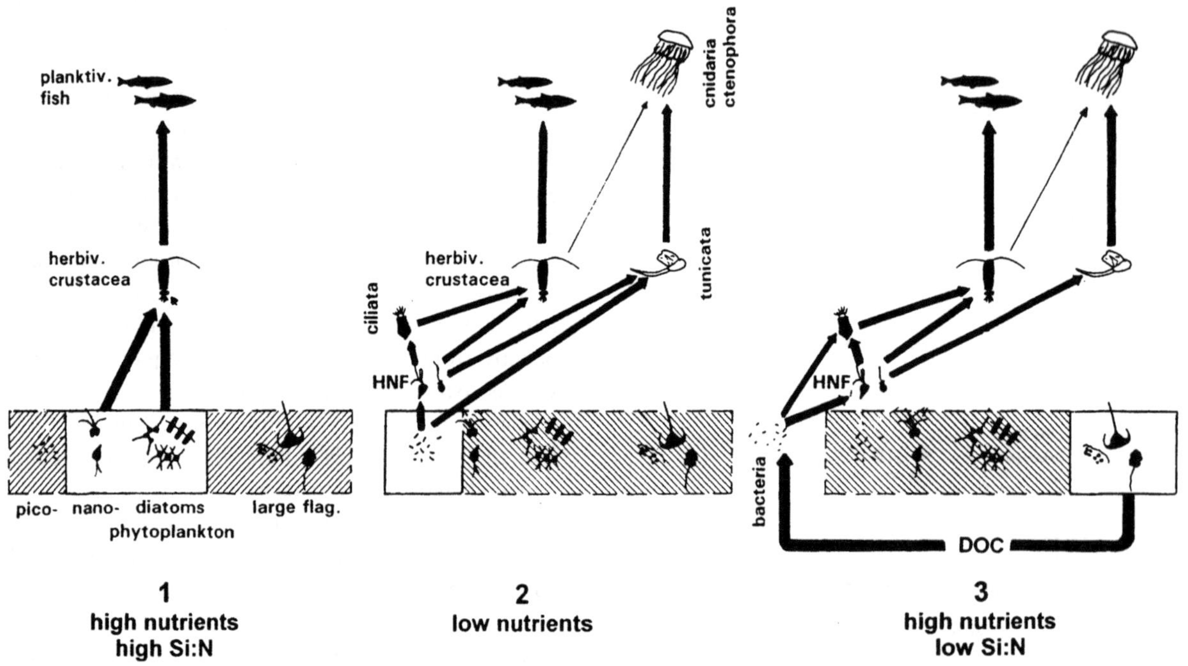

Figure 5. Main pathways of the energy flow to the primary carnivore trophicv levels unnder different nutrient scenarios; food web topography as in Figure 1.

food web, further energy will be lost by the feeding of tunicates on picoplankton and small nanoplankton, thus fuelling the jelly food chain at the expense of the muscle food chain.

Scenario 3 is characterised by high supply of N and P, but low supply of Si. This situation is characteristic for eutrophicated coastal seas during summer stratification, where nutrient supply to the euphotic zone is dominated by land-based sources. Cultural eutrophication has increased N and P emissions into the environment, while Si loading to the sea has not been affected or even been reduced by increased sedimentation (Billen & Garnier, 1997). Phytoplankton will be composed of large nuisance algae, which are not directly grazed. Primary production enters the energy flux via excretion and cell lysis (detritus food chain). DOC will be taken up by bacteria, which are fed upon by protozoa and tunicates. Similar to scenario 2, there is a strong microbial food web, but it is fuelled heterotrophically instead of autotrophically. Because of the additional bacterial link trophic levels of copepods and planktivorous fish are even higher: 4–5 for the copepods, 5–6 for the planktivorous fish.

The jellyfish–fish interaction

There is a growing concern that in many seas carnivorous, gelatinous megazooplankton (Cnidaria and Ctenophora) expand at the expense of fish in many seas (Legovic, 1987). Because of a shortage of reliable data, it is not yet clear whether this reflects only an increasing awareness or a real problem, except for the notorious invasion of the ctenophore *Mnemiopsis leidyi* in the Black Sea (Shushkina & Vinogradov, 1991). Its mass growth began in 1988 and led to massive declines in food mesozoopankton species and its planktonic competitors *Aurelia aurita* (Scyphozoa) and *Sagitta* (Chaetognatha) (Fig. 6). It is unclear whether the concomitant decline in fish stocks has facilitated the expansion of *Mnemiopsis* or whether *Mnemiopsis* has adversely affected fish. Neither is it clear whether the invasion success had been facilitated by some kind of environmental degradation, e.g. eutrophication.

In spite of the insufficient state of the art, there are several features in the in the jellyfish–fish interaction which give rise to the concern that the jellyfish-state might be an alternative stable state which is difficult to revert. Jellyfish act as competitors of juvenile and adult planktivorous fish and as predators on fish larvae (Fig. 7). High stocks of jellyfish might thus suppress fish recruitment and thereby release themselves from competition for the zooplankton food. Once such a state has been established, it might resist change and prevent fish stocks from recovery even if fishing pressure was reduced. It seems still less obvious but it is highly important from a managerial perspective, which mechanism would stabilise the fish state against shifts towards the jellyfish state.

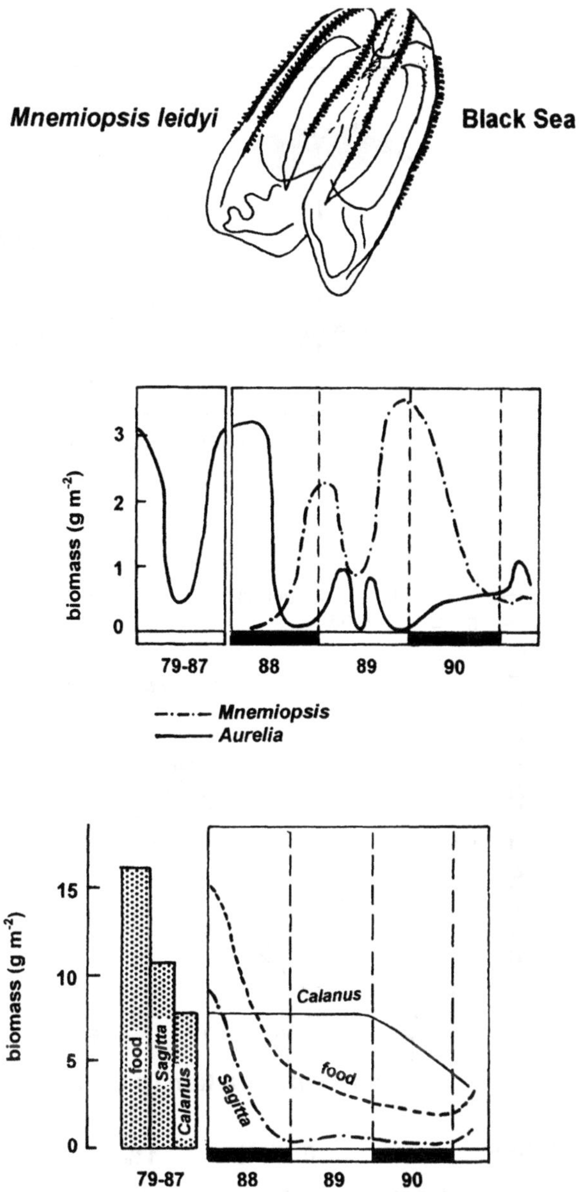

Figure 6. Impact of *Mnemiopsis leidyi* on the zooplankton community in the Black Sea: Middle: biomass of *Mnemiopsis* and its competitor *Aurelia*; bottom: biomass of food zooplankton, the competitor *Sagitta* and the unaffected copepod *Calanus*. Data source: Shushkina & Vinogradov, 1991.

Discussion

The food webs scenarios in Figure 4 overemphasise contrasts. In fact, picoplankton production, the microbial loop and protozoan feeding by copepods are by no means negligible under high Si – high N conditions. Fish feeding on tunicates might be relatively unimportant, but it is not absent and other crustaceans than copepods (cladocerans, krill) might short-cut the food chain when phytoplankton is unavailable for copepods. Therefore, it is no surprise that the nekton production: primary production relationship by Iverson (1990) implies a smaller trophic level difference of planktivorous fish than Figure 4 between the different scenarios. The low productivity end of Iverson's regression is compatible with a 15% ecological efficiency and fish at the 4th trophic level, while the high productivity end is compatible with a 10% ecological

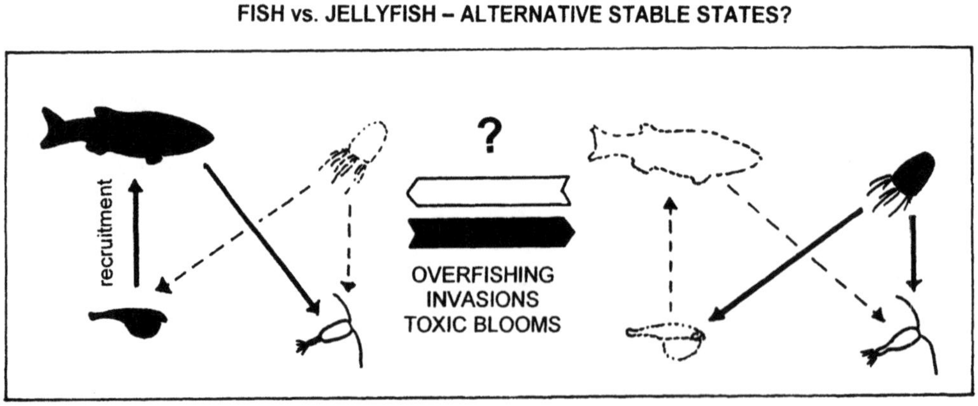

Figure 7. Interactions between fish and jellyfish; left: fish dominated state; right: jellyfish dominated state.

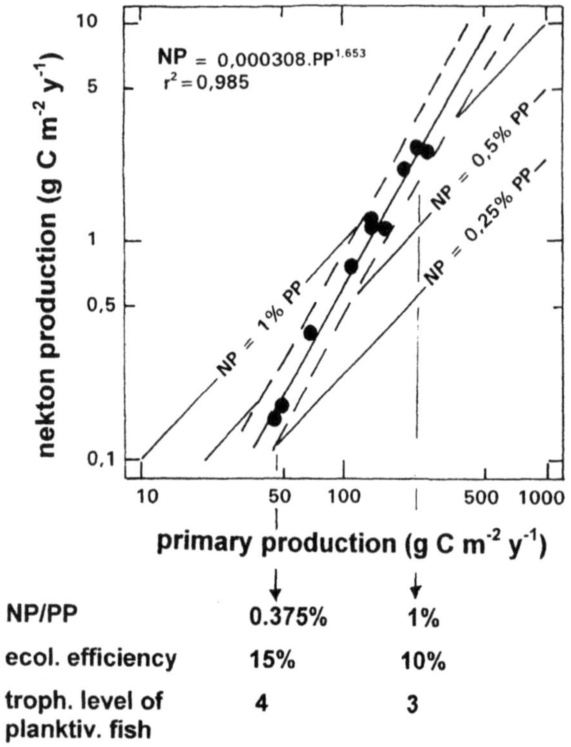

Figure 8. Relationship between nekton (fish and squid) production and primary production. Data source: Iverson, 1990.

efficiency and fish at the 3rd trophic level (Fig. 8). Other combinations with higher ecological efficiencies and longer food chains are similarly consistent with the data.

Most of the reasoning is still speculative, but a number of testable hypotheses and research questions can be derived from my food web consideration. Many of the most urgent questions can be studied along eutrophic-oligotrophic gradients, provided that the comparison includes both gradients starting from upwelling zones and gradients starting from eutrophicated river plumes. The questions are:

- Does the trophic level of copepods change along nutrient gradients?
- Does the protozoa:metazoa ratio in zooplankton change along nutrient gradients?
- Does the copepod:tunicate ratio change along nutrient gradients?
- To which extent are tunicates consumed by fish? How does their nutritional value compare with copepods?
- Does a shift from copepods to tunicates tilt the competitive balance between fish and jellyfish?

It should be kept in mind that the considerations here are based on a bottom-up transmission of effects in marine pelagic food webs. In lakes, strong top-down effects in the pelagic are well known (Carpenter et al., 1985), while it is still not clear whether such trophic cascades can be observed in the marine pelagic. A further obstacle for using our food web scenarios as a predictive tool for managerial purposes lies in the poor knowledge about the jelly food chain. While we know quite a lot about the phytoplankton, microbial loop and the copepods, increased research efforts are needed for tunicates, cnidaria and ctenophores.

Acknowledgement

This article is a contribution to the MAST III-project COMWEB funded by the EU.

References

Azam, F., T. Fenchel, J. G. Field, L. A. Meier-Reil & F. Thingstad, 1983. The ecological role of water column microbes in the sea. Mar. Ecol. Progr. Ser. 10: 257–263.

Barber, R. T. & R. L. Smith, 1981. Coastal upwelling ecosystems. In Longhurst, A. R. (ed.), Analysis of Marine Ecosystems. Academic Press, London: 33–68.

Billen, G. & J. Garnier, 1997. The Phison river plume: coastal eutrophication in response to changes in land use and water management in the watershed. Aquat. Microb. Ecol. 13: 3–17.

Cadée, G. C., 1986. Recurrent and changing seasonal patterns of phytoplankton in the westernmost inlet of the Wadden Sea, the Marsdiep, since 1973. In Lancelot, C., G. Billen & H. Bath (eds), Water Pollution Research Report 12. Commission of the European Community, Luxembourg: 105–112.

Cadée, G. C. & J. Hegeman, 1991. Historical phytoplankton data from the Marsdiep. Hydrobiol. Bull. 24: 111–119.

Carpenter, S. R., J. F. Kitchell & D. R. Hodgson, 1985. Cascading trophic interactions and lake productivity. BioScience 35: 634–639.

Coale, K. H., K. S. Johnson, S. E. Fitzwater, R. M. Gordon, S. Tanner, F. P. Chavez, L. Ferioli, C. Sakamoto, P. Rogers, F. Millero, P. Steinberg, P. Nightingale, D. Cooper, W. P. Cochlan, M. R. Landry, J. Constantinou, G. Rollwagen, A. Trasvina & R. Cudela, 1996. A massive phytoplankton bloom induced by an ecosystem-scale iron fertilisation experiment in the equatorial Pacific Ocean. Nature 383: 495–501.

Cushing, J. D. H., 1971. Upwelling and the productivity of fish. Adv. Mar. Biol. 9: 255–334.

DeMott, W. R., 1988. Discrimination between algae and artificial particles by freshwater and marine copepods. Limnol. Oceanogr. 33: 397–408.

Egge, J. K. & A. Jacobsen, 1997. Influence of silicate on particulate carbon production in phytoplankton. Mar. Ecol. Progr. Ser. 147: 219–230.

Escaravage, V., T. C. Prins, A. C. Smaal, J. C. H. Peeters, 1996. The response of phytoplankton communities to phosphorous input reductions in mesocosm experiments. J. Exp. Mar. Biol. Ecol. 198: 55–79.

Granéli, E., P. Carlsson, P. Olsson, B. Sundström, W. Granéli & O. Lindahl, 1989. From anoxia to fish poisoning: the last ten years of phytoplankton blooms in Swedish marine waters. In Cosper, E. M., V. M. Bricelj & E. J. Carpenter (eds), Novel Phytoplankton Blooms. Springer, New York: 407–427.

Ianora, A., A. Miralto & S. A. Poulet, 1999. Are diatoms good or toxic for copepods? Reply to comment by Jonasdottir et al. Mar. Ecol. Progr. Ser. 177: 305–308.

Ianora, A., S. A. Poulet & A. Miralto, 1995. A comparative study of the inhibitory effect of diatoms on the reproductive biology of the copepod *Temora stylifera*. Mar. Biol. 121: 533–539.

Iverson, R. L., 1990. Control of marine fish production. Limnol. Oceanogr. 35: 1593–1604.

James, A. G. & X. Chiappa-Carrara, 1990. A comparison of field studies on the trophic ecology of *Engraulis capensis* and *E. mordax*. In Barnes, M. & R. N. Gibson (eds), Trophic Relationships in the Marine Environment. Aberdeen Univ. Press: 208–221.

Jonasdottir, S. H., T. Kiørboe, K. W. Tang, M. St. John, A. W. Visser, E. Saiz & H. G. Dam, 1998. Role of diatoms in copepod production: good, harmless or toxic. Mar. Ecol. Progr. Ser. 172: 305–308.

Katechakis, A., 1999. Nischenüberlappung zwischen herbivorem gelatinösen und Crustaceen-Zooplankton im NW-Mittelmeer (Catalanisches Meer). Diploma thesis, Univ. Kiel.

Kiørboe, T., E. Saiz & M. Viitasalo, 1996. Prey switching behaviour of the planktonic copepod *Acartia tonsa*. Mar. Ecol. Progr. Ser. 143: 65–75.

Kleppel, G. S., 1993. On the diet of calanoid copepods. Mar. Ecol. Progr. Ser. 99: 183–195.

Legovic, T., 1987. A recent increase in jellyfish-populations: A predator-prey model and its implications. Ecol. Modell. 38: 243–256.

Li, W. K. W., T. Zohary, Y. Z. Jacobi & A. M. Wood, 1993. Ultraphytoplankton in the eastern Mediterranean Sea: towards deriving phytoplankton biomass from flow cytometric measures of abundance , fluorescence and light scatter. Mar. Ecol. Progr. Ser. 102: 79–87.

Lindell, D. & A. F. Post, 1995. Ultraphytoplankton succession is triggered by deep winter mixing in the Gulf of Aqaba (Eilat), Red Sea. Limnol. Oceanogr. 40: 1130–1141.

Nejstgaard, J. C., I. Gismervik & P. T. Solberg, 1997. Feeding and reproduction by *Calanus finnmarchicus*, and microzooplankton grazing during mesocosm blooms of diatoms and the coccolithophore *Emiliana huxleyi*. Mar. Ecol. Progr. Ser. 147: 197–217.

Pomeroy, L. R., 1974. The ocean foodweb, a changing paradigm. BioScience 24: 499–504.

Radach, G. & J. Berg, 1986. Trends in den Konzentrationen der Nährstoffe und des Phytoplanktons in der Deutschen Bucht. Ber. Biol. Anst. Helgoland 2: 1–165.

Raven, J. A., 1986. Physiological consequences of extremely small size for autotrophic organisms in the sea. Can. Bull. Fish. aquat. Sci. 214: 1–70.

Ryther, J. H., 1969. Photosynthesis and fish production in the Sea. Science 166: 72–78.

Schöllhorn, E. & E. Granéli, 1996. Influence of different nitrogen to silica ratios and artificial mixing on the structure of a summer phytoplankton community from the Swedish west coast (Gullmar Fjord). J. Sea Res. 35: 159–167.

Shushkina, E. A. & M. Y. Vinogradov, 1991. Long-term changes in the biomass of plankton in open areas of the Black Sea. Oceanology 31: 716–721.

Sieburth, J. M. & P. G. Davis, 1982. The role of heterotrophic nanoplankton in the grazing and nurturing of planktonic bacteria in the Sargasso and Caribbean Sea. Annls. Inst. oceanogr. Paris 58: 285–296.

Smayda, T. J., 1990. Novel and nuisance blooms in the sea: evidence for a global epidemic. In Granéli, E., B. Sundström, L. Edler & D. M. Anderson (eds), Toxic Marine Phytoplankton. Elsevier, Amsterdam: 29–41.

Sommer, F., H. Stibor, U. Sommer & B. Velimirov, 2000. Grazing by mesozooplankton from Kiel Bight, Baltic Sea, on different sized algae and natural seston size fractions. Mar. Ecol. Progr. Ser. 199: 43–53.

Sommer, U., 1983. Nutrient competition between phytoplankton species in multispecies chemostat experiments. Arch. Hydrobiol. 96: 399–416.

Sommer, U., 1994. The impact of light intensity and daylength on silicate and nitrate competition among marine phytoplankton. Limnol. Oceanogr. 39: 1680–1688.

Sommer, U., 1996a. Nutrient competition experiments with periphyton from the Baltic Sea. Mar. Ecol. Progr. Ser. 140: 161–167.

Sommer, U., 1996b. Plankton ecology: two decades of progress. Naturwiss. 38: 293–301.

Sommer, U., 1998. From algal competition to animal production: enhanced ecological efficiency of *Brachionus* with a mixed diet. Limnol. Oceanogr. 43: 1393–1396.

Sommer, U., 1999. A comment on the proper use of nutrient ratios in microalgal ecology. Arch. Hydrobiol. 146: 55–64.

Sommer, U., 2000. Scarcity of medium-sized phytoplankton in the northern Red Sea explained by strong bottom-up and weak top-down control. Mar. Ecol. Progr. Ser. 197: 19–25.

Stockner, J. G. & N. J. Antia, 1986. Algal picoplankton from marine and freshwater ecosystems: a multidisciplinary perspective. Can. J. Fish. aquat. Sci. 43: 2472–2503.

Suttle, C. A., J. G. Stockner & P. J. Harrison, 1987. Effects of nutrient pulses on community structure and cell size of a freshwater phytoplankton assemblage in culture. Can. J. Fish. Aquat. Sci. 44: 1768–1774.

Tilman, D., 1982. Resource Competition and Community Structure. Princeton Univ. Press.

Tilman, D., R. Kiesling, R. W. Sterner, S. S. Kilham & F. A. Johnson, 1986. Green, blue-green and diatom algae: taxonomic differences in competitive ability for phosphorous, silicon and nitrogen. Arch. Hydrobiol. 106: 473–485.

Turpin, D. H. & P. J. Harrison, 1980. Cell size manipulation in natural, marine , planktonic diatom communities. Can. J. Fish. aquat. Sci. 37: 1193–1195.

Verity, P. G. & V. Smetacek, 1996. Organism life cycles, predation, and the structure of marine pelagic systems. Mar. Ecol. Progr. Ser. 130: 277–293.

Hydrobiologia **484:** 21–31, 2002.
O. Vadstein & Y. Olsen (eds), Sustainable Increase of Marine Harvesting: Fundamental Mechanisms and New Concepts.

The importance of silicon for marine production*

Svein Kristiansen[1] & Espen Edward Hoell[2,3]
[1] *University of Tromsø, NFH, Breivika, N-9037 Tromsø, Norway*
Tel.: +47-77646000. Fax: +47-77646020. E-mail: sveinkr@nfh.uit.no
[2] *Hydro Research Centre Porsgrunn, P.O. Box 2560, N-3907 Porsgrunn, Norway*
[3] *Present address: SustainTech AS, Teknologisenteret, N-3918 Porsgrunn, Norway*
Tel.: +47-35574031. Fax: +47-35574010. E-mail: espen.hoell@sustaintech.no

Key words: Silicate, diatom, review, eutrophication, nutrient ratios

Abstract

A review of silicon, with emphasis on its forms, uptake, dissolution and role in marine primary production, is given. The importance of silicon in marine food webs is discussed, as well as the concentrations of silicon in various areas and the importance of changing N:Si:P ratios. The methodology for measuring silicate transformations has recently been improved by the introduction of the highly enriched ^{32}Si isotope. Results from uptake experiments using ^{32}Si in nutrient rich coastal water and in open ocean are presented. The uptake kinetic experiments showed that the silicate uptake usually is unsaturated. We propose that closer attention in the future should be paid to the importance of balanced nutrient composition as well as nutrient supply dynamics for the development of eutrophication versus efficient trophic transfer and fish production in nutrient enriched systems. Close attention should also be paid to the mechanisms that reduce the inputs of silicate to coastal waters.

Introduction

Diatoms have an absolute requirement for silicon (Lewin, 1962). Although other algal groups may require silicon (e.g. some chrysophytes), diatoms are by far the most important silicifying algal group in marine waters. They are the dominating primary producers in the high-production areas of the ocean, and silcon (Si) is therefore a key factor in the transfer of energy to higher tropic levels. Several reviews on diatoms and Si have been published (Lewin, 1962; Round et al., 1990). Here we will focus on the roles of Si in marine production.

Forms of silicon

The biological Si cycle in the marine environment is relatively simple (Dugdale et al., 1995). Diatoms take up silicate, which they mainly incorporate as biogenic silica (BSi) in their frustules. Siliceous particles (diatom frustules included) dissolve again to silicate (Werner, 1977). However, the frustules do not dissolve completely in the water column and diatom ooze does accumulate at the sea floor, especially in productive areas (Seibold & Berger, 1982). A recent budget indicates a silicon balance in the world oceans (Tréguer et al., 1995). The major inputs come from rivers, and these inputs are reasonably balanced by the net outputs of biogenic silica into the marine sediments (Conley, 1997). Thus, reduced inputs from rivers may reduce the primary production or alter the phytoplankton species composition.

Monomeric silicate ($Si(OH)_4$) accounts for nearly all dissolved silicon (97%). The remaining 3% is mainly present in the anionic form $SiO(OH)_3^-$ (Stumm & Morgan, 1996). These fractions of silicic acid are usually referred to as silicate. Particulate silica is summed biogenic particulate silica (BSi) and lithogenic particulate silica (LSi).

The analysis of silicate is simple and it is described in common manuals of seawater analysis. It

* This article is dedicated to Professor E. Paasche, University of Oslo, on his retirement in Autumn 2002.

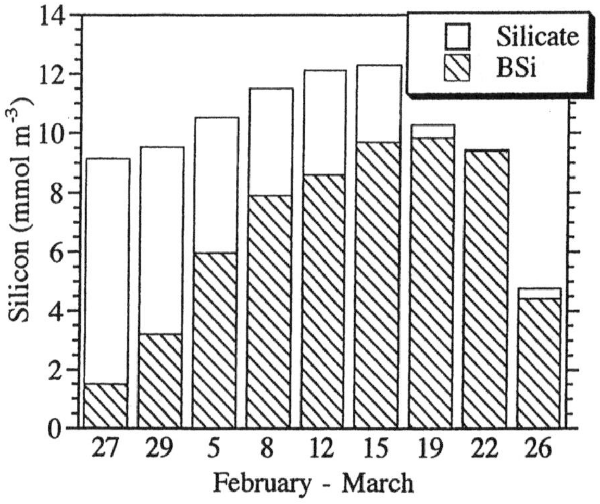

Figure 1. Surface concentrations of biogenic silica (BSi) and of silicate in mmol m^{-3} during the spring bloom in the Oslofjord, Southern Norway, in 1996.

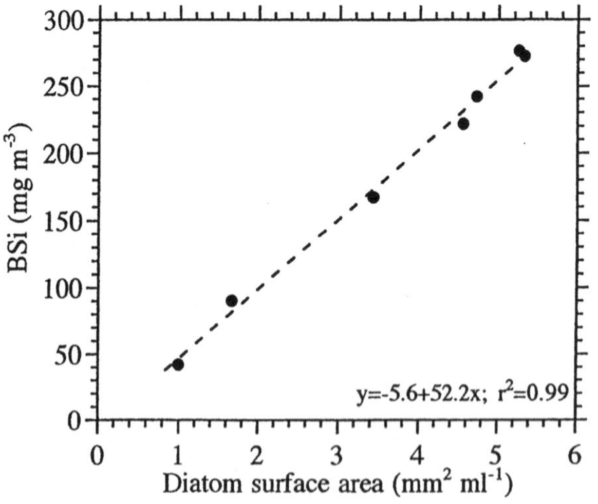

Figure 2. Concentration of biogenic silica (BSi in mg m^{-3}) *versus* diatom surface area (mm^2 ml^{-1}) during the spring bloom. The same samples as in Figure 1.

is a calorimetric method based on the production of a silicomolybdate complex. However, it should be kept in mind that silicate may polymerise and that not all polymers react in the silicomolybdate analysis. The term 'reactive silicate' is often used as the amount of silicate available for phytoplankton growth (Strickland & Parsons, 1972). The analyses of BSi and LSi are more complicated and several methods are available (Ragueneau & Tréguer, 1994). It may also be difficult to separate between the two forms, especially in samples from waters with high lithogenic content. Some of the BSi methods are not completely selective, and BSi may be overestimated because of leaching from LSiduring the analytical procedure. The latter may be a serious problem in samples from shallow areas and areas close to rivers that may have high LSi concentrations (Ragueneau & Tréguer, *op. cit.*). In some areas it is possible to measure BSi by soda hydrolysis of cells collected on proper filters followed by the traditional silicate analysis given above (Paasche, 1980b). BSi has been successfully measured using this method both in Norwegian coastal waters (Paasche & Østergren, 1980; Fig. 1) in open ocean (Kristiansen unpublished values). LSi can be measured on the same filters or on parallel filters (as the difference (particulate Si – BSi) using HF digestion followed by the traditional silicate analysis (Brzezinski & Nelson, 1995).

The concentration of silicate in natural seawater varies by more than 3 orders of magnitude within the euphotic zone of the world's oceans. The highest concentrations (110 mmol m^{-3}) are found in the northwest North Pacific and in the Southern Ocean whilst the concentration generally is lower in parts of the Atlantic Ocean, the Indian Ocean and the Pacific Ocean (Levitus et al., 1993). Since silicate is an essential nutrient for diatoms, it may be depleted during diatom blooms in coastal water (Fig. 1) and in the open ocean (Brzezinski & Nelson, 1989). Concentration of BSi is directly related to the diatom biomass, and Van Bennekom et al. (1974) found 40–110 mmol m^{-3} of BSi during a diatom bloom in the nutrient-rich Dutch Wadden Sea. Working with diatom cultures, Paasche (1980b) found that Si per unit cell surface area is a convenient measure of diatom silicification. This relationship can be used to sort out the dominating diatoms in natural samples (Fig. 2). The cell numbers and BSi in Figure 3 are from a spring bloom totally dominated by diatoms. The diatoms *Chaetoceros socialis* and *Thalassiosira nordenskioeldii* accounted for up to 40–50% of the total cell number. However, *T. nordenskioeldii* was the most important diatom and it accounted for up to 72% of the BSi. The smaller cell, *C. socialis*, accounted for < 11% of the BSi.

Uptake of Si

Marine diatoms mainly use silicate as a Si-source for growth, the anionic forms ($Si(OH)_3^-$) are of minor importance (Del Amo & Brzezinski, 1999). Until recently the ^{30}Si stable isotope tracer method (Nelson & Goering, 1977) was the only feasible way of measuring silicate uptake by natural diatom populations

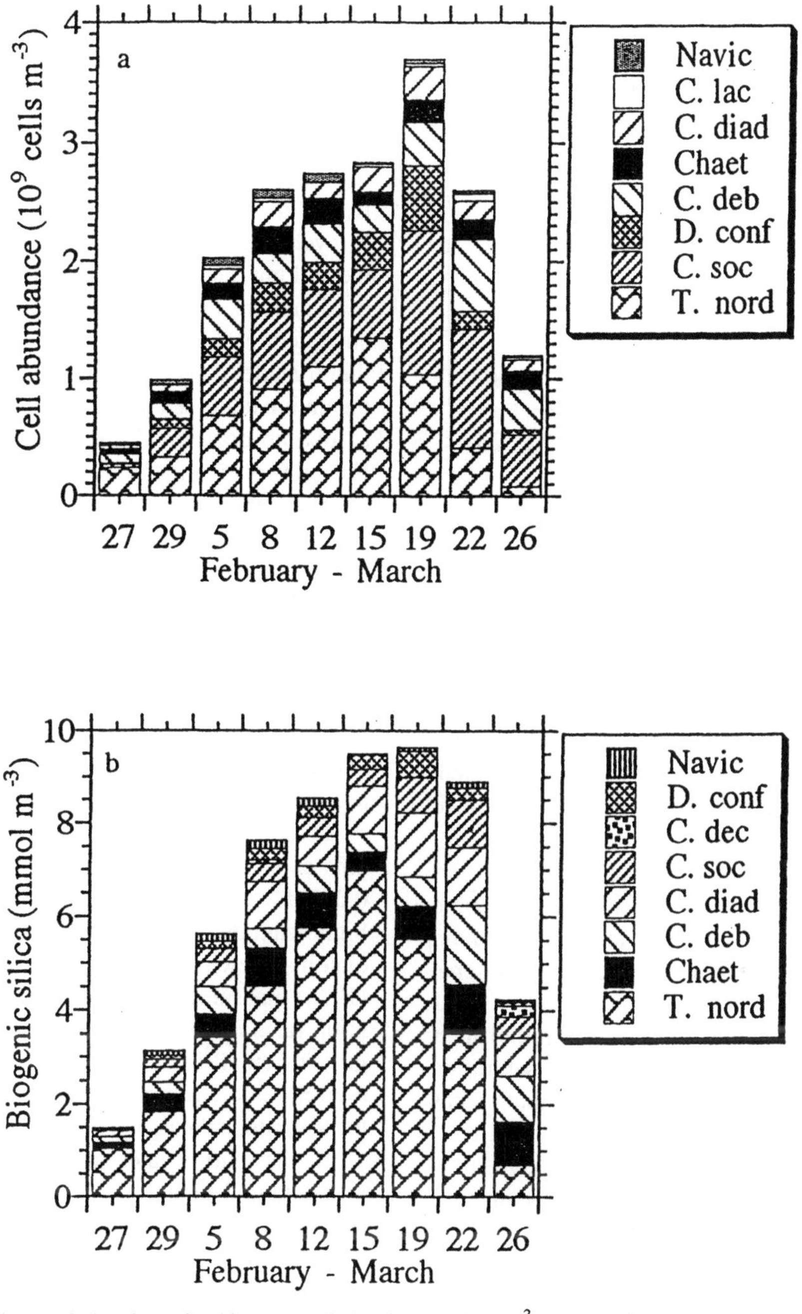

Figure 3. Dominating diatoms during the spring bloom as cell abundance (cells m^{-3}) in (a) and as biogenic silica (mmol m^{-3}) in (b). The same samples as in Figure. 1.

(Werner, 1977; Brzezinski & Phillips, 1997). The method is labour intensive and the samples have to be analysed by mass spectrometry, but it has been used successfully both in open ocean and in coastal systems (e.g. Brzezinski & Nelson, 1996; Nelson & Dortch, 1996). Tréguer et al. (1991) used the radioisotope ^{32}Si with low specific activity to measure silicate uptake rates in the silicate-rich Southern Ocean. ^{32}Si with high specific activity is now available (Brzezinski & Phillips, 1997), and it can be used to measure silicate uptake rates in most open ocean and coastal systems without perturbing the ambient silicate concentration (Fig. 4; Brzezinski et al., 1997). The highest uptake rate in Fig. 4 (82 μmol m^{-3} h^{-1} from the Oslof-

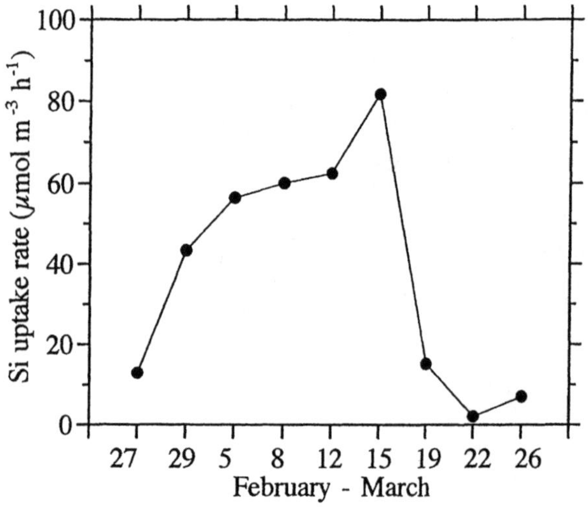

Figure 4. Silicate uptake rate (μmol m^{-3} h^{-1}) during the spring bloom. The same samples as in Figure 1.

jord) corresponds to a diatom growth rate of 0.2 d^{-1} which is a low though reasonable rate at the *in situ* temperature in the Oslofjord in March (ca. −1 °C).

Probably the most striking quality of silicate uptake is that it is unsaturated in most areas investigated. The corresponding half-saturation constant for uptake is high, usually 1–2 mmol m^{-3} or higher (Fig. 5), (Kristiansen et al., 2000). However, culture experiments have shown that the cellular BSi is flexible and that diatoms are able to grow at near maximum growth rates at low silicate concentrations (Paasche, 1980a). For example, half-saturation constant for silicate uptake in *T. nordenskioeldii* was 2.8 mmol m^{-3} in Figure 5, while the corresponding half-saturation constant for silicate-limited growth for this diatom has been found to be in the range 0.05–0.13 mmol m^{-3} (Paasche, 1975). In culture experiments, diatoms may show a decrease in cellular BSi by a factor up to 3 when grown under severe Si-limitation and by a factor of up to 5 because of cell size reduction during the cell cycle (Paasche, 1980a). In nature, silicate may be depleted to undetectable concentrations after diatom blooms, and the diatoms may be weakly silicified. Estimating diatom growth rates from silicate uptake rates must therefore be done with caution.

An interesting approach is using the ^{32}Si isotope in autoradiographic investigations. Shipe & Brzezinski (1999) used this technique to study silicate uptake by *Rhizosolenia* mats in the Pacific Ocean and revealed distinct labelled bands and/or valves. This technique can be used to study species-specific silicate uptake rates.

Dissolution of Si

Recycling of Si occurs by dissolution of silica (Lewin, 1962; Werner, 1977). In this respect, Si is fundamentally different from N and P which both are efficiently regenerated by biological activity (Harrison, 1993). Recycling of Si within the productive zone ('a short Si cycle') therefore depends on the dissolution rate, the sinking rate and the mixing depth. Frustules from dead diatoms dissolve at different speed and to different extent in the water column. Smaller diatoms with thin frustules can be completely dissolved during sedimentation to the sea floor while others will accumulate as diatom ooze on the sea floor (Werner, op. cit.). Frustules of living diatoms are protected from dissolution by an organic coating (Lewin, op. cit). Some diatoms also secrete macromolecules which have been suggested to protect against dissolution (Hoagland et al., 1993). Recently, bacterial activity has been shown to accelerate dissolution of Si by degrading the organic coating (Bidle & Azam, 1999). They found that the observed rates of bacteria-mediated Si dissolution could explain most of the reported upper-ocean Si dissolution and that this may control diatom productivity and the cycling of Si and carbon in the ocean.

Dissolution rate of silica from decomposing diatoms at 8–30 °C varies 2 orders of magnitude (0.2–0.002 d^{-1}; from Tables 2 and 3 in Kamatani, 1982). Significant dissolution also occurs in actively growing diatoms (Lewin, 1962; Paasche, 1973). Nelson et al. (1976) measured silicate uptake rates and Si dissolution rates in culture experiments with exponentially growing small centric diatoms at 20 °C. Their measured Si dissolution rate (0.05–0.2 d^{-1}) was 7–15% of the saturated silicate uptake rate.

Si dissolution rates have only been measured in a few natural plankton communities. Typically, high specific Si dissolution rates occur in areas with dead diatoms and low rates occur in productive waters dominated by active diatoms (Brzezinski & Nelson, 1989; Brzezinski & Nelson, 1995; Nelson et al., 1995). It has also been hypothesised that the specific Si dissolution rate generally exceeds 0.1 d^{-1} in surface waters that are 20 °C or warmer (Brzezinski & Nelson, 1995). Si dissolution rates during the spring bloom in the Oslofjord (Fig. 4) can be estimated from the seawater temperature using the equation for decomposing diatoms given in Kamatini (1982). However, the estimated rates will be potential Si dissolution rates because the equation used was made for decomposing diatoms. The seawater temperature during the bloom was

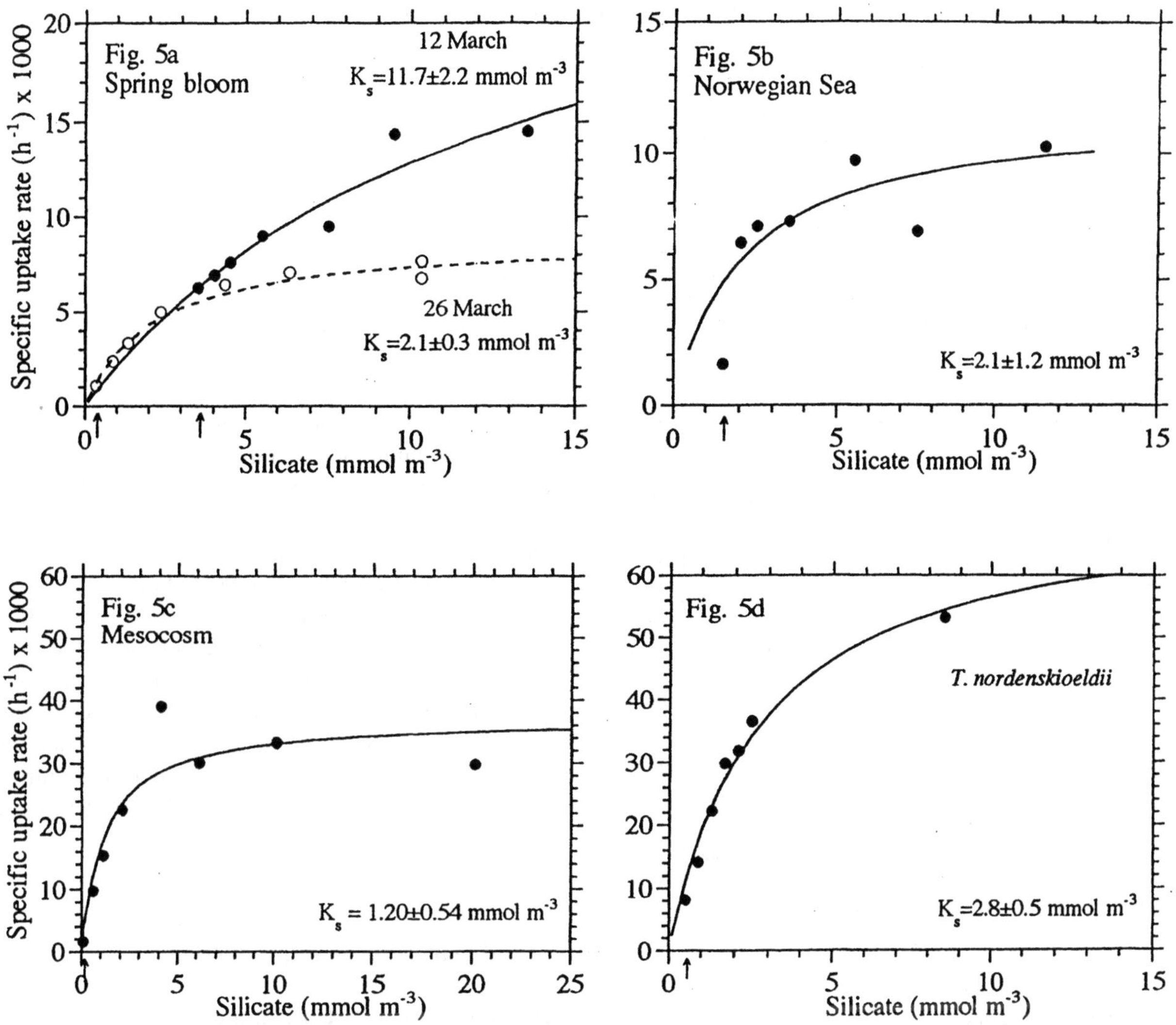

Figure 5. Silicate uptake kinetics. Specific silicate uptake rate (h^{-1}) plotted *versus* silicate concentration (mmol m^{-3}). Arrows indicate the ambient silicate concentrations. The values are from 3 different geographical areas and from a culture experiment. (a), (b) and (c) are from the spring bloom in the Oslofjord in March 1996, from the Norwegian Sea in July 1996 and from mesocosms experiments in Norwegian coastal water in May 1997, respectively. (d) is from a culture experiment with the diatom *Thalassiosira nordenskioeldii* isolated during the 1996 spring bloom in the Oslofjord. The half-saturation constants (K_S) are given ±1 SE. The solid lines were fitted to a Michaelis–Menten hyperbola by a non-linear procedure.

−1 °C which gives potential specific Si dissolution rates of <0.014 d^{-1} (Table 1). Specific Si dissolution rates calculated from the saturated silicate uptake rates were similar (0.007–0.015 d^{-1}) and thereby support the potential Si dissolution rates given in Table 1. The silicate uptake rates were high during the first 2 weeks of the bloom and then decreased abruptly as the bloom collapsed (Fig. 4). The spring bloom has therefore been divided into 2 periods; the first 2 weeks and the last week of the bloom. The average specific silicate uptake rate was high during the first 2 weeks (0.2 d^{-1}) and the potential specific dissolution rate was low (1–2 orders of magnitude lower; Table 1). Both the average specific silicate uptake rate and the potential specific Si dissolution rate were low during the last week of the bloom (<0.02 d^{-1}). Thus, Si dissolution was of minor importance for diatom growth during the intense part of the spring bloom in the fjord (the first 2 weeks). However, Si dissolution was important for maintaining the weak diatom growth during the last week of the bloom. Paasche & Østergren (1980) basically reached similar conclusion when discussing Si dissolution in

Table 1. Measured silicate uptake rates and estimated Si dissolution rates during the 1996 spring bloom in the Oslofjord

Rates	27 Feb–15 Mar	22–26 Mar
Average specific silicate uptake rate (d^{-1})	0.20	0.02
Average saturated specific silicate uptake rate[a] (d^{-1})	0.52	0.19
Si dissolution rate, estimate from the temperature[b] (d^{-1})	0.001–0.014	0.001–0.014
Si dissolution rate, estimate from the sat. uptake rate[b] (d^{-1})	0.007–0.015	–

[a] Saturated silicate uptake rates from kinetic experiments (Kristiansen et al., 2000).
[b] The Si dissolution rates were calculated from the *in situ* temperature (−1 °C) and from the measured saturated specific silicate uptake rate corrected for the *in situ* temperature (Nelson et al., 1976; Kamatini, 1982).

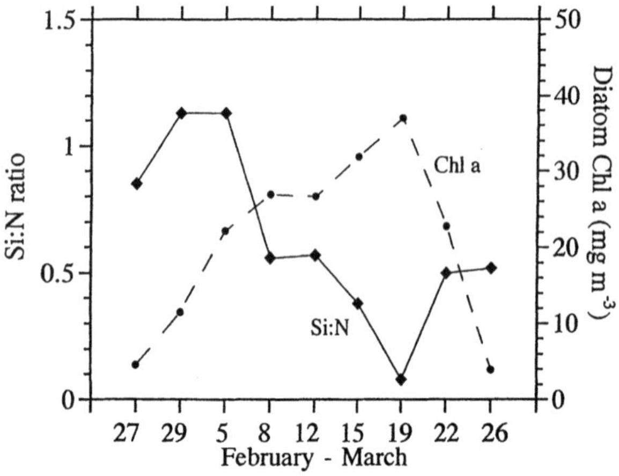

Figure 6. Time series of diatom abundance (as chlorophyll *a*) and the Si:N ratio from the spring bloom. Diatoms accounted for > 94% of the chlorophyll *a*. The same samples as in Figure 1.

the Oslofjord; Si dissolution may at times be important in the euphotic zone.

Abundance of diatoms

Diatoms will be an important algal group if silicate is available. This is well known from upwelling areas, during the spring bloom at temperate and moderately high latitudes and in freshwater influenced coastal areas (Smayda, 1980; Smetacek, 1986). It is illustrated in Figure 6 where time series of diatoms and the Si:N ratio from the Oslofjord are shown. The diatom bloom in the fjord collapsed when the Si:N ratio dropped below 0.4. If silicate is continuously supplied, diatoms may survive in apparently silicate-depleted water. This probably occurs during the extension of the spring bloom in the Trondheimsfjord that is fuelled by nutrients from the inflowing river water (Sakshaug & Myklestad, 1973). Diatoms will become weakly silicified during severe Si-limitation, and weakly silicified forms are often found in temperate coastal phytoplankton communities after the spring bloom (Paasche & Østergren, 1980). Already Gran (1912) discussed 'summer forms' of common diatoms that 'have usually thinner cell walls, and a more slender structure'. It may even be difficult to separate between some of the summer forms because of the weak silicification (Hasle & Smayda, 1960). By manipulating the N:Si ratio in mesocosms, it has been shown experimentally that small diatoms are favoured by low silicate concentrations (Carlsson & Granéli, 1999), and that the diatoms eventually are replaced by non-Si requiring phytoplankton species (Menzel et al., 1963; Egge & Aksnes, 1992).

Si and marine primary production

It has often been stated that nitrogen (N) limits total phytoplankton growth in the sea while phosphorus (P) limits growth in freshwater. During the last 10–15 years it has also become clear that grazing and trace metals (mainly concentration of soluble iron) may regulate phytoplankton production in oceans where inorganic nitrogen and phosphate are in excess throughout the year (see Falkowski & Raven, 1997). In freshwater influenced areas, however, phosphorus may be limiting (Thingstad et al., 1993). Si plays an equally important role in development of diatom blooms. Diatoms require N and Si in a ratio of about 1:1 (Redfield et al., 1963; Brzezinski, 1985). Diatom blooms may be initiated experimentally by adding balanced ratios of nutrients to samples from the oligotrophic open ocean (Menzel et al., 1963) and from coastal waters (Parsons et al., 1978; Doering et al., 1989; Egge & Aksnes, 1992).

In fresh water, competition for resources (N, P and Si) has successfully been used to explain the structure of phytoplankton communities (Tilman et al., 1982). The same success has not been achieved in marine waters. There is, however, experimental evidence for dominance of diatoms at high Si:N ratios and dominance of non-siliceous flagellates at low Si:N ratios in marine waters (Sommer, 1994).

The silicate uptake rate seems to be substrate-limited in most environments. This does not imply that diatom growth rates will be reduced because silicate may be reduced to undetectable concentrations following diatom blooms (see above). Diatom blooms may also collapse before silicate is depleted. Several explanations for this have been offered, including shortage of other nutrients (e.g. nitrate; Brzezinski & Nelson, 1995), iron (Martin & Fitzwater, 1988), diatom aggregation and sedimentation (Jackson, 1990), grazing (Miller, 1993) and deep mixing (Nelson & Smith, 1991). Direct proof of Si-limited diatom growth in the sea is scarce. It is complicating that the nutrient concentrations generally decrease during diatom blooms, and it therefore may be difficult to identify one limiting nutrient in the Liebig sense. However, Si has been found to control diatom growth during the spring bloom and also to control new production in upwelling areas (Dugdale & Wilkerson, 1998; Kristiansen et al., 2000).

Extraordinary flood events will supply unusual amounts of nutrients (also silicate) to the influenced coastal areas. Cloern (1996) reviewed phytoplankton blooms in coastal ecosystems, focusing on the San Francisco Bay. He found large diatom dominated blooms during years of exceptionally high river flow. Similar diatom dominated blooms have also been observed in Norwegian coastal water after extraordinary high river flow (Kristiansen, 1998). Damming and flood prevention efforts may have the opposite effect. They will reduce the LSi and silicate concentrations in the river water, which again may reduce diatom abundance in the influenced coastal areas. Building of the Iron Gate Dam reduced the silicate transport to the Black Sea to 1/3 of the pre-dam levels (Humborg et al., 1997; Milliman, 1997). Similar effect has also been pointed out for other major rivers which discharge water into the Baltic Sea and the Black Sea (Humborg et al., 2000).

Silicate is important in structuring the phytoplankton communities (see above). However, the effects of changed silicate input on marine production are not well understood. Recent experimental studies in Norwegian coastal water resulted in increased carbon production in mesocosms enriched with silicate. Nevertheless, the increase in carbon production was not followed by a increased accumulation of particulate carbon in the same mesocosms (Egge & Jacobsen, 1997; Svensen et al., 2001).

Si and eutrophication

Nixon (1995) emphasises that coastal marine eutrophication is a process, defined as 'an increase in the rate of supply of organic matter to an ecosystem'. We use this definition below. Vollenweider (1976) developed a model for lake eutrophication based on phosphorus availability, a model which successfully has been used in management of freshwater systems. According to Gray (1992), no simple relationships of the Vollenweider type are likely in coastal marine areas. Eutrophication, as defined, also occurs in natural systems (e.g. during upwelling or spring blooms). Here we will concentrate on cultural eutrophication (i.e. associated with anthropogenic activities) in stabilised water masses after the spring bloom has culminated. Already, Officer & Ryther (1980) suggested that during eutrophication Si may be the controlling nutrient in altering a diatom dominated community into a flagellate dominated community.

Justić et al. (1995) analysed the stoichiometry of dissolved nutrients in 10 large world rivers, and in two river-dominated coastal ecosystems (Adriatic Sea and Gulf of Mexico). They found that pristine rivers delivered Si to the coastal ocean in great excess over N and P, and thus contributed to the N- or P-deficiency in coastal waters. Moreover, rivers influenced by cultural eutrophication tend to have changed nutrient ratios to such an extent that the excess of silicate diminished in the coastal waters influenced by those rivers. Jickells (1998) reviewed nutrient biogeochemistry of the coastal zone and reached similar conclusions: Inputs of nutrients to the coastal zone are usually rich in N and P and low in Si resulting in high N:Si and P:Si ratios in the influenced coastal waters. Studies have also shown long-term declining silicate concentrations in some rivers (Conley et al., 1993; Turner & Rabalais, 1994; Humborg et al., 2000). The reduction in silicate transport is especially pronounced where large dams have been built on some of the world's major rivers, and it has been termed the 'artificial lake effect' (Van Bennekom & Salomons, 1980; Humborg et al., 2000). Well documented case studies include the Nile River (Wahby & Bishara, 1980), the Danube River (Humborg et al., 1997) and the Colorado River (Mayer & Gloss, 1980).

Another concern is that the loaded nutrients will increase the diatom production and thereby increase the diatom sedimentation, which again may result in decreased dissolved silicate in the water column. The silica depletion hypothesis has been questioned (see Conley et al., 1993). However, after reviewing the effects of eutrophication on the biogeochemical cycle of Si, Conley and co-workers concluded that the reduction of water column silicate occurs in many fresh water and marine systems throughout the world.

The effects of the anthropogenic nutrient inputs will necessarily be more pronounced in systems with long residence times. Therefore, the decline in Si is especially prevalent in large enclosed marine systems. The Baltic Sea is probably the best investigated system of this kind. Wulff et al. (1990) found significant changes in the dissolved nutrient pools of the Baltic, with a clear trend towards increased nitrogen and phosphorus concentrations and reduced silicate concentrations over several years (1971–1981). The trend in nutrient concentrations in the Baltic continued into the 1990s (Humborg et al., 2000). Phytoplankton production and biomass have also increased in the Baltic since the late 1960s (Cederwall & Elmgren, 1990). Increased phytoplankton production and changes in N:P:Si ratios may lead to changed species composition. Such changes have been reported from some areas of the Baltic (Elmgren, 1989; Wasmund et al., 1998). Other case studies include the German Bight (Radach et al., 1990), the Black Sea (Humborg et al., 1997), and the Mississippi River shelf system (Turner & Rabalais, 1994). Increased N:Si and P:Si ratios and reduced abundance of diatoms were reported from these studies.

High N:Si and P:Si ratios in coastal areas is a world-wide phenomenon which has caused major concerns because it has been related to novel and nuisance phytoplankton blooms (Smayda, 1990). However, the suggested link between cultural eutrophication and nuisance blooms is controversial (Gray, 1992; Richardson & Jørgensen, 1996). Probably the most conspicuous harmful bloom in Scandinavian waters is the outburst of the prymnesiophyte *Chrysochromulina polylepis* in 1988. It had a profound effect on marine life, but the exposed communities recovered surprisingly fast (Dahl et al., 1989; Gjøsæter et al., 2000). Economically, however, the dinoflagellate *Gyrodinium aureolum* probably is the most serious harmful algae in Scandinavian waters. It has regularly bloomed and caused severe economic losses to the aquaculture industry since 1966 (Dahl & Tangen, 1993). The initial populations of these blooms were associated with the pycnocline in the Skagerrak/Kattegat, and the blooms were probably not initiated by unusually high N:Si and P:Si ratios (Dahl & Tangen, 1993; Heilmann et al., 1994; Richardson & Jørgensen, 1996).

It is important to realise that harmful algal blooms are natural phenomena that have occurred for a long time, though the frequency of these blooms seems to have increased during the last 2–3 decades (Hallegraeff, 1993). Most harmful species are flagellates or dinoflagellates. Though diatoms may damage or clog fish gills in fish farms, and some diatoms (*Pseudo-nitzschia* spp.) may produce toxins (Smayda, 1990; Hallegraeff et al., 1995). The mechanisms behind harmful blooms are not yet completely understood (Richardson & Jørgensen, 1996). It is, however, becoming increasingly clear that they cannot be explained by high nutrient ratios alone (Hallegraeff, 1993; Anderson, 1994; Smayda, 1997).

Silicate and the marine food web

The great fisheries of the world depend on phytoplankton blooms dominated by diatoms in upwelling areas or during spring and autumn in temperate waters (Ryther, 1969). Diatom blooms are often followed by peak production of herbivorous zooplankton and then fish (the traditional food chain). Copepods have for long been recognised as the major grazers in the traditional food chain. However, the quality of the food is important (Kleppel & Burkart, 1995), and copepods do not necessarily prefer diatoms as diet (Harris, 1996). In a recent word-wide study, Ban et al. (1997) showed that the majority of the diatoms in their experiments could not support high egg production rates and hatching rates in copepods. Later investigations have reached similar conclusions, that diatoms may reduce the hatching success of copepod eggs (Miralto et al., 1999; Kang & Poulet, 2000). Copepods do, however, feed on a mixed diet at sea, and it is therefore difficult to evaluate the effect of diatoms on copepods in nature. Kiørboe & Nielsen (1994) studied seasonal distributions of biomass egg production and production rates in the pelagic copepod community in the Kattegat. They observed a peak in mortality during the spring bloom that coincided with high rates of egg production. However, they concluded that copepod production depended mainly on net phytoplankton blooms. It is still generally accepted that copepods feed on diatoms. A negative effect of diatoms on copepods is disputable, and it has not yet been tested in an ecological meaningful way (Jonasdottir et al., 1998).

Systems dominated by the traditional food chain usually also show microbial activity (Pierce & Turner, 1992). For example, does ciliate biomass in the Kattegat follow phytoplankton biomass with a distinct spring peak (Nielsen & Kiørboe, 1994), and the ciliates are grazed by copepods (Kiørboe & Nielsen, 1994). A general food web may therefore be viewed as continuum of trophic pathways, the traditional food chain and the microbial web being the 2 extremes (Legendre & Rassoulzadegan, 1995). In the South African upwelling system Benguela, Probyn (1992) found that the system was altering between these pathways. For periods of 4–8 days after an upwelling event, the traditional food chain dominated. For periods of more that 8 days after an upwelling event, the microbial web was the important one. Duarte et al. (2000) tested experimentally the effects of nutrient supply on Mediterranean plankton communities. They found a shift in the biomass structure towards a relatively higher autotrophic fraction with increasing nutrient inputs.

Coastal areas exposed to cultural eutrophication are characterised by high supplies of N and P and low supply of Si (see above). Sommer et al. (this volume) compared phytoplankton responses to different nutrient regimes with food size spectra of herbivorous zooplankton. They found that crustaceans occupied trophic level 2 in upwelling systems. Large nuisance algal cells often dominate in coastal areas exposed to cultural eutrophication. In such a scenario they found a strong microbial food web. The large nuisance algal cells where not grazed upon, and the crustaceans occupied trophic level as high as 4–5. Upwelling systems are therefore characterised by efficient trophic transfer while culturally eutrophicated systems often are characterised by inefficient trophic transfer.

Conclusion

Silicate is an important regulating factor for marine production. It is important in structuring the phytoplankton communities. Diatoms usually dominate the phytoplankton community in silicate-rich waters while flagellates usually dominate in silicate-poor water. Changes in the nutrient ratios (N:Si and P:Si) typically occur in coastal areas exposed to cultural eutrophication. These changes probably cause increased frequency of novel and nuisance algal blooms and inefficient trophic transfer of the energy.

References

Anderson, D.M., 1994. Red tides. Scientific American 271: 52–58.

Ban, S., C. Burns, J. Castel, Y. Chaudron, E. Christou, R. Escribano, S. F. Umani, S. Gasparini, F. G. Ruiz, M. Hoffmeyer, A. Ianora, H.-K. Kang, M. Laabir, A. Lacoste, A. Miralto, X. Ning, S. Poulet, V. Rodriguez, J. Runge, J. Shi, M. Starr, S.-i. Uye & Y. Wang, 1997. The paradox of diatom-copepod interactions. Mar. Ecol. Prog. Ser. 157: 287–293.

Bidle, K. D. & F. Azam, 1999. Accelerated dissolution of diatom silica by marine bacterial assemblages. Nature 397: 508–512.

Brzezinski, M. A., 1985. The Si:C:N ratio of marine diatoms: interspecific variability and the effect of some environmental variables. J. Phycol. 21: 347–357.

Brzezinski, M. A. & D. M. Nelson, 1989. Seasonal changes in the silicon cycle within a Gulf Stream warm-core ring. Deep-Sea Res. 36: 1009–1030.

Brzezinski, M. A. & D. M. Nelson, 1995. The annual silica cycle in the Saragasso Sea near Bermuda. Deep Sea Res. 42: 1215–1237.

Brzezinski, M. A. & D. M. Nelson, 1996. Chronic substrate limitation of silicate uptake rates in the western Sargasso Sea. Deep-Sea Res. II 43: 437–453.

Brzezinski, M. A. & D. R. Phillips, 1997. Evaluation of 32Si as a tracer for measuring silica production rates in marine waters. Limnol. Oceanogr. 42: 856–865.

Brzezinski, M. A., D. R. Phillips, F. P. Chavez, G. E. Friederich & R. C. Dugdale, 1997. Silica production in the Monterey, California, upwelling system. Limnol. Oceanogr. 42: 1694–1705.

Carlsson, P. & E. Granéli, 1999. Effects of N:P:Si ratios and zooplankton grazing on phytoplankton communities in the northern Adriatic Sea. II. Phytoplankton species composition. Aquat. Microb. Ecol. 18: 55–65.

Cederwall, H. & R. Elmgren, 1990. Biological effects of eutrophication in the Baltic Sea, particularly the coastal zone. Ambio 19: 109–112.

Cloern, J. E., 1996. Phytoplankton bloom dynamics in coastal ecosystems: A review with some general lessons from sustained investigations of San Francisco Bay, California. Rev. Geophys. 34: 127–168.

Conley, D. J., 1997. Riverine contribution of biogenic silica to the oceanic silica budget. Limnol. Oceanogr. 42: 774–777.

Conley, D. J., C. L. Schelske & E. F. Stoermer, 1993. Modification of the biogeochemical cycle of silica with eutrophication. Mar. Ecol. Prog. Ser. 101: 179–192.

Dahl, E. & K. Tangen, 1993. 25 years experience with *Gyrodinium aureolum* in Norwegian waters. In Smayda, T. J. & Y. Shimizu (eds), Toxic Phytoplankton Blooms in the Sea. Elsevier, Amsterdam: 15–21.

Dahl, E., O. Lindahl, E. Paasche & J. Throndsen, 1989. The *Chrysochromulina polylepis* bloom in Scandinavian waters during spring 1988. In Cosper, E. M., V. M. Bricelj & E. J. Carpenter (eds), Novel Phytoplankton Blooms. Springer-Verlag, Berlin: 383–405.

Del Amo, Y. & M. A. Brzezinski, 1999. The chemical form of dissolved Si taken up by marine diatoms. J. Phycol. 35: 1162–1170.

Doering, P. H., C. A. Oviatt, L. L. Beatty, V. F. Banzon, R. Rice, S. P. Kelly, B. K. Sullivan & J. B. Frithsen, 1989. Structure and function in a model coastal ecosystem: Silicon, the benthos and eutrophication. Mar. Ecol. Prog. Ser. 52: 287–299.

Duarte, C. M., S. Agustí, J. M. Gasol, D. Vaqué & E. Vazquez-Dominguez, 2000. Effect of nutrient supply on the biomass structure of planktonic communities: an experimental test on a Mediterranean coastal community. Mar. Ecol. Prog. Ser. 206: 87–95.

Dugdale, R. C. & F. P. Wilkerson, 1998. Silicate regulation of new production in the equatorial Pacific upwelling. Nature 391: 270–273.

Dugdale, R. C., F. P. Wilkerson & H. J. Minas, 1995. The role of a silicate pump in driving new production. Deep Sea Res. 42: 697–719.

Egge, J. K. & D. L. Aksnes, 1992. Silicate as regulating nutrient in phytoplankton competition. Mar. Ecol. Prog. Ser. 83: 281–289.

Egge, J. K. & A. Jacobsen, 1997. Influence of silicate on particulate carbon production in phytoplankton. Mar. Ecol. Prog. Ser. 147: 219—230.

Elmgren, R., 1989. Man's impact on the ecosystem of the Baltic Sea: Energy flows today and at the turn of the century. Ambio 18: 326-332.

Falkowski, P. G. & J. A. Raven, 1997. Aquatic Photosynthesis. Blackwell, Malden, 375 pp.

Gjøsæter, J., K. Lekve, N. C. Stenseth, H. P. Leinaas, H. Christie, E. Dahl, D. S. Danielssen, B. Edvardsen, F. Olsgard, E. Oug & E. Paasche, 2000. A long-term perspective on the *Chrysochromulina* bloom on the Norwegian Skagerrak coast 1988: a catastrophe or an innocent incident? Mar. Ecol. Prog. Ser. 207: 201–218.

Gran, H. H., 1912. Pelagic plant life. In Murray, J. & J. Hjort (eds), The Depths of the Ocean. MacMillian, London: 307–386.

Gray, J. S., 1992. Eutrophication in the sea. In Colombo, G., I. Ferrari, V. U. Ceccherellii & R. Rossi (eds), Marine Eutrophication and Population Dynamics. Olsen and Olsen, Fredensborg: 3–15.

Hallegraeff, G. M., 1993. A review of harmful algal blooms and their apparent global increase. Phycologia 32: 79–99.

Hallegraeff, G. M., D. M. Anderson & A. D. Cembella, 1995. Manual on harmful marine microalgae. UNESCO, Paris, 551 pp.

Harris, R. P., 1996. Feeding ecology of Calanus. Ophelia 44: 85–109.

Harrison, W. G., 1993. Nutrient recycling in production experiments. ICES mar. Sci. Symp. 197: 149–158.

Hasle, G. R. & T. J. Smayda, 1960. The annual phytoplankton cycle at Drøbak, Oslofjord. Nytt Mag. Bot. 8: 53–75.

Heilmann, J. P., K. Richardson & G. Ærtebjerg, 1994. Annual distribution and activity of phytoplankton in the Skagerrak/Kattegat frontal region. Mar. Ecol. Prog. Ser. 112: 213–223.

Hoagland, K. D., J. R. Rosowski, M. R. Gretz & S. C. Roemer, 1993. Diatom extracellular polymeric substances: function, fine structure, chemistry, and physiology. J. Phycol. 29: 537–566.

Humborg, C., V. Ittekkot, A. Cociasu & B. v. Bodungen, 1997. Effect of Danube River dam on Black Sea biogeochemistry and ecosystem structure. Nature 386: 385–388.

Humborg, C., D. J. Conley, L. Rahm, F. Wulff, A. Cociasu & V. Ittekkot, 2000. Silicon retention in river basins: Far-reaching effects on biogeochemistry and aquatic food webs in coastal marine environments. Ambio 29: 45–50.

Jackson, G. A., 1990. A model of the formation of marine algal flocs by physical coagulation processes. Deep Sea Res. 37: 1197–1211.

Jickells, T. D., 1998. Nutrient biogeochemistry of the coastal zone. Science 281: 217–222.

Jonasdottir, S. H., T. Kiørboe, K. W. Tang, M. St. John, A. W. Visser, E. Saiz & H. G. Dam, 1998. Role of diatoms in copepod production: good, harmless or toxic? Mar. Ecol. Prog. Ser. 172: 305–308.

Justić, D., N. N. Rabalais & R. E. Turner, 1995. Stoichiometric nutrient balance and origin of coastal eutrophication. Mar. Pollut. Bull. 30: 41–46.

Kamatani, A., 1982. Dissolution rates of silica from diatoms decomposing at various temperatures. Mar. Biol. 68: 91–96.

Kang, H.-K. & S. A. Poulet, 2000. Reproductive success in *Calanus helgolandicus* as a function of diet and egg cannibalism. Mar. Ecol. Prog. Ser. 201: 241–250.

Kiørboe, T. & T. G. Nielsen, 1994. Regulation of zooplankton biomass and production in a temperate, coastal ecosystem. 1. Copepods. Limnol. Oceanogr. 39: 493–507.

Kleppel, G. S. & C. A. Burkart, 1995. Egg production and the nutritional environment of *Acartia tonsa*: the role of food quality in copepod nutrition. ICES J. mar. Sci. 52: 297–304.

Kristiansen, S., 1998. Impact of increased river discharge on the phytoplankton community in the outer Oslofjord, Norway. Hydrobiologia 363: 169–177.

Kristiansen, S., T. Farbrot & L.-J. Naustvoll, 2000. Production of biogenic silica by spring diatoms. Limnol. Oceanogr. 45: 472–478.

Legendre, L. & F. Rassoulzadegan, 1995. Plankton and nutrient dynamics in marine waters. Ophelia 41: 153–172.

Levitus, S., M. E. Conkright, J. L. Reid, R. G. Najjar & A. Mantyla, 1993. Distribution of nitrate, phosphate and silicate in the world oceans. Progr. Oceanogr. 31: 245–273.

Lewin, J. C., 1962. Silicification. In Lewin, R. A. (ed.), Physiology and Biochemistry of Algae. Academic, New York: 445–455.

Martin, J. H. & S. E. Fitzwater, 1988. Iron deficiency limits phytoplankton growth in the north-east Pacific subarctic. Nature 331: 341–343.

Mayer, L. M. & S. P. Gloss, 1980. Buffering of silica and phosphate in a turbid river. Limnol. Oceanogr. 25: 12–22.

Menzel, D. W., E. M. Hulburt & J. H. Ryther, 1963. The effects of enriching Sargasso Sea water on the production and species composition of the phytoplankton. Deep Sea Res. 10: 209–219.

Miller, C. B., 1993. Pelagic production processes in the subarctic Pacific. Progr. Oceanogr. 32: 1–15.

Milliman, J. D., 1997. Blessed dams or damned dams? Nature 386: 325–327.

Miralto, A., G. Barone, G. Romano, S. A. Poulet, A. Ianora, G. L. Russo, I. Buttino, G. Mazzarella, M. Laabir, M. Cabrini & M. G. Glacobbe, 1999. The insidious effect of diatoms on copepod reproduction. Nature 402: 173–176.

Nelson, D. M. & Q. Dortch, 1996. Silicic acid depletion and silicon limitation in the plume of the Mississippi River: evidence from kinetic studies in spring and summer. Mar. Ecol. Prog. Ser. 136: 163–178.

Nelson, D. M. & J. J. Goering, 1977. A stable isotope tracer method to measure silicate uptake by marine phytoplankton. Anal. Biochem. 78: 139–147.

Nelson, D. M. & W. O. Smith, Jr., 1991. Sverdrup revisited: Critical depths, maximum chlorophyll levels, and the control of Southern Ocean productivity by the irradiance-mixing regime. Limnol. Oceanogr. 36: 1650–1662.

Nelson, D. M., J. J. Goering, S. S. Kilham & R. R. L. Guillard, 1976. Kinetics of silicic acid uptake and rates of silica dissolution in the marine diatom *Thalassiosira pseudonana*. J. Phycol. 12: 246–252.

Nelson, D. M., P. Tréguer, M. A. Brzezinski, A. Leynaert & B. Quéguiner, 1995. Production and dissolution of biogenic silica in the ocean: Revised global estimates, comparison with re-

gional data and relationship to biogenic sedimentation. Global Biogeochem. Cycle 9: 359–372.
Nielsen, T. G. & T. Kiørboe, 1994. Regulation of zooplankton biomass and production in a temperate, coastal ecosystem. 2. Ciliates. Limnol. Oceanogr. 39: 508–519.
Nixon, S. W., 1995. Coastal marine eutrophication: A definition, social causes, and future concerns. Ophelia 41: 199–219.
Officer, C. B. & J. H. Ryther, 1980. The possible importance of silicon in marine eutrophication. Mar. Ecol. Prog. Ser. 3: 83–91.
Paasche, E., 1973. Silicon and the ecology of marine plankton diatoms. II. Silicate-uptake kinetics in five diatom species. Mar. Biol. 19: 262–269.
Paasche, E., 1975. Growth of the plankton diatom *Thalassiosira nordenskioeldii* Cleve at low silicate concentrations. J. exp. mar. Biol. Ecol. 18: 173–183.
Paasche, E., 1980a. Silicon. In Morris, I. (ed.), The Physiological Ecology of Phytoplankton. Blackwell, Oxford: 259–284.
Paasche, E., 1980b. Silicon content of five marine plankton diatom species measured with a rapid filter method. Limnol. Oceanogr. 25: 474–480.
Paasche, E. & I. Østergren, 1980. The annual cycle of plankton diatom growth and silica production in the inner Oslofjord. Limnol. Oceanogr. 25: 481–494.
Parsons, T. R., P. J. Harrison & R. Waters, 1978. An experimental simulation of changes in diatom and flagellate blooms. J. exp. mar. Biol. Ecol. 32: 285–294.
Pierce, R. W. & J. T. Turner, 1992. Ecology of planktonic ciliates in marine food webs. Rev. Aquat. Sci. 6: 139–181.
Probyn, T. A., 1992. The inorganic nitrogen nutrition of phytoplankton in the southern Benguela: New production, phytoplankton size and implications for pelagic foodwebs. S. Afr. J. mar. Sci. 12: 411–420.
Radach, G., J. Berg & E. Hagmeier, 1990. Long-term changes of the annual cycles of meteorological, hydrographic, nutrient and phytoplankton time series at Helgoland and at LV ELBE 1 in the German Bight. Cont. Shelf Res 10: 305–328.
Ragueneau, O. & P. Tréguer, 1994. Determination of biogenic silica in coastal waters: applicability and limits of the alkaline digestion method. Mar. Chem. 45: 43–51.
Redfield, A. C., B. H. Ketchum & F. A. Richards, 1963. The influence of organisms on the composition of sea-water. In Hill, M. N. (ed.), The Sea. Vol. 2. Interscience Publishers, New York: 26–77.
Richardson, K. & B. B. Jørgensen, 1996. Eutrophication: Definition, history and effects. In Jørgensen, B. B. & K. Richardson (eds), Eutrophication in Coastal Marine Ecosystems. American Geophysical Union, Washington: 1–19.
Round, F. E., R. M. Crawford & D. G. Mann, 1990. The Diatoms. Biology & Morphology of the Genera. Cambridge University Press, Cambridge, 477 pp.
Ryther, J. H., 1969. Photosynthesis and fish production in the sea. Science 166: 72–76.
Sakshaug, E. & S. Myklestad, 1973. Studies on the phytoplankton ecology of the Trondheimsfjord. III. Dynamics of phytoplankton blooms in relation to environmental factors, bioassay experiments and parameters for the physiological state of the populations. J. exp. mar. Biol. Ecol. 11: 157–188.
Seibold, E. & W. H. Berger, 1982. The Sea Floor. Springer, Berlin, 288 pp.
Shipe, R. F. & M. A. Brzezinski, 1999. A study of Si deposition synchrony in *Rhizosolenia* (Bacillariophyceae) mats using a novel ^{32}Si autoradiographic method. J. Phycol. 35: 995–1004.
Smayda, T. J., 1980. Phytoplankton species succession. In Morris, I. (ed.), The Physiological Ecology of Phytoplankton. Blackwell, Oxford: 493–570.
Smayda, T. J., 1990. Novel and nuisance phytoplankton blooms in the sea: evidence for a global epidemic. In Granéli, E., B. Sundström, L. Edler & D. M. Anderson (eds), Toxic Marine Phytoplankton. Elsevier, New York: 29–40.
Smayda, T. J., 1997. Harmful algal blooms: Their ecophysiology and general relevance to phytoplankton blooms in the sea. Limnol. Oceanogr. 42: 1137–1153.
Smetacek, V. S., 1986. Impact of freshwater discharge on production and transfer of materials in the marine environment. In Skreslet, S. (ed.), The Role of Freshwater Outflow in Coastal Marine Ecosystems. Springer-Verlag, Berlin: 85–106.
Sommer, U., 1994. Are marine diatoms favoured by high Si:N ratios? Mar. Ecol. Prog. Ser. 115: 309–315.
Sommer, U., H. Stibor, A. Katechakis, F. Sommer & T. Hansen, 2002. Pelagic food web configurations at different levels of nutrient richness and their implications for the ratio fish production:primary production. Hydrobiologia 484/Dev. Hydrobiol. 167: 11–20.
Strickland, J. D. H. & T. R. Parsons, 1972. A practical handbook of seawater analysis. 2nd edn. Bull. Fish. Res. Bd Can. 167: 1–310.
Stumm, W. & J. J. Morgan, 1996. Aquatic Chemistry. 3rd edn. Wiley, New York, 1022 pp.
Svensen, C., J. K. Egge & J. E. Stiansen, 2001. Can silicate and turbulence regulate the vertical flux of biogenic matter? A mesocosm study. Mar. Ecol. Prog. Ser. 217: 67–80.
Thingstad, T. F., E. F. Skjoldal & R. A. Bohne, 1993. Phosphorus cycling and algal-bacterial competition in Sandsfjord, western Norway. Mar. Ecol. Prog. Ser. 99: 239–259.
Tilman, D., S. S. Kilham & P. Kilman, 1982. Phytoplankton community ecology: The role of limiting nutrients. Ann. Rev. Ecol. Syst. 13: 349–372.
Tréguer, P., L. Lindner, A. J. Van Bennekom, A. Leynaert, M. Panouse & G. Jacques, 1991. Production of biogenic silica in the Weddell-Scotia Seas measured with ^{32}Si. Limnol. Oceanogr. 36: 1217–1227.
Tréguer, P., D. M. Nelson, A. J. Van Bennekom, D. J. DeMaster, A. Leynaert & B. Quéguiner, 1995. The silica balance in the world ocean: A reestimate. Science 268: 375–379.
Turner, R. E. & N. N. Rabalais, 1994. Coastal eutrophication near the Mississippi river delta. Nature 368: 619–621.
Van Bennekom, A. J., E. Krijgsman-van Hartingsveld, G. C. M. van der Veer & H. F. J. van Voorst, 1974. The seasonal cycles of reactive silicate and suspended diatoms in the Dutch Wadden Sea. Neth. J. Sea Res. 8: 174–207.
Van Bennekom, A. J. & W. Salomons, 1980. Pathways of nutrients and organic matter from land to ocean through rivers. In Martin, J.-M., J. D. Burton & D. Eisma (eds), River Inputs to Ocean Systems. UNEP/UNESCO: 33–56.
Vollenweider, R. A., 1976. Advances in defining critical loading levels for phosphorus in lake eutrophication. Mem. 1st. Ital. Idrobiol. 33: 53–84.
Wahby, S. D. & N. F. Bishara, 1980. The effect of the river Nile on Mediterranean water, before and after the construction of the High Dam at Aswan. In Martin, J.-M., J. D. Burton & D. Eisma (eds), River Inputs to Ocean Systems. UNEP/UNESCO: 311–315.
Wasmund, N., G. Nausch & W. Matthäus, 1998. Phytoplankton spring blooms in the southern Baltic Sea – spatio-temporal development and long-term trends. J. Plankton Res. 20: 1099–1117.
Werner, D., 1977. Silicate metabolism. In Werner, D. (ed.), The Biology of Diatoms. University of California Press, Berkeley: 110–149.
Wulff, F., A. Stigebrandt & L. Rahm, 1990. Nutrient dynamics of the Baltic Sea. Ambio 19: 126–133.

Hydrobiologia **484:** 33–48, 2002.
O. Vadstein & Y. Olsen (eds), Sustainable Increase of Marine Harvesting: Fundamental Mechanisms and New Concepts.

Silicate availability, vertical mixing and grazing control of phytoplankton blooms in mesocosms

V. Escaravage[1]* & T. C. Prins[2]
[1]*Netherlands Institute of Ecology – Centre for Estuarine and Coastal ecology / NIOO-CEMO, Korringaweg 7, 4401 NT Yerseke, The Netherlands*
[2]*National Institute for Coastal and Marine Management / RIKZ, The Netherlands*
*(*Author for correspondence) E-mail: v.escaravage@rikz.rws.minvenw.nl*

Key words: diatoms, *Phaeocystis*, *Cerataulina marina*, stratification, ciliates, copepods

Abstract

The present paper explores (i) the way silicate concentrations may control the switch from diatom to non-diatom dominated phytoplankton, (ii) the consequence of mixing intensity and nutrient loading on phytoplankton composition (and potentially harmful species), (iii) to what extent trophic interactions may interfere with this abiotic control. The data used for this paper were selected from experiments run in 3-m^3 land-based mesocosms (SW Netherlands) since 1993. Various nutrient and/or mixing regimes were applied to seawater incubated for periods varying from 1 to five months. The termination of the initial diatom bloom was induced by silicate concentrations below 2 μM, whereas the subsequent flagellate bloom was limited by N or P availability. Combined high nitrogen loadings and stratification events promoted the development of potentially harmful dinoflagellates but other factors such as grazing and temperature played a decisive role in the appearance of such blooms. Despite their relatively low intrinsic growth rate, copepods were able to control diatoms and microzooplankton development. Microzooplankton efficiently incorporated blooms of small phytoplankton species. These results, combined with previous experiments where nitrogen enrichment induced increases in small phytoplankton biomasses, stress the positive effect of N-enrichment on the microbial food web. Conversely, blooms of large diatom species exemplified the trophic mismatch leading to the accumulation of algal biomass without transfer to higher trophic levels.

Introduction

Silicate availability has been abundantly described in the literature as a crucial factor in determining whether flagellates or diatoms will dominate phytoplankton assemblages. It has also been discussed that anthropogenic eutrophication and the associated high N/Si ratio, could be the cause of the increase in toxic phytoplankton blooms observed in marine coastal waters during recent decades (Smayda, 1990; Egge & Asknes, 1992; Schøllhorn & Granéli, 1993; Sommer, 1994; Schöllhorn & Granéli, 1996).

An important notion brought up by these studies is that the development of non-diatom phytoplankton depends on the quantity of nitrogen and phosphorus left over by diatoms after the spring bloom becomes Si-limited. In the same vein, Egge & Asknes (1992) observed in mesocosm experiments that diatoms needed at least 2 μmol silicate l^{-1} to successfully develop; at lower Si concentrations, non-diatom species developed and numerically dominated phytoplankton (at least temporarily). However, laboratory experiments have shown the flexibility of diatoms with respect to their silicon requirements (Harrison et al., 1977) since they appear to be able to modulate their internal Si pool according to the availability of this nutrient in the cell's environment.

Nutrient ratios, however, cannot solely account for phytoplankton composition since other factors also play a significant role. In a well-mixed environment, both diatoms and flagellates experience the same nutrient/light conditions. In calm environments, diatoms have to lower their density to escape from sedimentation (Gradmann & Boyd, 1995; Moore & Villareal, 1996), whereas flagellates may freely migrate between the nutrient-rich bottom layer and the well-illuminated upper layer (Eppley et al., 1968; Edler & Olsson, 1985; Olsson & Granéli, 1991). Following Cushing (1989), the ratio of respiration to maximal photosynthesis is three times higher for dinoflagellates than for diatoms, as a consequence of which the former need waters that are strongly stratified, whereas the latter can survive in well-mixed environments. Kiørboe et al. (1990) observed that the central stratified part of the Skagerrak was characterised by small flagellates, while large and chain-forming diatoms dominated along the periphery of this area. In the same vein, calm, sunny weather with low wind speed appears to promote summer blooms of the harmful dinoflagellate *Dynophysis acuminata* off the Dutch coast (Peperzak et al., 1996).

Besides the abiotic controls (nutrient, mixing), biotic factors such as resource competition and grazing may complicate the features described above. These processes are closely related to the size of the organisms: small species show higher growth rates than large ones due to more efficient light harvesting and nutrient uptake kinetics. Small algae are subject to a high grazing pressure by microzooplankton, whereas larger phytoplankton may escape size-selective microzooplankton grazing, and reach large biomass. The efficiency of microzooplankton in controlling (small-sized) phytoplankton biomass follows from its high growth rate equalling or possibly exceeding that of phytoplankton (Banse, 1982). Conversely, the trophic mismatch for high resource availability can be ascribed to the fact that potential grazers for large phytoplankton (mainly copepods) have generation times in the order of weeks (Klein Breteler et al., 1982) that do not allow them to check the phytoplankton biomass (Kiørboe, 1998; Wassmann, 1998).

The present paper aims at exploring (i) the way silicate concentrations may control the switch from diatom to non-diatom dominated phytoplankton, (ii) the consequence of mixing intensity and nutrient loading on the balance between diatoms and flagellates (and potentially harmful species), (iii) to what extent trophic interactions may interfere with this abiotic control.

Results of mesocosm experiments were used to investigate the way nutrient ratios and/or mixing intensity affected the development of phytoplankton blooms. Besides phytoplankton, zooplankton (micro- and meso-) was also monitored to account for trophic interactions.

Materials and methods

Mesocosm description

The land-based mesocosms are located at the field-station of RIKZ near the mouth of the Oosterschelde estuary (SW Netherlands). The mesocosms consist of black, solid polyethylene tanks (height 3 m, diameter 1.2 m, volume 3000 l). A central rotating mixer mounted with horizontal paddles mixed the water column. A scraper, coated with a strip of abrasive material, was used to remove fouling organisms from the walls of the tanks. Additional manual removal of fouling organisms was carried out when necessary. Above the mesocosms an optical diffuser of structured Plexiglas (Groenendijk, Pl 20070 TK) was installed to ensure a homogeneous light gradient in the water column. In all experiments the mesocosms were filled with water from the nearby Oosterschelde estuary (SW Netherlands) after filtration on a 1-mm plankton net. Inorganic nutrients were continuously added to the mesocosms from stock solutions with a peristaltic pump. The mesocosms were continuously flushed with sea water at a rate of 100 l day^{-1}.

The mixing regime used in the mesocosms was adjusted by changing the number and position of the paddles and/or the rotation speed of the mixer. In 'well-mixed' experiments, the mixer speed was set at 10 rpm and three paddles were mounted on the central axis. In 'stratified' experiments, the mixer speed was set to 3 rpm with two paddles mounted on the central axis.

Stratification of the water column was obtained by establishing a sharp salinity gradient between the upper (0–1.5 m) and the lower (1.5–3 m) layers. The gradient was induced by injecting 150 l demineralised water at 1.50 m depth. In order to maintain the saline stratification, the top layer was flushed with seawater diluted with demineralised water to a final salinity of about 27‰. The bottom layer was flushed with seawater spiked with artificial seawater salt to a final salinity of about 38‰. In both the Top and the Bottom compartments, the flushing rate induced a 30-day water renewal time. Vertical profiles of salinity were recorded each day with a portable conductivity meter (WTW™). Heating of the mesocosms by solar radiation was diminished by spraying sea water on the outer wall of the tanks and by insulating the tanks with foil. An extensive description of the mesocosms is given in Prins et al. (1995).

Table 1. Description of the mesocosm experiments, periods and treatments. The codes used throughout this paper are given here. The average nutrient loadings are expressed as mmol of nutrient added per day to each mesocosm

Exp. no.	Experimental period (day no.)	Stratification period(s)	Treatment code	N-loading (mmol day^{-1})	P-loading (mmol day^{-1})	Si-loading (mmol day^{-1})
Exp93	81-116	–	93MixH1	7.00	1.32	6.30
Exp94 (Spring)	136-164	–	94MixH1	8.43	0.54	4.22
			94MixH2	8.45	0.54	4.22
Exp94 (Summer)	173-201	–	94MixH1	9.00	0.59	2.62
			94MixH2	8.97	0.58	2.61
Exp97	188-234	204-234	97StratH1	7.28	0.25	1.54
			97StratH2	6.84	0.25	1.33
			97StratL1	2.84	0.25	1.73
			97StratL2	2.90	0.26	1.20
		–	97MixH1	7.37	0.25	1.18
			97MixH2	7.47	0.26	1.20
Exp98	90-230	140-167 and 184-209	98StratH1	6.94	0.27	1.48
			98StratH2	6.99	0.27	1.49
			98StratL1	2.20	0.27	1.50
			98StratL2	2.18	0.27	1.50

Mesocosm experiments

The data used for this paper were selected from experiments run in mesocosms since 1993 (listed in Table 1). The effect of silicate concentrations on the diatoms/flagellates succession was documented with four separate experiments: spring experiment in 1993 (Escaravage et al., 1996), spring and summer experiments in 1994 (Escaravage et al., 1999) and a summer experiment run in 1997. In all these experiments, the initial diatom bloom was primarily resource-limited. Only the mesocosms with the highest nutrient loadings (N or P) were considered here, since excess of these nutrients is expected to promote the shift in species composition. The study was furthermore restricted to the initial period when the diatom/flagellate succession occurred and when N and P concentrations were above 1 and 0.1 μM, respectively, considered as potential threshold levels for nutrient limitation of phytoplankton development (Peeters & Peperzak, 1990).

The study on the combined effect of stratification and nutrients on the balance between flagellates and diatoms was based on results from the experiment run in 1997 (Table 1). The experiment in 1997 started with an initial period (Period I, day no. 188–203) when the water column of the six mesocosms was kept well mixed, four of them receiving a high nitrogen loading (H) while a low nitrogen loading (L) was applied to the two other systems (Table 1). During the following period (Period II, day no. 204–234), a stratification event was simulated in four mesocosms (two from each nitrogen treatment: 97Strat-H1, 97Strat-H2, 97Strat-L1, 97Strat-L2) while the two other mesocosms with high N-loading remained well mixed (97Mix-H1, 97Mix-H2).

Besides the mixing and nutrient-treatments, a dinoflagellate (*Prorocentrum micans*) was added to the mesocosm water and served as a 'model species' to identify environmental conditions that could be profitable to potential toxic dinoflagellates. This dinoflagellate is often found in considerable concentrations during blooms of the DSP producing *Dinophysis acuminata* (Peperzak et al., 1996). A batch culture of *P. micans* was run in 8-l polyethylene bags with f/2 medium without silicate. Each week, 6 l of culture medium (about 1000–2000 cells l^{-1} *P. micans*) was distributed over the six mesocosms, the rest was diluted with fresh medium and the culture was left to grow again for another week.

The question of to what extent trophic interactions may interfere with abiotic controls was studied on the basis of additional observations made during the experiments described above and an experiment run in 1998 (Table 1). This last experiment used a design similar to the 1997 experiment but showed clear indications of a tight top-down control of the plankton dynamics.

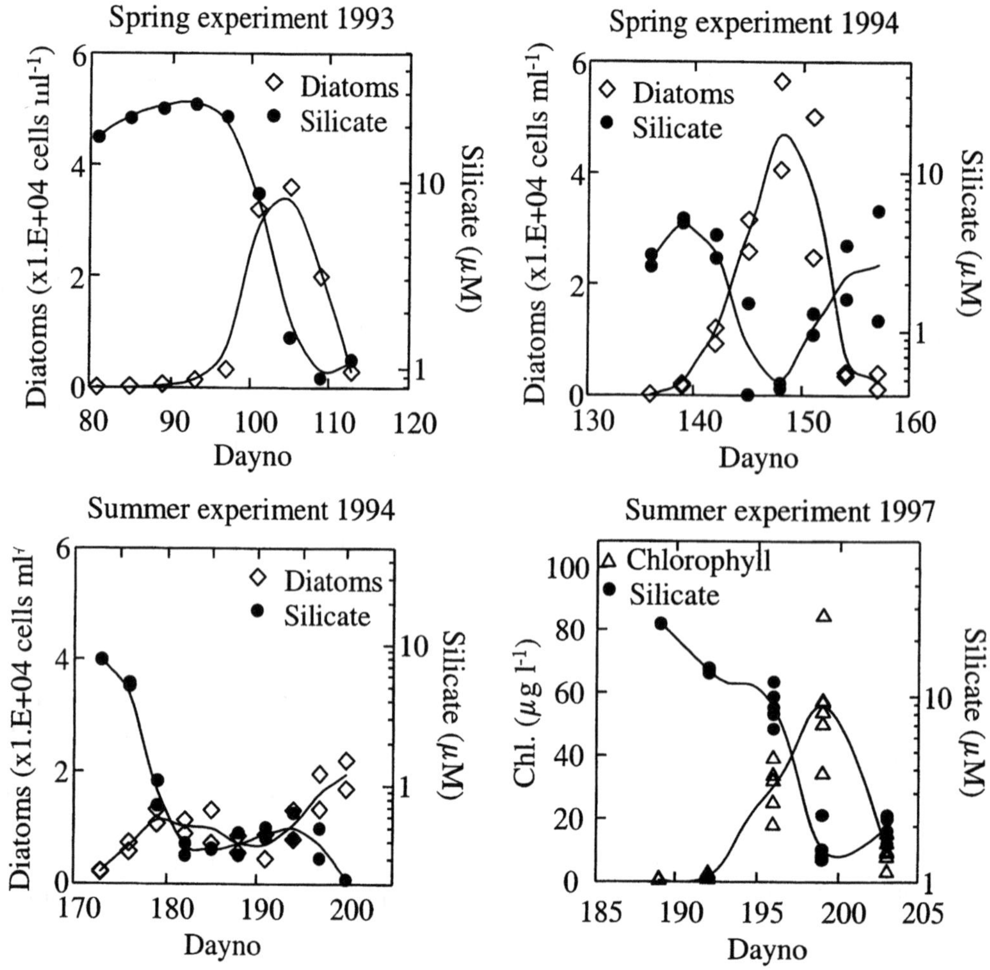

Figure 1. Silicate concentrations against diatom concentrations (1993, 1994) and chlorophyll-*a* concentrations (1997) in mesocosms with the highest N or P loading.

Sampling and analytical procedures

Dissolved inorganic nutrients DIN (NH_4^+, NO_3^-, NO_2^-), PO_4^{3-} and H_4SiO_4 and chlorophyll-*a* were determined in bi-weekly samples. Nutrients were analysed with an Autoanalyzer. Chlorophyll-*a* was extracted on GF/C filters according to Gieskes & Kraay (1984) and analysed with a HPLC method using a 85–100% acetone–water–water gradient and a reversed phase RP18 Novopack column (Waters) in a Spectra Physics Chromatography station. Chlorophyll-*a* was detected with a Perkin-Elmer LS-2B fluorometer (excitation, 410–430 nm; emission, >530 nm). The observed chlorophyll-*a* values were used to convert the continuously measured fluorescence data to chlorophyll-*a*.

Samples for phytoplankton analysis were taken bi-weekly in 1993 and 1994 and weekly in 1997 and 1998. Microzooplankton and mesozooplankton were analysed in weekly samples. Phytoplankton and microzooplankton (20–200 µm, mainly heterotrophic dinoflagellates and ciliates) abundances and species composition were determined by the Utermöhl technique from the same water sample fixed with acid Lugol's iodine solution (Utermöhl, 1958). Samples for mesozooplankton were collected at two depths (0.75 and 2.75 m) with a Zenit submersible pump (stratified experiments) or with a 2.5-m PVC-tube, provided with a ball valve at the lower end for depth-integration (Kuiper, 1977). After filtration on a 55-µm plankton-net, samples, were fixed with borax-buffered formalin (Steedman, 1976) to a final concentration of ca. 4%. After staining with Bengal rose, zooplankton species composition and abundance were determined with a stereomicroscope. Carbon biomass was estimated

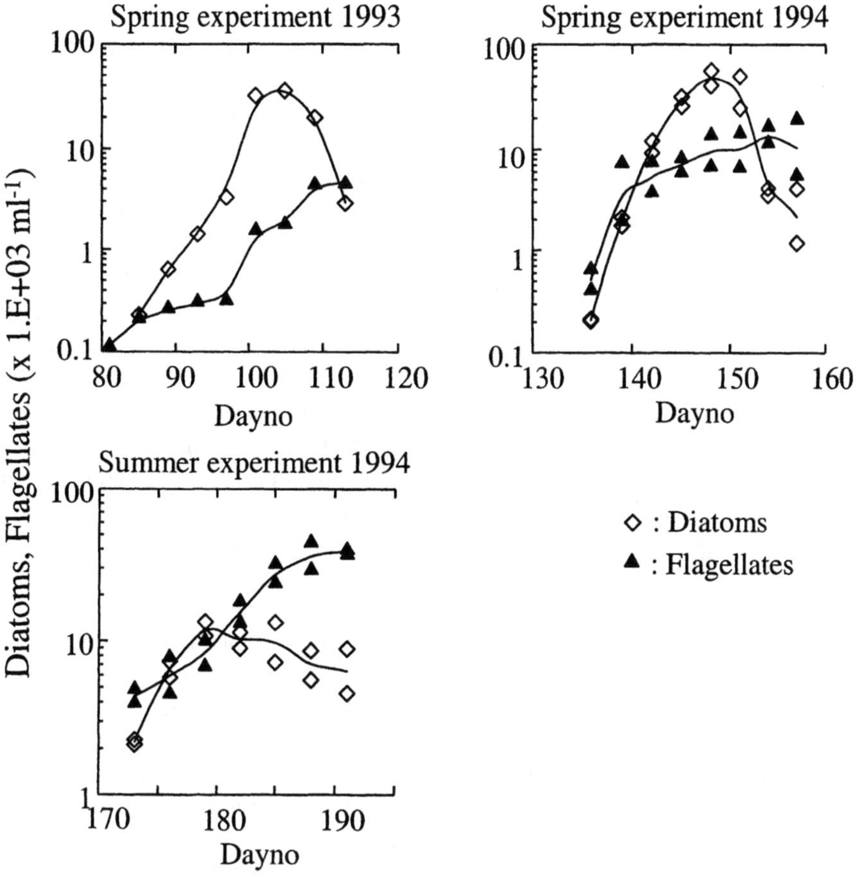

Figure 2. Diatom and flagellate concentrations in three mesocosm experiments run in 1993 and 1994.

from appropriate geometric approximation applied to measured dimensions and conversion formula's from literature: phytoplankton (Smayda, 1978), ciliates (Putt & Stoecker, 1989), dinoflagellates (Lessard, 1991) and copepods (Klein-Breteler et al., 1982).

Results

Silicate control of diatom dynamics

In most mesocosm experiments, the first week of incubation was characterised by an intense diatom bloom whose decline coincided with the exhaustion of the silicate concentrations. To what extent silicate concentrations may control diatom development was studied in experiments run in 1993, 1994 and 1997 (Fig. 1). Silicate and diatom concentrations were followed during the course of the initial diatom bloom development in mesocosms with the highest N or P loadings, where nutrient molar ratios N:Si and P:Si were higher than the optimum ratio for diatoms (Gillbricht, 1988), suggesting that silicate should be the first potential limiting nutrient for diatoms. Since no phytoplankton samples were taken during the first week of the experiment in 1997, chlorophyll concentrations instead of numerical abundance were used to describe the phytoplankton development. Phytoplankton samples from the second week of this experiment were highly dominated by diatoms, suggesting that diatoms were responsible for the initial chlorophyll peak.

In all examples shown here, a decline in the silicate concentrations preceded the collapse of the diatom bloom that developed during the first weeks of the experiments (Fig. 1). The diatom concentrations decreased after silicate concentrations had dropped below 2–1 μM.

Diatom–flagellate succession

The succession between flagellates and diatoms was studied by following concentrations of both flagellates and diatoms from the experiment onset until flagellates

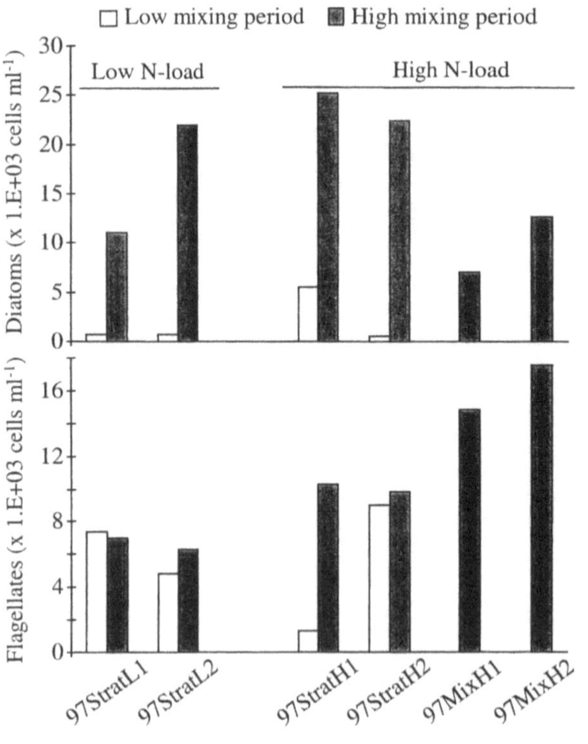

Figure 3. Diatom and flagellate concentrations measured during low mixing (white bars) and high mixing periods (shaded bars) in mesocosms from Exp97. Mesocosms (names as in Table 1) on the x-axis are grouped according to their nutrient treatment.

reached their highest concentrations. The same experiments as in the previous section were considered, except Exp97, since no phytoplankton sample was taken during the first week of this experiment. In none of the cases was a time-lag observed between the growth periods of diatoms and flagellates (Fig. 2). Both groups grew from the beginning of the experiments but, when diatom development was limited by the silicate concentrations (see previous section), flagellates continued to grow after the silicate pool was exhausted. Previous analyses showed that the level eventually reached by the flagellates (dominated by *Phaeocystis*) was controlled by the availability of P in 1993 (Escaravage et al., 1996) or N in 1994 (Escaravage et al., 1999).

Flagellate–diatom development vs mixing rate and nutrient loading

The effect of the treatments on phytoplankton was studied on mean water column concentrations of flagellates and diatoms calculated per mesocosm over periods characterised by a distinct combination of treatments: high or low N-load combined with high or low mixing (Fig. 3). The effect of the treatments was evaluated with a Bonferroni adjusted t-test after normalisation (log-transformation) of the data. Flagellate and diatom concentrations were generally higher in the well-mixed conditions when compared with the stratification periods but this relation was found significant only for the diatoms ($P = 0.022$). When the comparison was restricted to Period II between well-mixed (97MixH1-2) and stratified (97StratH1-2) mesocosms, flagellate concentrations were also found to be significantly higher in the well-mixed mesocosms (nested ANOVA, $P < 0.01$).

Nitrogen loadings did not significantly explain the variation observed in diatom or flagellate concentrations. As a consequence of the atomic ratios in the nutrient loading ($4.7<N/Si<1.5$ and $Si/P=11.5$) silicon-using phytoplankton may have been limited rather by silicate than by nitrogen (Gillbricht, 1988). That could explain the lack of treatment effect observed for the diatoms. The lack of resource control for flagellates resulted mainly from low flagellate concentrations measured in mesocosm 97StratH1 during the low-mixing period (see Fig. 3). When this point is excepted, flagellate concentrations appeared to be significantly controlled by the nitrogen loading ($P = 0.014$). A more detailed examination of the data revealed that the dynamics and composition of the flagellate community in mesocosm 97StratH1 during the stratification period were completely different from what was observed in the replicate mesocosm (97StratH2) (Fig. 4). Whereas a *Phaeocystis* bloom developed in 97StratH2 during the stratification period, *Phaeocystis* concentrations remained low in 97StratH1. Instead of *Phaeocytis* sp., *Prorocentrum micans* developed in mesocosm 97StratH1.

Whereas phytoplankton counts equivocally responded to nutrient addition during the stratification periods, chlorophyll-*a* biomass showed a clearer response to the treatment with marked biomass peaks in the upper layer of the water column between 40 and 25 $\mu g\ l^{-1}$ in mesocosms 97Strat-H1 and 97Strat-H2, whereas it did not exceed 10 $\mu g\ l^{-1}$ during the stratification period in mesocosms 97Strat-L1 and 97Strat-L2 (Fig. 4). In both cases, the increase in chlorophyll-*a* concentrations coincided with an increase in flagellate concentrations: *Phaeocystis* from day 210 to 224 in 97Strat-H2 and *Prorocentrum micans* from day 217 to 224 in 97Strat-H1 (Fig. 4).

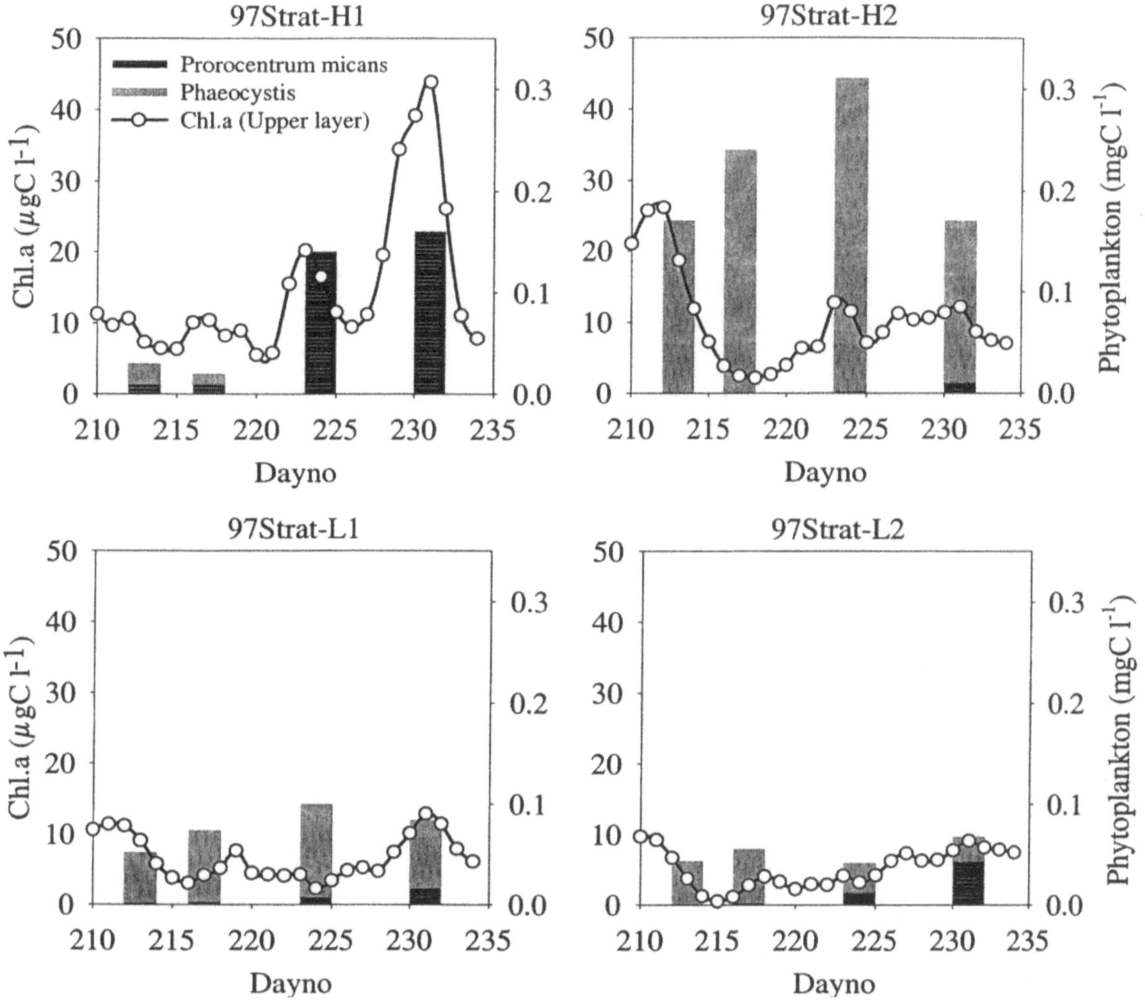

Figure 4. Chlorophyll-*a* concentrations (dots and line) in the lower and upper layer of mesocosm during the stratification period in 1997. *Phaeocytis* and *P. micans* biomass are plotted (bars) on the same graph.

Trophic interactions evidenced from mesocosm experiments

The experiment run in 1998 had the same design as in 1997, with two stratification periods (spring/summer) and two distinct N-loadings (**H**igh and **L**ow) (Table 1). Nutrient concentrations showed a similar evolution in all mesocosms, with exhaustion of the initial pools (values below detection limits) of DIP and Silicate after days 110–130. The initial DIN-pool became exhausted in mesocosm STRAT98-L1 and -L2 only after day 180, whereas DIN concentrations remained high (>15 μM) in mesocosms STRAT98-H1 and -H2. As a result, phytoplankton as a whole may have become P-limited from day 110 on, whereas diatoms may have encountered silicate limitation around day 130. This hypothesis was asserted by the chlorophyll-*a* concentrations, which showed a termination of the initial increase in all mesocosms after day 120.

Exp.98 ran from April to August, covered most of the spring phytoplankton/mesozooplankton growth period and showed a dynamic succession in ecological groups, as illustrated in Figure 5, where the concentrations of diatoms, *Phaeocystis*, microzooplankton and mesozooplankton in each mesocosm have been plotted against time. Three main periods could be distinguished during this experiment, each presenting contrasting trophic structures.

The first mixed period (days 90–139) was characterised by the development of a copepod community coincidentally with the initial diatom development (Fig. 5). A significant negative relation was found

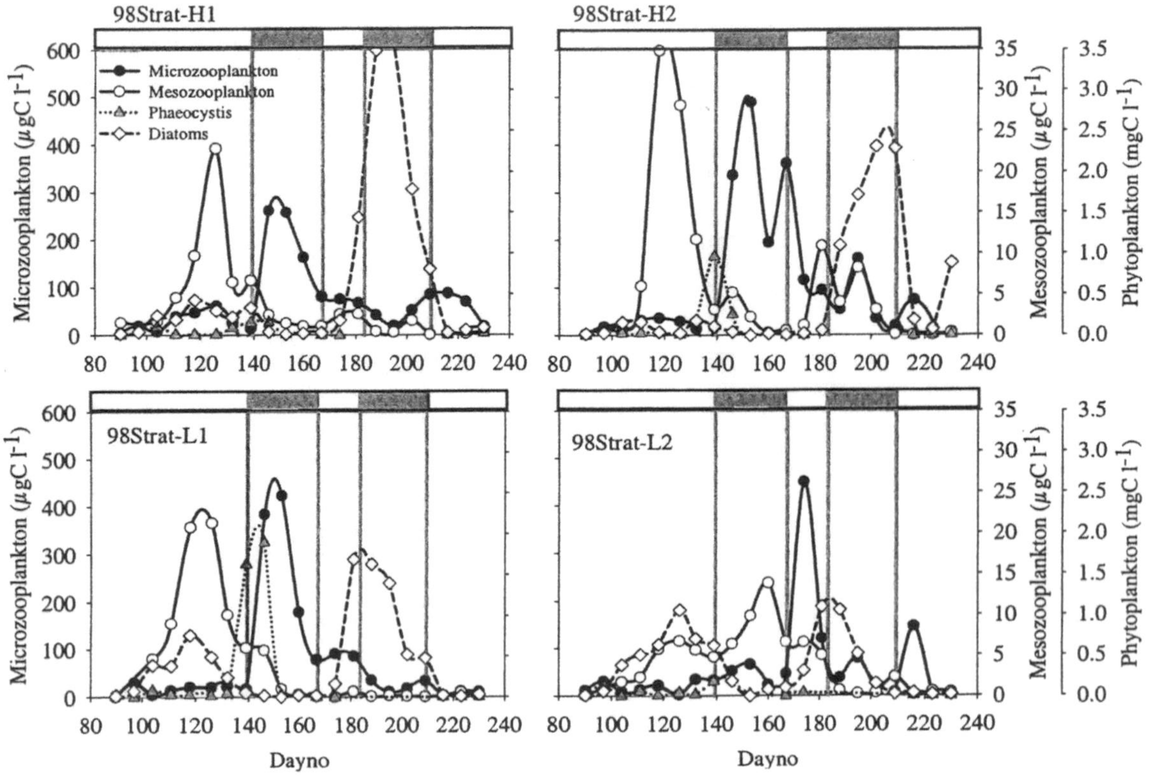

Figure 5. Microzooplankton, copepod and phytoplankton biomass (diatoms and *Phaeocystis* are distinguished) in the four mesocosms in 1998. The shading on top of the graphs indicates periods when mesocosm water column was stratified.

between the average biomass of copepods and diatoms during the mixing period (Fig. 6a).

The decrease in diatom biomass was followed with a sharp increase in *Phaeocystis* biomass around day 139 (Fig. 5), due to the development of a colonial bloom (colonies were reported but not quantified). The maximum level of biomass reached by *Phaeocystis* in all mesocosms at day 139 was inversely related to the diatom biomass measured on the same day (Fig. 6b).

The *Phaeocystis* colonial bloom culminated at the beginning of the stratification period and declined rapidly from that time onwards. This decline was accompanied by an increase in ciliate biomass that culminated about 1 week after the *Phaeocystis* bloom (Fig. 5). The maximum biomass levels of *Phaeocystis* and microzooplankton were compared (Fig. 6c). Our data were completed with observations from a previous mesocosm experiment (Summer 1994) (Wetsteyn & Vink-Lievaart, 1995; Escaravage et al., 1999) and from the field, Marsdiep (1990, 1992) (van Boekel et al., 1992; Brussard et al., 1995). *Phaeocystis* biomass was not supplied by the latter authors in their publication but was calculated by applying a cell biomass of 20 pg C (from our own database). This comparison showed a linear relation between the maximum biomass values of consecutive peaks of*Phaeocystis* and of ciliates (Fig. 6c).

The last period that was distinguished corresponded with the second stratification period and the final mixed period (Fig. 5). This period was characterised by the decrease of the microzooplankton community biomass and the increasing dominance of diatoms almost entirely represented by the large chain-forming species *Cerataulina pelagica.* A second microzooplankton development showed up in the last weeks of the experiment but did not reach the levels of the first bloom. Mesozooplankton also remained rather low during this period (Fig. 5).

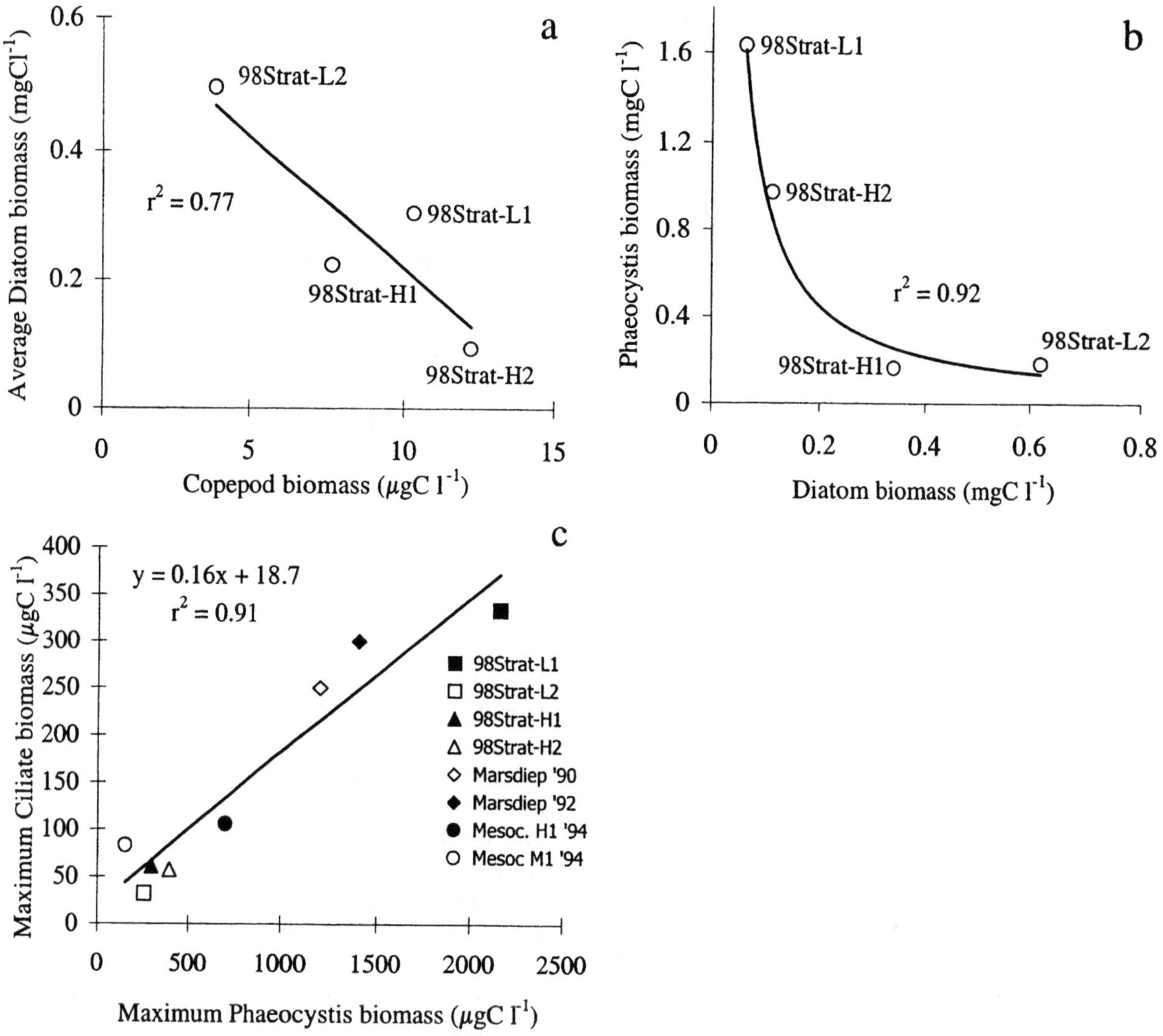

Figure 6. (a) Average biomass of diatoms and copepods during the initial mixing period in 1998. (b) Relation between diatom and *Phaeocystis* biomass on day 139 in 1997 (during the colonial bloom). (c) Maximum ciliate biomass as function of the preceding *Phaeocystis* bloom (all in μg C l^{-1}) in 1998, the Marsdiep (1990, 1992) (van Boeckel et al., 1992; Brussart et al., 1995) and previous mesocosm experiments (Mesocosm M1, H1 in 1994 in Wetsteyn & Vink-Lievaart, 1995; Escaravage et al., 1999).

Discussion

Silicate control of diatom dynamics

Twelve independent observations from mesocosm experiments described the tight control by silicate concentrations on the diatom spring bloom. In all cases, concentrations reaching values lower than 2 μM induced a termination of the diatom bloom (Fig. 1). This was in accordance with enclosure experiments by Egge & Asknes (1992), where silicate concentrations above 2 μM were required for the development of diatom-dominated phytoplankton. This is only verified, however, when silicate is the limiting factor. In earlier papers we showed that with N or P becoming limiting, a second generation of diatoms may develop and dominate the phytoplankton at low silicate concentrations (Escaravage et al., 1996, 1999). Species that develop during the initial bloom, when nutrients are available in excess, differ from species that develop when nutrients become limiting. During our mesocosm experiments, species from the first group were: *Rhizosolenia delicatula, Thalassiosira fallax, T. rotula, T. noordenskioldii, Chaetoceros socialis* and *Skeletonema costatum.* Diatoms from the same species or genus are described in the literature as typical species for spring bloom conditions. For example, *Rhizosolenia* and *Thallassiosira* in the North Sea (Mills et al., 1994), *Chaetoceros socialis* in the Funka Bay (Kudo et al., 2000) and in the Barents Sea (Rey & Skjoldal, 1987), *Skeletonema, Thalassiosira* and *Chaetoceros* in the mouth of the Yellow Sea (Cho et al., 1994).

The second group of diatoms that appeared in our experiments after N- or P-pools were exhausted, was formed by *Leptocylindrus danicus, Eucampia zoodiacus* and *Asterionella glacialis*. These diatoms are usually found in the mouth of the Oosterschelde estuary in late-spring/summer (Bakker et al., 1990). This also agreed with Egge & Asknes (1992) who observed that the diatom community, dominated by *Skeletonema* in N,P-enriched enclosures, became dominated by *Leptocylindrus danicus* when no nutrients were added. As a consequence, whereas silicate concentration below 2 μM may induce the termination of the diatom spring bloom, it may not exclude the development of typical summer diatom species when other nutrients (N or P) are also close to limiting values for phytoplankton growth.

Diatom–flagellate succession

Egge & Asknes (1992) described the 2-μM level as a threshold where a switch occurs between diatom- and flagellate-dominated systems. Instead of such a clear cut between the two kinds of systems, our data suggested that diatoms and flagellates initially grew simultaneously (Fig. 2). The apparent sharp increase of flagellate abundance at the end of the diatom bloom resulted from the exponential nature of concentration increase. Moreover, the decline in diatom concentrations did not influence flagellate development, suggesting that both groups grew independently. As far as *Phaeocystis* is concerned (the dominant non-diatom component in the present experiments) Peperzak (1993) identified high irradiance as a prerequisite for the build-up of the colonial bloom, whereas diatoms are relatively better adapted for growth at low light intensity (Langdon, 1988). This differential light requirement may be part of the mechanism inducing the time-lag observed in the field between the spring diatom and *Phaeocystis* blooms (Peperzak et al., 1998).

Flagellate-diatom development versus mixing rate and nutrient loading

Experiment 97

The positive response of both *Phaeocystis* and diatoms to mixing (Fig. 3) may have been due not only to the decrease of sedimentation losses but also to an increased nutrient availability in the highly productive zone located in the upper water layer. Under stratified conditions, nutrients added to the bottom water layer were not directly available for phytoplankton in the upper layer and frequently accumulated in the bottom layer.

Higher nitrogen addition resulted in increased phytoplankton biomass (Chlorophyll-*a*), mainly consisting of *Phaeocystis*. This resembled previous results (Lancelot, 1983; Escaravage et al., 1999) where increased nitrogen loadings induced higher *Phaeocystis* concentrations.

Experiment 98

The diatom/*Phaeocystis* succession occurring in Exp.98 was also affected by meso- and microzooplankton grazing processes (see below). The most striking feature in this experiment was the massive diatom bloom developing through the second stratification period (Fig. 5). The *C. pelagica* bloom developed, whereas silicate concentrations were below 2 μM. Microscopic observations showed that *C. pelagica* in the mesocosms had very thin walls (van der Voorn, pers comm.). This feature illustrates the finding that diatoms may reduce their Si cellular content when subject to silicate limitation (Harrison et al., 1977). One interesting aspect of this bloom was that this diatom was able to accumulate a large biomass in the water column despite the low mixing rate. Some studies suggest that diatoms may not be as heavy as is intuitively assumed. The voltage recordings experiments by Gradmann & Boyd (1995) support the idea that marine diatoms can adjust their buoyancy by the same machinery used by plant guard cells for the control of their volume by turgor. Direct observations of positive buoyancy and associated vertical migration ability have been made on numerous marine diatoms (Moore & Villareal, 1996). There is strong evidence from field observations that low mixing rate may not unequivocally promote flagellates to the detriment of diatoms (Taylor et al., 1985). These authors reported a 3-month lasting bloom of the diatom *C. pelagica* off the northeast coast of New Zealand in the summer of 1983. *C. pelagica* surface concentrations were one order of

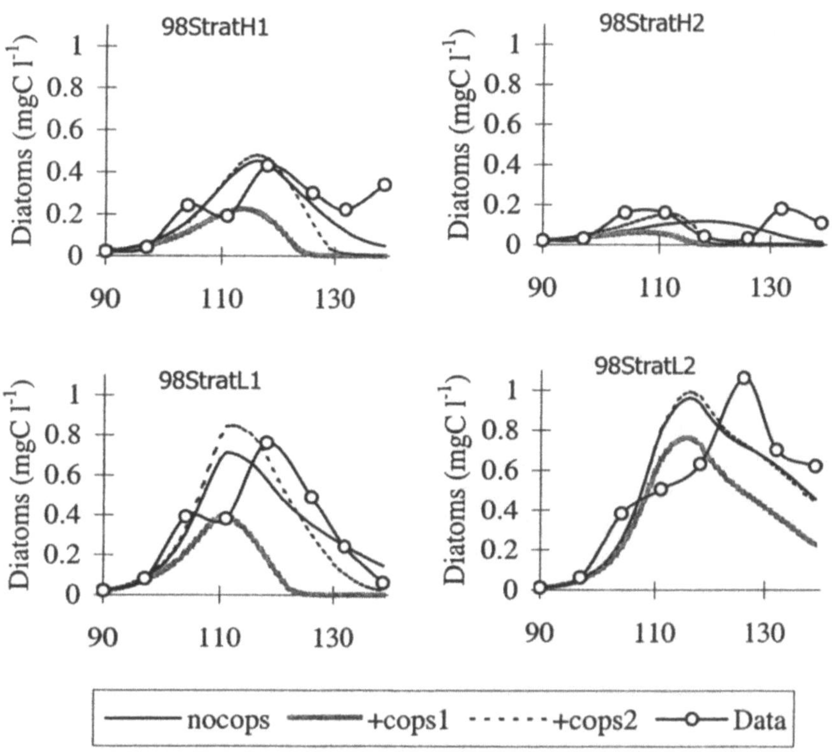

Figure 7. Simulation of copepod grazing on diatoms in Exp98. Actual diatom concentrations (Data) are compared with three simulations. In 'nocops', no copepod is present in the model (purely nutrient driven). In '+cops1', the copepod grazing is visualised by adding observed copepod biomass to the previous run (phytoplankton parameters unchanged). In '+cops2', the phytoplankton growth parameters are fitted after introduction of copepods in the model (see text for discussion).

magnitude higher than the bottom concentrations, suggesting that this diatom was actively maintaining itself in the well-illuminated upper part of the water column. The magnitude of this outburst resulted in high mortality of benthic shellfish and fish, mainly due to anoxia. This bloom was related to unusually calm conditions and temperatures 2 °C lower than the average summer value in this region. In our experiment, the *C. pelagica* bloom developed at temperatures between 16 and 18 °C and irradiance mostly below 30 E m^{-2} day^{-1}. In contrast, when *Prorocentrum micans* bloomed in mesocosm 97Strat-H1 in Exp. 97, temperature was around 23 °C and daily irradiance between 35 and 45 E m^{-2} day^{-1}. We know from our own cultures that *Prorocentrum micans* growth rate becomes optimal above 20 °C. Peperzak et al. (1996) identified stratification and high temperature (>19 °C) as crucial factors for the development of *Dinophysis acuminata* blooms off the Dutch coast. In a statistical analysis on phytoplankton species succession in the Oosterschelde estuary, Bakker et al. (1990) described *C. pelagica* as an early summer species, and *P. micans* as a summer species, which confirms their occurrence in the mesocosm under respectively low temperature, low light and high temperature, high light conditions.

Trophic interactions

Experiment 98

The ecological succession during the first mixed period illustrated how quickly food webs may switch from Bottom-up to Top-down controls. The negative relation between the intensity of the first diatom bloom and the copepod biomass (Fig. 6a) could be interpreted as a control of diatoms by copepod grazing (Top-down). A mathematical simulation, incorporating the daily interpolated data (nutrients, zooplankton) from the four mesocosms between days 90 and 139, was used to ascertain this hypothesis. Diatom biomass was allowed to grow following a nutrient-controlled exponential function $e^{r-d(1-Fn)}$ with r, intrinsic growth rate; d, max mortality rate (0.4 day^{-1}); Fn, $\min_{Si,P}(S/(K_s+S))$, K_s=2 and 1 μM for silicate and phosphorus, respectively. Copepod daily clearance rate was set as 0.02 l per μg C copepod (Hansen, 1995) without exceeding a biomass specific rate of 1.5 day^{-1} (Klein Breteler et al., 1990; Hansen & Boekel, 1991). Initial diatom biomass was measured on day 90. For the following dates of the series, data were iteratively simulated from the biomass calculated the day before and from the copepod biomass and nutrient concentration measured on the same day. The model was allowed to vary the phytoplankton intrinsic growth rates r in such a way as to minimise (Microsoft-Excel® Solver) the differences between the measured and the simulated diatom biomass. Three simulations were executed (Fig. 7): (1) calibration with no copepod grazing in the model (nocops); (2) simulation of copepod grazing applied after the calibration step (+cops1); (3) calibration with copepods in the model (+cops2). The first run estimated the average phytoplankton intrinsic growth rate at 0.26 day^{-1} and produced a good fit between the predicted and the observed data (Fig. 8). When copepods were allowed to graze on this community (without changing the growth parameter), the diatom biomass decreased by 25–50%. When the model with copepods was calibrated, the average phytoplankton intrinsic growth rate became 0.28 day^{-1} and the goodness of fit obtained with the first run was restored. These results support the hypothesis that copepods present in the mesocosms may have significantly controlled diatom biomass, as suggested by the negative relation between copepod and diatom biomass (Fig. 6a).

The termination of the copepod development coincided with decreasing diatom biomass that may have become resource-controlled after day 120 due to the exhaustion of the Si-pool (Bottom-up control).

Phosphorus and silicate pools became exhausted after days 110 and 130, respectively. Consequently, Si-limitation of diatoms occurred simultaneously with phosphorus limitation of non-diatom phytoplankton. As a result, *Phaeocystis* development was limited by the amount of phosphorus left over after diatom uptake, as suggested by the negative relation between the maximum biomass level of *Phaeocystis* (day 139) and the diatom biomass (Fig. 6b). The copepod control on diatom biomass combined with the *Phaeocystis* limitation by competition for P with diatoms, suggests that the development of *Phaeocystis* may have been promoted by copepod grazing on diatoms (potential competitor for *Phaeocystis*). Hansen et al. (1994) also described a positive effect of the copepod activity on the *Phaeocystis* blooming but in their case the involved mechanism was the copepod predation on microzooplankton that released the grazing pressure on *Phaeocystis*.

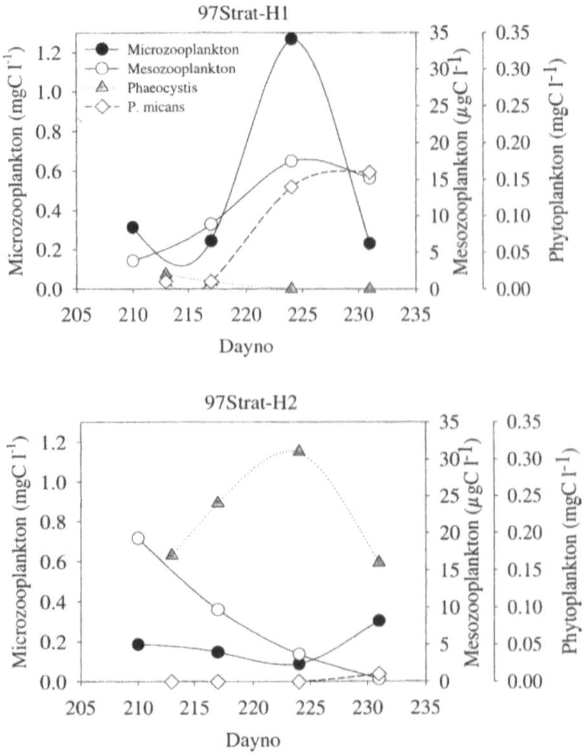

Figure 8. Microzooplankton, mesozooplankton and phytoplankton biomass (*Phaeocystis* and *P. micans* are distinguished) during the stratification period in mesocosms 97Strat-H1 and 97Strat-H2.

An exemplary pattern of a trophic cascade was given by the *Phaeocystis*/microzooplankton succession (Fig. 5). A similar *Phaeocystis*/ciliate trophic succession was described in the Marsdiep in 1990 by van Boekel et al. (1992) and in 1992 by Brussard et al. (1995). When biomass is considered, both Marsdiep measurements and results from the present experiment showed ciliate development reaching one quarter of the biomass of the preceding *Phaeocystis* bloom (Fig. 6c).

The fact that the ciliate peak generally occurred during the decline of the *Phaeocystis* bloom was attributed by van Boekel et al. (1992) to the fact that ciliates may not be able to graze on healthy colonies but rather well on the disrupted colony membranes and cells coming free at the end of the *Phaeocystis* bloom. A consequence of this delayed mode of grazing is that the ciliate grazing does not reduce *Phaeocystis* production, which could explain the positive relation between ciliate and *Phaeocystis* biomass. Consequently, enhanced *Phaeocystis* yield due to increased nutrient loadings should translate into increased ciliate biomass. This was confirmed from previous mesocosm experiments where microzooplankton biomass responded positively to nitrogen addition (Wetsteyn & Vink-Lievaart, 1995; Prins et al., 1999).

The systematic appearance of a microzooplankton bloom after the decline of the mesozooplankton community strongly suggested a trophic relation between the two groups (Fig. 5). In the first stratification period, three mesocosms showed a sharp increase in microzooplankton biomass, whereas it remained low in mesocosm 98Strat-L2. This was also the period when copepods showed a second biomass increase in 98Strat-L2, while already declined in the other mesocosms. Copepods developing exponentially in 98Strat-L2 may have exerted a direct predation or competitive exclusion on microzooplankton.

In order to estimate whether this grazing pressure could have been responsible for the low microzooplankton concentrations in 98Strat-L2, a comparison was made with mesocosm 98Strat-H1 that showed similarly low *Phaeocystis* biomass but where copepod biomass was also very low during the stratification period. Microzooplankton biomass increased between days 139 and 146 from 16 to 265 μg C l^{-1} in 98Strat-H1, corresponding to a specific growth rate of 0.40 day^{-1}. Copepod biomass at the beginning of the stratification period was 5 μg C l^{-1} in 98Strat-L2 and increased exponentially during the following days. Assuming that growing copepods may daily ingest up to 275% of their own biomass in the form of ciliates (Hansen et al., 1993), corresponds to a potential daily grazing pressure of 15 μg C l^{-1} for mesocosm 98Strat-L2, or about 50% of the microzooplankton biomass at day 139 (32 μg C l^{-1}). When it is assumed that microzooplankton, free of copepod grazing, should have grown in 98Strat-L2 at the same rate as in 98Strat-H1 (0.4 day^{-1}), it seems plausible to assume that copepod grazing (0.5 day^{-1}) may have controlled microzooplankton development in 98Strat-L2.

Although microzooplankton concentrations were lower in 98Strat-L2 than in the other mesocosms, small phytoplankton species did not respond positively to this decrease in microzooplankton grazing pressure. Many copepods are able to prey on both phytoplankton (filter feeding) and microzooplankton (ambush feeding) and may switch between the two feeding types depending on the availability of prey in the environment (Kiørboe, 1998). It is probable that copepods in 98Strat-L2 resorted to both food sources, acting at the same time as predators and food competitors of microzooplankton.

The massive bloom of *Cerataulina pelagica* developing after the first stratification period highlighted the hypothesis by Riegman et al. (1993) on the ability of large algae (mostly diatoms) to escape from grazing. Indeed the development of this bloom coincided with a decrease of the microzooplankton biomass which suggests the low edibility of such large (chain forming) phytoplankton for micro- and mesozooplankton.

Experiment 97

When the evolution of the main ecological groups identified in this study is plotted for the stratification period in 97Strat-H1 and Strat-H2, comparable processes as those described for Exp.98 seem to be responsible for the observed successional pattern (Fig. 8):

(1) low copepod biomass (<5 μg C l^{-1}) at the beginning of the stratification period in 97Strat-H1 could not control the microzooplantkon biomass, whereas the high copepod biomass (>15 μg C l^{-1}) present in 97Strat-H2, maintained the microzooplankton maintained at low levels;

(2) high microzooplankton biomass developing in 97Strat-H1 prevented *Phaeocystis* to bloom, thus increasing resource availability for *Prorocentrum micans*;

(3) large *Phaeocystis* biomass (colonial bloom) in 97Strat-H2 monopolized nutrient resources (P) and prevented the development of *Prorocentrum micans*.

The first question to be answered is whether the difference in copepod biomass between 97Strat-H1 and 97Strat-H2 could explain the difference in microzooplankton development. This problem was tackled with a simulation of the microzooplankton measured in mesocosms 97Strat-H1 and 97Strat-H2 between days 217 and 224. Starting from the biomass measured on day 217, simulated populations of microzooplankton were allowed to vary (daily time-step) depending on an intrinsic exponential growth process and a copepod-dependent loss function. Copepods (interpolated biomass from measured series) were allowed to graze on the microzooplankton biomass at a maximum rate of 2.75 day^{-1} (Hansen et al., 1993). The model was calibrated by adjusting the parameters for microzooplankton growth and mesozooplankton grazing, aiming at reproducing the microzooplankton biomass in the mesocosms on day 224. The experimental biomass of microzooplankton at day 217 in mesocosm 97Strat-H1 and 97Strat-H2, were fully recovered by applying 0.15 and 2.75 day^{-1} for the microzooplankton increase rate and the copepod grazing rate respectively. This rather large value for the copepod grazing corresponds to the upper range of experimental values presented in Hansen et al. (1993) for *Temora longicornis* feeding on microzooplankton. This result suggests that copepod grazing may have been one of the main factors explaining the different microzooplankton development observed in mesocosms 97Strat-H1 and 97Strat-H2.

The second question is whether *Phaeocystis* was controlled by microzooplankton, which could explain the low concentrations in 97Strat-H2 compared with 97Strat-H1. This was investigated by introducing the initial (on day 217) biomass of *Phaeocystis* in the model used previously and optimising the parameters for *Phaeocystis* growth and microzooplankton grazing in order to recover the experimental biomass of *Phaeocystis* measured on day 224 in the mesocosms. The experimental values of biomass were fully recovered by applying 0.56 day^{-1} for the *Phaeocystis* increase rate and 1.41 day^{-1} for the microzooplankton grazing rate. Microzooplankton biomass in 97Strat-H1 and 97Strat-H2 was mostly composed of the heterotrophic dinoflagellate *Oxyrrhis marina.* This species is known to graze very efficiently on autotrophic flagellates with maximum specific ingestion rates close to the present estimates: from 0.7 day^{-1} in Hansen et al. (1993) to 1.3 day^{-1} in Goldman et al. (1989), 1.9 day^{-1} in Hansen et al. (1996) and 2.7 day^{-1} in Hansen et al. (1997). This result supports the hypothesis that lower *Phaeocystis* concentrations in 97Strat-H1, compared with 97Strat-H2, could have been induced by the large microzooplankton biomass in this mesocosm. The removal of *Phaeocystis* sp. by overgrazing may have benefited *P. micans*, which is probably a less edible item due to its armoured shell and relatively large size (ca. 30 μm) when compared to the naked *Phaeocystis* cells (ca. 8 μm).

The third question that is motivated from the measurements in mesocosms 97Strat-H1 and 97Strat-H2 is, why did *P. micans* not develop in mesocosm 97Strat-H2 together with *Phaeocystis?* The stratification period was characterised by low phosphate concentrations (below 0.5 μM). Laboratory cultures have shown that the growth rate of *P. micans* was significantly lower than by *Phaeocystis*: up to 1 day^{-1} (Costas, 1990) and up to 2 day^{-1} (Peperzak, 1993) respectively. This lower growth rate presents a disadvantage for *P. micans* against *Phaeocystis* in a period of potential nutrient limitation and may explain its successful development in a mesocosm where its competitor was removed by microzooplankton grazing.

Conclusion

Diatom spring development appeared to be strictly controlled by silicate availability. Conversely, diatom summer species have shown their ability to compete successfully against flagellates for N or P at low silicate concentrations (<2 μM).

When grazing is not considered, the diatom/flagellate succession may be relatively well explained by this simple three-step scenario: (1) early spring, at low light and high nutrient availability, diatoms grow faster than the light-limited flagellates; (2) spring , light has increased but silicate is exhausted, diatom growth rate declines whereas light allows flagellates to grow faster; (3) late spring, flagellate growth may continue until exhaustion of either N and P.

Combined high nitrogen loadings and stratification events are favourable conditions for potentially harmful dinoflagellates (*Prorocentrum micans* here) but other factors, such as in our examples, grazing and temperature may play a decisive role in the appearance of such blooms. Despite their relatively low intrinsic growth rate, copepods were able to control diatoms and microzooplankton development. Microzooplankton efficiently incorporated blooms of small phytoplankton species. These results, combined with previous experiments where nitrogen enrichment induced increases in small phytoplankton biomass, stress the positive effect of N-enrichment on the microbial food web. Conversely, the bloom of the diatom *C. pelagica* exemplified the notion of trophic mismatch leading to the accumulation of algal biomass without trophic transfer to higher trophic levels.

Acknowledgements

This research was funded by the RIKZ/National Institute for Coastal and Marine Management under projects BEON*EUTRO*ECO, PenP-Eutrofiëring en productiviteit, and the EU under the MAST-III programme (ELOISE, MAS3-CT96-0053-PHASE). We gratefully aknowledge the following persons for their contribution to the present paper. The daily maintenance of the mesocosms was ensured by Arjen Pouwer. The lab analyses were performed by Wim Schreurs and staff (Analytical Laboratory of RIKZ-Middelburg). Microscopic work was done by Elmira Boikova and staff (Latvia University-Salaspils) / (Phytoplankton), Peter de Vries (NIOO-CEMO)/ (Microzooplankton) and Marjo Vink (RIKZ) / (Mesozooplankton). The database management was ensured by Cynthia van der Voorn (NIOO-CEMO). We also thank Louis Peperzak and two referees for comments on earlier versions of this paper. This is publication number 3042 of the NIOO Centre for Estuarine and Coatal Ecology, Yerseke, The Netherlands.

References

Bakker, C., P. M. J. Herman & M. Vink, 1990. Changes in seasonal succession of phytoplankton induced by the storm-surge barrier in the Oosterschelde (S.W. Netherlands). J. Plankton Res. 12: 947–972.

Banse, K., 1982. Cell volumes, maximal growth rates of unicellular algae and ciliates in the marine pelagial. Limnol. Oceanogr. 27: 1059–1071.

van Boekel, W. H. M., F. C. Hansen, R. Riegman & R. P. M. Bak, 1992. Lysis-induced decline of a *Phaeocystis* spring bloom and coupling with the microbial food web. Mar. Ecol. Prog. Ser. 81: 269–276.

Brussard, C. P. D., R. Riegman, A. A. M. Noordeloos, G. C. CadJe, H. Witte, A. J. Kop, G. Nieuwland, F. C. van Duyl & R. P. M. Bak, 1995. Effects of grazing, sedimentatiom and phytoplankton cell lysis on the structure of a coastal pelagic food web. Mar. Ecol. Prog. Ser. 123: 259–271.

Cho, B. C., J. K. Choi, C. S. Chang & G. H. Hong, 1994. Uncoupling of bacteria and phytoplankton during a spring diatom bloom in the mouth of the Yellow Sea. Mar. Ecol. Prog. Ser. 115: 181–190.

Costas, E., 1990. Genetic variablity in growth rates of marine dinoflagellates. Genetica 83: 99–102.

Cushing, D. H., 1989. A difference in structure between ecosystems in strongly stratified waters and in those that are only weakly stratified. J. Plankton Res. 11: 1–13.

Edler, L. & P. Olsson, 1985. Observations on diel migration of *Ceratium furca* and *Prorocentrum micans* in a stratified bay on the Swedish west coast. In: Anderson, White & Baden (eds), Toxic Dinoflagellates. Elsevier, Amsterdam: 195–200

Egge, J. K. & D. L. Asknes, 1992. Silicate as regulating nutrient in phytoplankton competition. Mar. Ecol. Prog. Ser. 83: 281–289.

Eppley, R. W., O. Holm-Hansen & J. D. H. Strickland, 1968. Some observations on the vertical migration of dinoflagellates. J. Phycol. 4: 333–340.

Escaravage, V., T. C. Prins, A. C. Smaal & J. C. H. Peeters, 1996. The response of phytoplankton communities to phosphorus input reduction in mesocosm experiments. J. Exp. Mar. Biol. Ecol. 198: 55–79.

Escaravage, V., T. C. Prins, C. Nijdam, A. C. Smaal & J. C. H. Peeters, 1999. Response of phytoplankton communities to nitrogen input reduction in mesocosm experiments. Mar. Ecol. Prog. Ser. 179: 187–199.

Gieskes, W. W. C. & G. W. Kraay, 1984. Phytoplankton, its pigments, and primary production at a central North Sea station in May, July and September 1981. Neth. J. Sea Res. 18: 51–70.

Gillbricht, M., 1988. Phytoplankton and nutrients in the Helgoland region. Helgol. Meeresunters. 42: 435–467.

Goldman, J. C., M. R. Dennett & H. Gordin, 1989. Dynamics of herbivorous grazing by the heterotrophic dinoflagellate *Oxyrrhis marina*. J. Plankton Res. 11: 391–407.

Gradmann, D. & C. M. Boyd, 1995. Membrane voltage of marine phytoplankton, measured in the diatom *Coscinodiscus radiatus*. Mar. Biol. 123: 645–650.

Hansen, B., P. G. Verity, T. Falkenhaug, K. S. Tande & F. Norrbin, 1994. On the trophic fate of *Phaeocystis pouchetii* (Harriot). V. Trophic relationships between *Phaeocystis* and zooplankton: an assessment of methods and size dependence. J. Plankton Res. 16: 487–511.

Hansen, F. C., 1995. Trophic interactions between zooplankton and *Phaeocystis cf. globosa*. Helgol. Meeresunters. 49: 283–293.

Hansen, F. C. & W. H. M. van Boekel, 1991. Grazing pressure of the calanoid copepod *Temora longicornis* on a *Phaeocystis* dominated spring bloom in a Dutch tidal inlet. Mar. Ecol. Prog. Ser. 78: 123–129.

Hansen, F. C., M. Reckermann, W. C. M. Klein Breteler & R. Riegman, 1993. *Phaeocystis* blooming enhanced by copepod predation on protozoa: evidence from incubation experiments. Mar. Ecol. Prog. Ser. 102: 51–57.

Hansen, F. C., H. J. Witte & J. Passarge, 1996. Grazing in the heterotrophic dinoflagellate *Oxyrrhis marina*: size selectivity and preference for calcified *Emiliania Huxleyi* cells. Aquat Microb Ecol 10: 307–313.

Hansen, P. J., P. K. Bjørnsen & B. W. Hansen, 1997. Zooplankton grazing and growth: Scaling the 2–2,000-μm body size range. Limnol. Oceanogr. 42: 687–704.

Harrison, P. J., H. L. Conway, R. W. Holmes & C. O. Davis, 1977. Marine diatoms grown in chemostats under silicate or ammonium limitation III. Cellular chemical composition and morphology of *Chaetoceros debilis*, *Skeletonema costatum*, and *Thalassiosira gravida*. Mar. Biol. 43: 19–31.

Kiørboe, T., 1998. Population regulation and role of mesozooplankton in shaping marine pelagic food webs. Hydrobiologia 363: 13–27.

Kiørboe, T., H. Kaas, B. Kruse, F. Mrhlenberg, P. Tiselius & G. Ertejerg, 1990. The structure of the pelagic food web in relation to water column structure in the Skagerrak. Mar. Ecol. Prog. Ser. 59: 19–32.

Klein Breteler, W. C. M., H. G. Fransz & S. R. Gonzalez, 1982. Growth and development of four calanoid copepod species under experimental and natural conditions. Neth. J. Sea Res. 16: 195–207.

Klein Breteler, W. C. M., N. Shogt & S. R. Gonzalez, 1990. On the role of food quality in grazing and development of life stages, and genetic change of body size during cultivation of pelagic copepods. J. Exp. Mar. Biol. Ecol. 135: 177–189.

Kudo, I., T. Yoshimura, M. Yanada & K. Matsunaga, 2000. Exhaustion of nitrate terminates a phytoplankton bloom in Funka bay, Japan: change in SiO_4:NO_3 consumption rate during the bloom. Mar. Ecol. Prog. Ser. 193: 45–51.

Kuiper, J., 1977. Development of North Sea coastal plankton communities in separate plastic bags under identical conditions. Mar. Biol. 44: 97–107.

Lancelot, C., 1983. Factors affecting phytoplankton extracellular release in the Southern Bight of the North Sea. Mar. Ecol. Prog. Ser. 12: 115–121.

Langdon, C., 1988. On the cauess of interspecific differences in the growth-irradiance relationship for phytoplankton. II A general review. J. Plankton Res. 10: 1291–1312.

Lessard, E. J., 1991. The trophic role of heterotrophic dinoflagellatesin diverse marine environments. Mar. Microb. Food Webs 5: 49–58.

Mills, D. K., P. B. Tett & G. Novarino, 1994. The spring bloom on the South Western North Sea in 1989. Neth. J. Sea Res. 33: 65–80.

Moore, J. K. & T. A. Villareal, 1996. Buoyancy and growth characteristics of three positively buoyant marine diatoms. Mar. Ecol. Prog. Ser. 132: 203–214.

Olsson, P. & E. Granéli, 1991. Observations on diurnal vertical migration and phased cell deivision for three coexisting marine dinoflagellates. J. Plankton Res. 13: 1313–1324.

Peeters, J. C. H. & L. Peperzak, 1990. Nutrient limitation in the North Sea: a bioassay approach. Neth. J. Sea Res. 26: 61–73.

Peperzak, L., 1993. Daily irradiance governs growth rate and colony formation of *Phaeocystis* (Prymnesiophyceae). J. Plankton Res. 15: 809–821.

Peperzak, L., G. J. Snoeijer, R. Dijkema, W. W. C. Gieskes, J. Joordens, J. C. H. Peeters, C. Schol, E. G. Vrieling & W. Zevenboom, 1996. Development of a *Dinophysis acuminata* bloom in the River Rhine plume (North Sea). In Yasumoto T., Y. Oshima & Y. Fukuyo (eds), Harmful and Toxic Algal Blooms. Unesco, Sendai: 273–276

Peperzak, L., F. Colijn, W. W. C. Gieskes & J. C. H. Peteers, 1998. Development of the diatom-*Phaeocystis spring* bloom in the Dutch coastal zone of the North Sea: the silicon depletion versus the daily irradiance threshold hypothesis. J. Plankton Res. 20: 517–537.

Prins, T. C., V. Escaravage, A. J. Pouwer, L. P. M. J. Wetsteyn & H. A. Haas, 1995. Description of mesocosms, methods and comparison with North Sea conditions. In Peeters, J. C. H., A. C. Smaal, H. A. Haas & C. H. R. Heip (eds), The Impact of Marine Eutrophication on Phytoplankton and Benthic Suspension Feeders. Progress Report II: Results of Mesocosm Experiments with Reduced N-load and Increased Grazing Pressure. Report RIKZ-95.048, NIOO-CEMO 1995-3, Middelburg, The Netherlands: 10–31.

Prins, T. C., V. Escaravage, L. P. M. J. Wetsteyn, J. C. H. Peeters & A. C. Smaal, 1999. Effects of different N- and P-loading on primary and secondary production in an experimental marine ecosystem. Aquat. Ecol. 33: 65–81.

Putt, M. & D. K. Stoecker, 1989. An experimentally determined carbon:volume ratio for marine 'oligotrichous' ciliates from estuarine and coastal waters. Limnol. Oceanogr. 34: 1097–1103.

Rey, F. & H. R. Skjoldal, 1987. Consumption of silicic acid below the euphotic zone by sedimenting diatoms in the Barents Sea. Mar. Ecol. Prog. Ser. 36: 307.

Riegman, R., B. R. Kuipers, A. A. M. Noordeloos & H. J. Witte, 1993. Size-differential control of phytoplankton and the structure of plankton communities. Neth. J. Sea Res. 31: 255–265.

Schøllhorn, E. & E. Granéli, 1993. Is the increase of flagellates in coastal waters caused by changes in ratios of N, P and Si? In Smayda, T. J. & Y. Shimizu (eds), Toxic Phytoplankton Blooms in the Sea. Elsevier, Amsterdam: 811–817

Schøllhorn, E. & E. Granéli, 1996. Influence of different nitrogen to silica ratios and artificial mixing on the structure of a summer phytoplankton community from the Swedish west coast (Gullmar fjord). J. Sea Res. 36: 159–167.

Smayda, T. J., 1978. from phytoplankters to biomass. In Sournia, A. (ed.), Phytoplankton Manual. Unesco Press, Paris: 273–279

Smayda, T. J., 1990. Novel and nuisance phytoplankton blooms in the sea: evidence for a global epidemic. In Granéli, E., B. Sundstrøm, L. Edler & D. M. Anderson (eds), Toxic Marine Phytoplankton. Elsevier, Amsterdam: 29–40

Sommer, U., 1994. Are marine diatoms favoured by high Si:N ratios. Mar. Ecol. Prog. Ser. 115: 309–315.

Steedman, H. F., 1976. General and applied data on formaldehyde fixation and preservation of marine zooplankton. In Steedman, H. F. (ed.), Zooplankton Fixation and Preservation. Monographs on Oceanographic Methodology 4. Unesco Press, Paris: 103–154

Taylor, F. J., N. J. Taylor & J. R. Walsby, 1985. A bloom of the planktonic diatom, *Cerataulina pelagica*, off the coast of northeastern New Zealand in 1983, and its contribution to an associated mortality of fish and benthic fauna. Int. Rev. ges. Hydrobiol. 70: 773–795.

Utermøhl, H., 1958. Zur vervollkommung der quantitativen Phytoplankton Methodik. Mitt. int. Verein. Limnol. 9: 1–39.

Wassmann, P., 1998. Retention versus export food chains: processes controlling sinking loss from marine pelagic systems. Hydrobiologia 363: 29–57.

Wetsteyn, L. P. M. J. & M. Vink-Lievaart, 1995. Zooplankton dynamics in N-manipulated mesocosms. In Peeters, J. C. H., A. C. Smaal, H. A. Haas & C. H. R. Heip (eds), The Impact of Marine Eutrophication on Phytoplankton and Benthic Suspension Feeders. Progress Report II: Results of Mesocosm Experiments with Reduced N-load and Increased Grazing Pressure. Report RIKZ-95.048, NIOO/CEMO-1995-3, Middelburg: 82–113.

Hydrobiologia **484:** 49–59, 2002.
O. Vadstein & Y. Olsen (eds), Sustainable Increase of Marine Harvesting: Fundamental Mechanisms and New Concepts.

The influence of turbidity on growth and survival of fish larvae: a numerical analysis

Øyvind Fiksen, Dag L. Aksnes, Magni H. Flyum & Jarl Giske
Department of Fisheries and Marine Biology, University of Bergen, P.O. Box 7800, N-5020 Bergen, Norway
E-mail: oyvind.fiksen@ifm.uib.no

Key words: larval feeding, model, phytoplankton concentration, recruitment, turbidity, vision

Abstract

Growth and survival through the early larval phase probably limit the production potential of many commercially important fish stocks. Attempts to increase the production of these stocks by restocking of juveniles have generally failed. Here, we analyse how enhanced concentrations of phytoplankton and zooplankton affect the survival of fish larvae during their early life stages. The analysis is developed for larvae feeding on copepod eggs and nauplii, with fish and invertebrates as major predators. A model of feeding and growth of fish larvae is applied to assess the benefit of enhanced phytoplankton and zooplankton abundance. The analysis shows that the shading effect of higher phytoplankton concentration may reduce predation rates on fish larvae substantially. This 'top-down' effect may be more important for the cohort survivorship than the 'bottom-up' mechanism in situations when larval food is sufficiently abundant. However, while increased algal biomass will improve recruitment at high zooplankton concentration, it may also reduce recruitment at low zooplankton concentrations and shallow mixing depths. Both the larvae and their vertebrate predators are dependent on light to detect their prey, and the longer reactive distance of the predators make them more susceptible than the larvae to reduced light levels and increased turbidity. We discuss the implications of reduced predation and increased zooplankton abundance on recruitment and production of fish larvae, and point at environmental conditions where changing algal biomass is likely to affect recruitment success.

Introduction

The issue of how the production of food from the ocean can be increased has gained interest as the world fish catch has levelled out (Cushing, 1996). Improved management of exploited stocks is an obvious solution to increase yields and avoid recruitment overfishing. Typically, fish are food- or density-regulated at a young age, while carrying capacities of exploited age classes are often above current stock levels. Consequently, the number of recruits that are supplied to the stock frequently limits the production of fish biomass and yield to fisheries. To accommodate this, large-scale rearing of juvenile fish and extensive sea ranching projects have been launched. However, these attempts have frequently failed, as the juveniles are expensive to produce and seem to have low survival rates after release (Svåsand et al., 2000).

An alternative approach to facilitate the supply of recruits to exploited stocks may be to improve the conditions for growth and survival through critical periods, such as the first feeding and larval stages, when larval fish are particularly vulnerable. The growth (and therefore survival) of fish larvae is often limited by the availability of food organisms (e.g., Ellertsen et al., 1989; Fossum, 1996). The single most important food item for many fish larvae in the North Atlantic is the nauplii of the calanoid copepod *Calanus finmarchicus*. There seems to be at least some correlation between growth and production of these nauplii and the abundance of autotrophs in the sea (e.g., Richardson et al., 1999). According to this reasoning, it may be possible to increase the number of recruits by artificially enhancing the primary production in the main nursery or other selected areas of these fish stocks.

We acknowledge that this is a question that requires multiple approaches and extensive scientific effort to be answered. More recruits will not increase production of fish if strong density-dependent regulation operates at juvenile or adult stages (Giske & Salvanes, 1999). Neither will nutrient enrichment have any effect on zooplankton and fish unless the ecosystem response to enrichment is such that energy can be transferred to higher trophic levels (Hulot et al., 2000).

We anticipate and analyse three mechanisms by which increased phytoplankton concentration affects growth and survival of fish larvae: positively through (1) the potential for higher food abundance and (2) reduced predation pressure from visually searching predators and negatively by (3) reduced feeding and growth rates due to more turbid water. We use an individual-based model of growth of herring larvae as a function of environmental parameters (Fiksen et al., 1998; Fiksen & Folkvord, 1999). Then we add predation risk and model the effects of prey, predator and chlorophyll concentrations on the recruitment success of the fish larvae cohort.

The model

Feeding and growth of fish larvae

The individual based model (IBM) of herring larvae is presented by Fiksen & Folkvord (1999). Based on theoretical and experimental studies, they developed a model of prey encounter, prey capture, growth and metabolism for larval herring. All these processes were size-dependent and tied to environmental variables such as temperature, turbulence, light, turbidity, prey concentration and prey size structure. Here, this model is extended to include predation risk from visually searching planktivores such as fish, and the focus will be on how growth, predation risk and starvation of fish larvae are affected by changes in the concentration of phytoplankton.

Both larvae and their predators are modelled as visual, raptorial foragers that detect their prey by sight (Fig. 1). This implies that the optical properties of the water column are important not only for the ability of the larvae to feed, but also for the probability that predators will detect it. The encounter rate between a herring larva and its prey is modelled by the cylinder equation

$$e = \frac{1}{2}\pi R_1^2 N V, \qquad (1)$$

where R_l is the reactive distance (visual range) of the larvae, N is prey abundance and V is the relative velocity (including swimming and small scale turbulent motion) between the larvae and the prey. The reactive distance were calculated by use of a model first derived by Aksnes & Giske (1993) and later elaborated by Aksnes & Utne (1997):

$$R_1^2 \exp(cR_1) = CA_p E' \frac{E_b}{K_e + E_b}. \qquad (2)$$

Here, E' is the size-specific sensitivity of the visual system of the larvae (Fiksen et al., 1998), which specifies the ratio of the weakest and the strongest difference between the reflected light from the prey and the local background irradiance (E_b) that can be recognised by the larvae. The satiation parameter (K_e) defines how visual range increase with light, A_p is the area of the prey and C is the inherent contrast (visibility) of the prey. The local illumination level E_b at depth z is a function of surface irradiance E_s and absorption in the water column:

$$E_b = E_s \exp(-Kz). \qquad (3)$$

The diffuse attenuation (extinction) coefficient K will normally vary through the water column, but here we assume a homogenous, well-mixed upper layer where K is constant. We also assume a fixed relationship between K and the beam attenuation coefficient (c), such that $c = 3K$ (Kirk, 1980). The coefficients c and K thus specify the overall turbidity of the water column, and the extinction and scattering of light both before and after reflection from the prey (s and a in Fig. 1). The main source of turbidity in oceanic environments is phytoplankton. From Riley (1956), we found the following relationship between K and chlorophyll a (Chl$a$, mg m^{-3}):

$$K = k_0 + 0.054\text{Chl}a^{(2/3)} + 0.0088\text{Chl}a. \qquad (4)$$

In the absence of Chl$a$, the extinction coefficient is k_0 (= 0.1 m^{-1}). The details of the larval growth and foraging processes are described in Fiksen & Folkvord (1999). The relation between growth and ingestion is taken from Kiørboe et al. (1987), and the prey capture success is modelled as in Heath (1993). Small scale turbulence is accounted for in the encounter rate (V in Equation (1)) and in the capture success calculation (MacKenzie et al., 1994). The larva starves (dies) if its

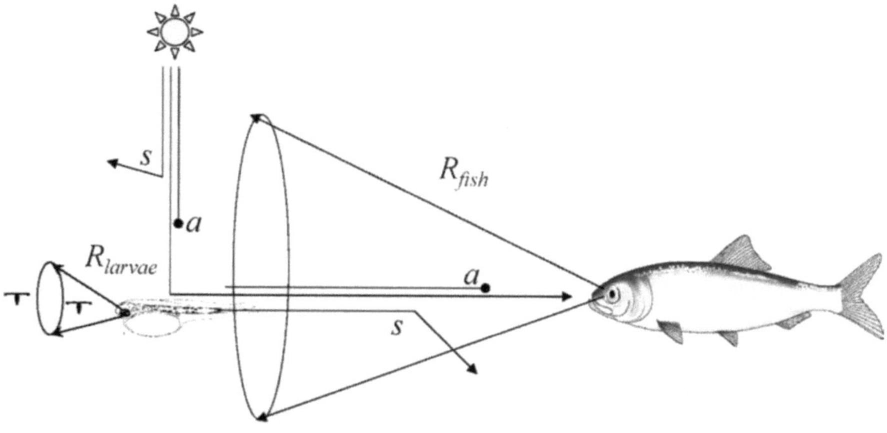

Figure 1. The basic principles of the encounter rate kernel in the model. Light is absorbed (*a*) and scattered (*s*) as it penetrates the water column, depending on the amount and type of particles in the water. Light is reflected (and absorbed) at the surface of prey (copepods or fish larvae), and this reflected light may be absorbed or scattered before reaching the eye of the larvae or the fish. Scattering and absorption depend on the properties of the particles; materials such as clay has a high scattering/absorption ratio; while, e.g., mud absorb rather than scatter light. Due to more sensitive eyes and larger prey, the visual range of the fish (R_{fish}) is generally much larger than the visual range of the larvae (R_{larva}). An encounter occurs when the predator is able to separate the prey image from the background irradiance. Modified from Aksnes & Giske (1993).

body mass is less than 80% of the expected body mass at any specific length.

Predation risk

Both invertebrates and fish are important predators of larval fish. The vulnerability of the larvae depends on the ability these predators have to detect and capture their prey (Bailey & Houde, 1989). Generally, as the larvae grow, both the encounter rate with predators and the ability to escape an attack increase. For most ambush or entangling invertebrate predators (or filtering fish) the prey encounter rate changes little with prey size, while the capture success drops rapidly as prey gets larger. For visually feeding fish, the encounter rate increases considerably with prey size, and the capture success remains quite high, although it will also decline with prey size (Folkvord & Hunter, 1986; Bailey & Houde, 1989).

The feeding efficiency of fish is highly dependent on reactive distance (Aksnes & Giske, 1993). As in Giske et al. (1994), we assume predation from fish to be proportional to visual range squared:

$$\mu_{\text{f}} \propto (1 - P_{\text{e}})R^2{}_{\text{f}} = k(1 - P_{\text{e}})R^2{}_{\text{f}}. \qquad (5)$$

Here, P_{e} is the probability of escape after an encounter and k is a constant of proportionality comprising all other factors such as predator density, efficiency and swimming speed. The probability of escaping an encounter with fish is a function of larval size (L, in mm) and the relative size of the predator and the prey (Bailey & Houde, 1989). From experiments on adult anchovies feeding on larval anchovies (Folkvord & Hunter, 1986), we let

$$P_{\text{e}} = \frac{0.92}{1 + \exp(\frac{L - 16.3}{4.13})}. \qquad (6)$$

The visual range of the planktivore is calculated as in Equation (2), only replacing A_{p} with the area of the larvae (0.1 L^2) and E' with a constant (10^4, Fig. 2). Since the area increases with larval size, the detection distance and encounter rate also increase with the size of the larvae (Fig. 2). On the other hand, the susceptibility to planktivores decreases with length. With the assumptions made above, the vulnerability of herring larvae to fish predation will increase up to about 17 mm, and then decrease. The predation risk as a function of larval size for different values of k is shown in Figure 3.

Estimates of mortality rates in the sea demonstrate a strong reduction with increased body size (Peterson & Wroblewski, 1984; McGurk, 1986; Bailey & Houde, 1989). A number of causes (e.g., developmental malfunction, infections, parasites, and size spectra of predators) could produce this pattern. Since predation risk of small and undeveloped fish larvae may not show strong size-dependence (Leggett & deBlois, 1994), we added a term to include all such effects:

$$\mu_n = aL^b, \qquad (7)$$

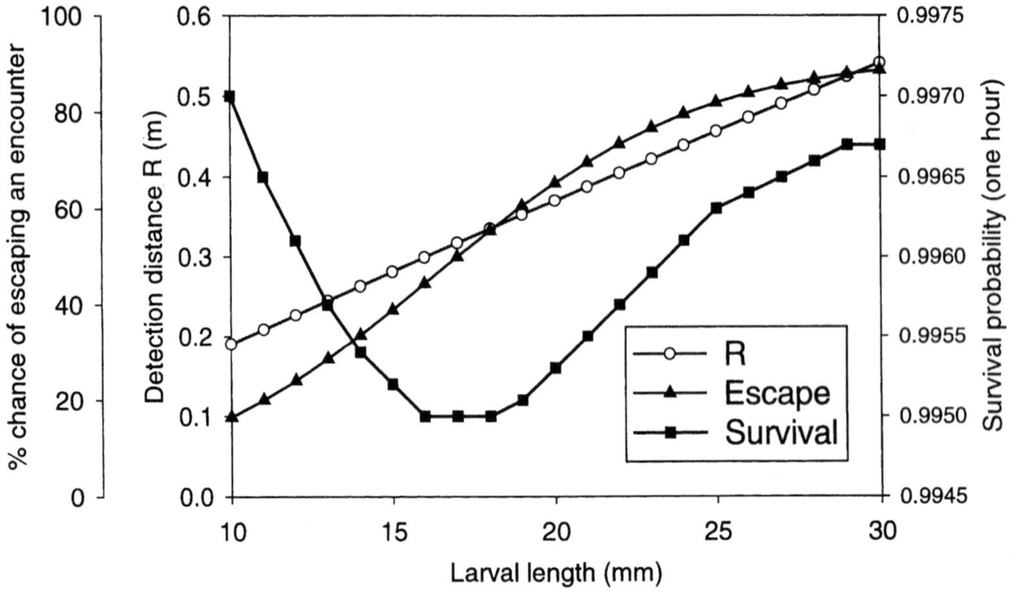

Figure 2. The relationship between the detection distance of the planktivorous predator, the probability of escape, and the probability of survival for a range of larval lengths. In this example, invertebrate predation rate is zero ($a = 0$), $k = 0.1$, depth is 10 m and time is midday (irradiance at surface was 300 μmol m^{-2} s^{-1}).

where a and b are constants (see McGurk (1986) or Bailey & Houde (1989) for reviews of mortality rates in larval fish). At their default values, the parameters $a = 0.1$ h^{-1} and $b = -1.3$ will cause visual and invertebrate predation rates to be similar (and reasonable; 0.01 h^{-1} or ~0.2 d^{-1}) in environments where fish are efficient (Fig. 3). The survival probability P_s is then calculated each hour from the exponential probability distribution:

$$P_s = \exp(-\mu_n - \mu_f). \quad (8)$$

If the body mass of the larvae are 80% or less of the expected value at any length, then the larvae starves and $P_s = 0$. The probability of $P_i(\Omega)$ individual i to have survived Ω h after hatching is

$$P_i(\Omega) = \prod_{h=0}^{H=\Omega} P_s. \quad (9)$$

Results

Vertical profiles of growth and predation risk

Equipped with these equations and assumptions, we can model the multiple effects of increased chlorophyll a levels in the water column on growth and predation risk of fish larvae. First, we present some static, vertical profiles of growth and survival of a 15-mm long larvae (Fig. 4). The profiles show that growth is temperature-limited (food is available in excess) near the surface, while food is increasingly limiting in deeper waters. At higher algal biomass the visual field of the larvae is reduced so that the profitable part of the water column (where growth is positive) is severely reduced. For first-feeding larvae, which is too small to search for benign conditions, this may increase the risk of starvation. At the same time, predation risk from planktivores is also reduced, and depending on food abundance, predation pressure and vertical distribution of the larvae the level of standing stock of algae may be beneficial or detrimental to recruitment. More turbid waters will reduce ambient light at depth, but visibility will be reduced even at similar levels of diffuse light (see Fig. 1). The longer reactive distance of piscivorous predators relative to their larval prey makes them more susceptible to scattering and absorbency within their reactive volume. The volume effectively scanned per unit time by larger fish will therefore drop faster than the volume scanned by fish larvae as turbidity increase (Fig. 5). If, however, the contrast between prey and the background also improve with turbidity (e.g., Utne-Palm, 1999), then the search volume will be less reduced, or even increased, with turbidity under a given ambient light level (Fig. 5).

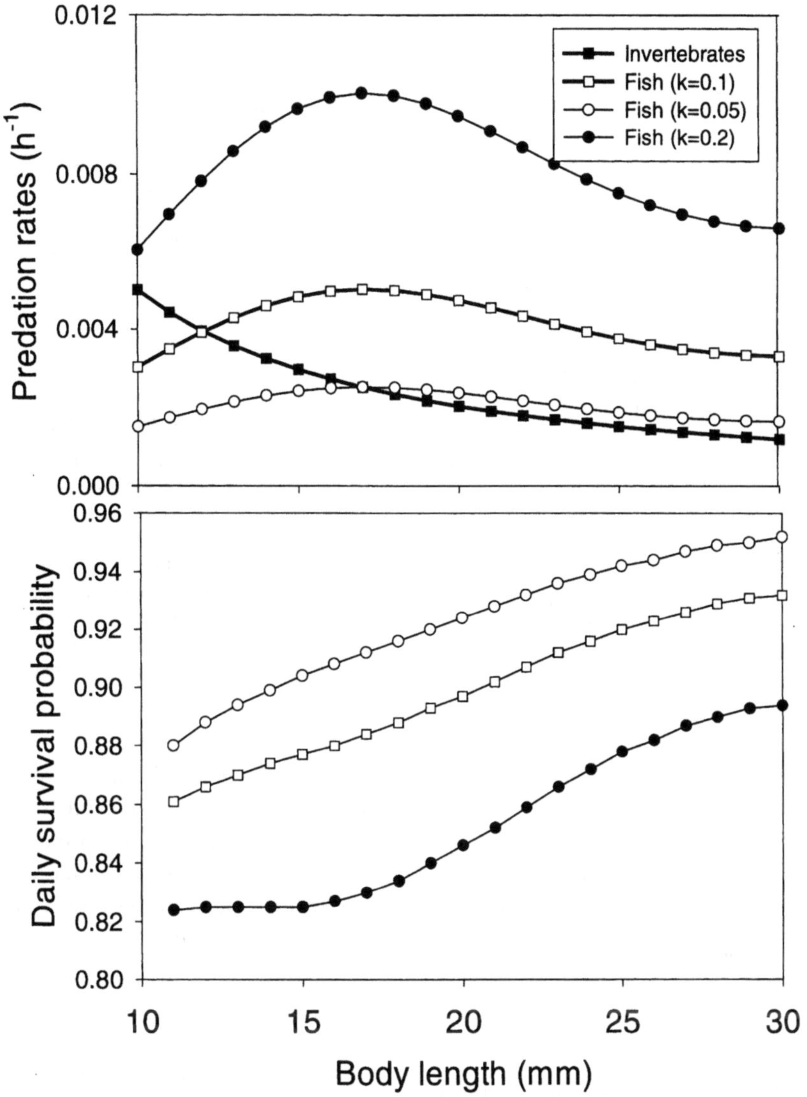

Figure 3. Predation rates at midday when visual predation is at its maximum level (upper panel) and integrated daily survival probabilities (vulnerabilities) under different levels of fish predation pressure and larval sizes (lower panel). Parameters for invertebrate predation rates are: $a = 0.1$ and $b = -1.3$, and depth = 10 m. Surface irradiance fluctuate from near zero at night to 300 μmol m^{-2} s^{-1} at midday.

The effect on predation risk from a layer of highly turbid water close to the surface is shown in Figure 6. For larger fish, the scattered light in the turbid water creates a refuge with low detection risk. The risk of being detected by predators for smaller fish is less dependent on turbidity. The same reasoning also applies to foraging: large fish searching for large prey will encounter fewer prey in the turbid layer compared to the deeper and clearer water, while small fish will find less food purely as a function of ambient light (Fig. 6). Giske et al. (1994) found the same general pattern using an earlier version of the model for the visual range (Aksnes & Giske, 1993). The revised model of visual range (Aksnes & Utne, 1997), which includes a term for satiation of the eye as light increases, shows however that the vertical differences in predation risk are less prominent.

Survival of a cohort in different mixing depths

To assess the combined effects of reduced growth and predation risk in a homogeneous water column under increasing chlorophyll concentrations, we simulated a cohort of 500 newly hatched larvae drifting passively with currents within a surface mixed layer of 10, 20 and 30 m depth (Fig. 7). These larvae are subject to

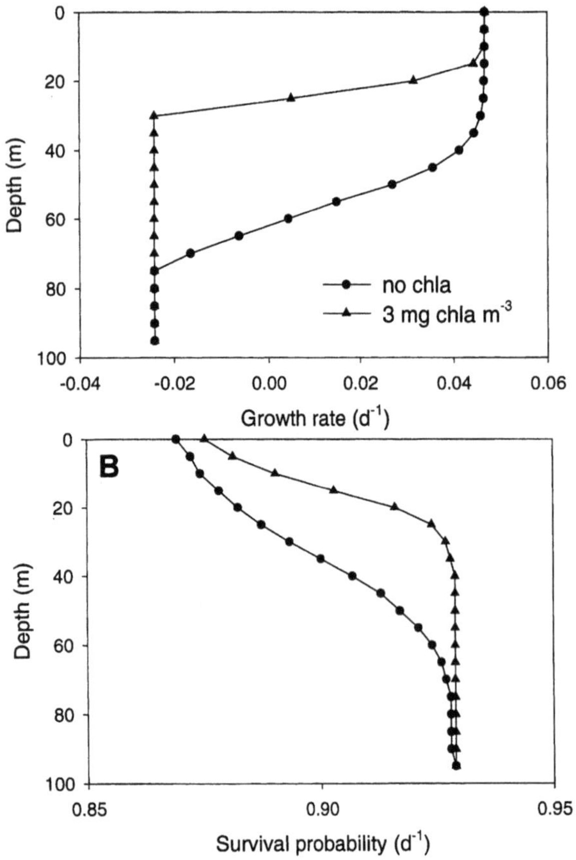

Figure 4. The profiles of (A) growth and (B) survival probability during 1 day for a herring larva (15 mm) staying at different depths in clear (no Chl*a*) and turbid (3 mg Chl*a* m^{-3}) water. The simulation use $N = 5$ mg dry weight nauplii and copepodite prey m^{-3} and 10 m s^{-1} wind stress at surface (see Fiksen & Folkvord, 1999, for details).

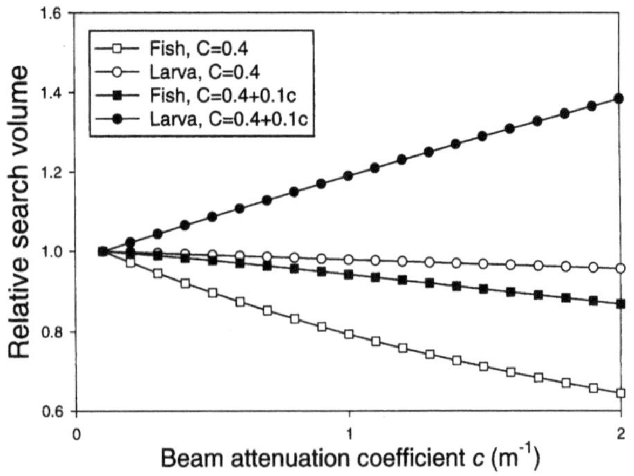

Figure 5. The effect of the beam attenuation coefficient c (turbidity) on the search volume of a herring larva (15 mm) eating nauplii and a fish eating the larva. When contrast C is not affected by turbidity (open symbols), both predator and prey experience reduced feeding abilities, although the impact is mainly on the predator. If the contrast is only slightly improved (assuming there is a statistical relation between C and c so that $C = 0.4 + 0.1c$) as turbidity increase (black symbols), the larvae will improve its search volume, while the predator's search volume is still reduced by turbidity.

mortality as expressed above and grow according to the amount of food they are able to eat each day. After 50 days, the population is evaluated by the cohort survival index (S):

$$S = \frac{1}{N} \sum_{i=1,N} w_i P_i, \qquad (10)$$

where w_i is the body mass (μg) and P_i is the survival probability (zero if starvation occurs) of individual i after 50 days. This index will return high values when growth and survival are high. An additional penalty is put on slowly growing individuals, since they also suffer higher risk of predation (Fig. 3).

The results show that at prey densities of 5 mg dw m^{-3} increased turbidity will cause lower recruitment (S) for all mixing depths (Fig. 7A). At this resource level, the larvae will be food limited during their first feeding period, and at deep mixing depths many larvae starve when turbidity is high. As food availability increase to 7.5 (Fig. 7B) and 10 (Fig. 7C) mg dw m^{-3}, the effect of lowered predation rates from fish (due to higher turbidity) becomes more important to the survivorship index. At these prey concentrations, higher turbidity (from the increased chlorophyll content) is predicted to be beneficial to recruitment provided the mixing depth does not exceed ~20 m (as long as the larvae cannot search for the optimal depth). Some combinations of prey and mixing depth seems to give maximum S at intermediate levels of turbidity. The actual change in S as a function of turbidity will depend in part on the intensity of visual predation on the cohort (Fig. 8). The relative change in larval growth and survival may be quite opposite in high and low predation regimes.

The latter prediction is made under the assumption that no extra zooplankton biomass is produced from the enhanced algal concentration. However, a striking feature of Figure 9 is that increasing prey abundance from 5 to 10 mg dw m^{-3} may increase S with about an order of magnitude. At high chlorophyll *a* levels (4 mg m^{-3}), the expected effect of satiation with increasing prey concentration is prevented (Fig. 9). The vertical aspects of prey encounter rate in the pelagic can explain this feature (Fig. 4). In clear water, the larvae in the bottom part of the mixed layer will still be encountering prey at reasonable rates, but under more turbid

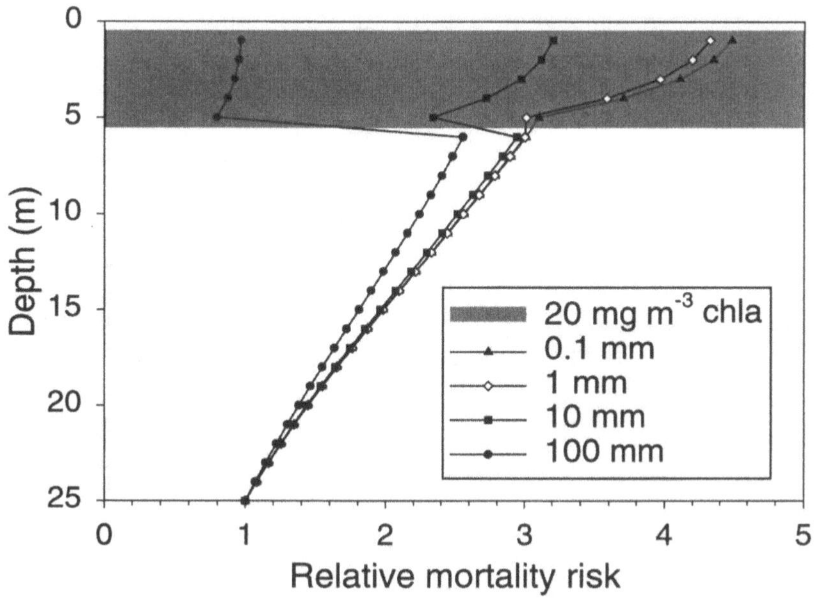

Figure 6. Chlorophyll as a refugium. Recalculation of Figure 6 in Giske et al. (1994) using the revised model of visual range (Aksnes & Utne 1997). The general picture is the same as the original, but the profiles are considerably less sensitive to depth (diffuse light).

conditions the growth of these larvae will be limited by prey encounter. Therefore, more phytoplankton or turbidity will augment the effect of higher food concentrations (Fig. 9). More zooplankton will ensure that the larvae grow more rapidly through the critical first feeding stages and the turbidity will provide shelter against predators as they grow larger. However, this effect will only have significant impact on recruitment success at quite high prey concentrations (>8 mg m^{-3}).

Discussion

Some limitations of the model

The model presented here is a simplified representation of the foraging process of fish larvae and their fish predators. We have analysed how growth and survival of a cohort of fish larvae (herring) respond to nutrient enrichment and stimulated production by autotrophs. In contrast to traditional food-web analyses, our attention was on how the altered optical properties of the water column modified predator–prey relationships. This is obviously an important aspect of the interaction between visually searching foragers such as fish larvae and planktivorous (or piscivorous) fish and their prey. The model is not intended to explore the trophic nutrients–phytoplankton–zooplankton–fish larvae dynamics. The model does suggest, however, that fish larvae with developing brain and nervous system will benefit from increasing food concentration during the first feeding phase (Figs 7 and 9, Fiksen & Folkvord, 1999). This is due to the low prey encounter and capture rates during the first feeding stages, and the effect of more food would persist even at quite high concentrations (Fig. 9). Larger larvae will become satiated at far lower prey densities, as they are more efficient both in finding and capturing prey.

Another limitation of the model is the lack of behavioural responses, and diel vertical migration in particular. The simulations shown in Figure 7 assume that the fish larvae drift passively with convection in the mixed layer. If the larvae were allowed to select their optimal position in the vertical, they would stay nearer the surface as turbidity increased, and deeper as they grew older. This behaviour would greatly offset the negative influence of the deeper mixed layer, and also reduce the probability of encountering planktivorous fish during daytime. On the other hand, only the slightest diel vertical migration of prey organisms (nauplii) or variation in prey contrast over the day (diel feeding patterns) would greatly reduce food availability of the larvae (Giske et al., 1994). In a model including optimal diel vertical migration of zooplankton, Fiksen & Giske (1995) found that increased resource levels would reduce the exposure of zooplankton to predat-

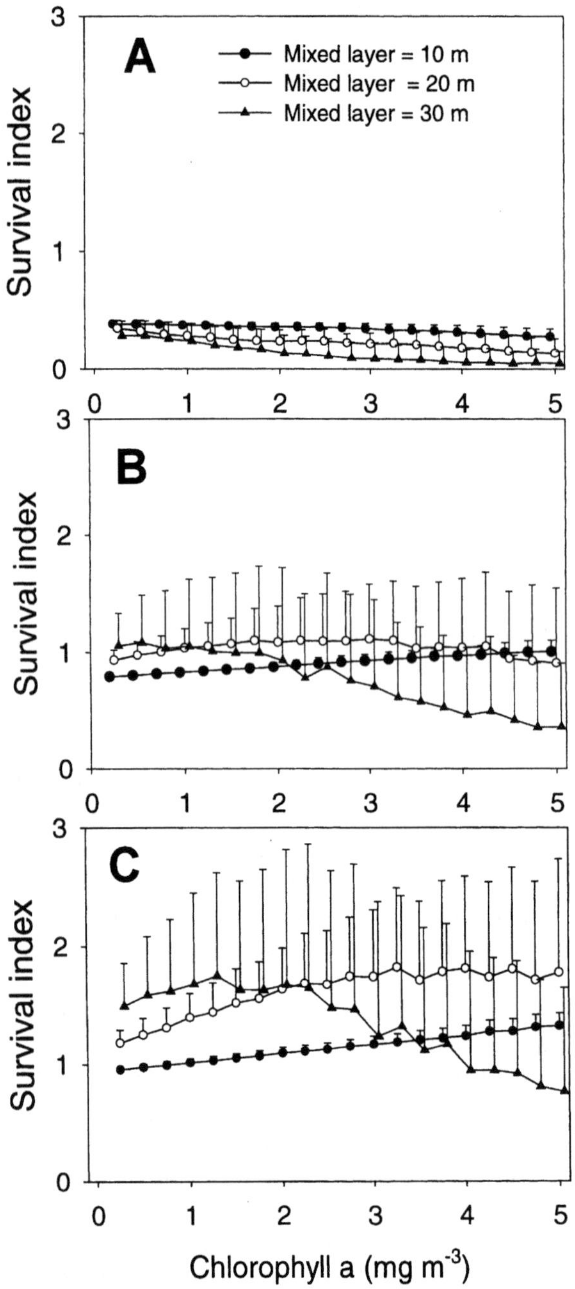

Figure 7. The effect of increasing chlorophyll levels on survival index *S* (Equation (10)) at (A) 5, (B) 7.5 and (C) 10 mg dry mass of prey m^{-3} for three different mixed layer depths, other parameters are as in the previous simulations. A concentration of 1 mg dry mass of prey m^{-3} would correspond to about one *Calanus* nauplii (stage 4) per l, and in the nursery areas of Norwegian spring spawning herring the density of nauplii would typically vary between 0 and ~10 per l at the time of first feeding (Fossum, 1996). The simulation is initiated with 500 identical, newly hatched larvae that are tracked for 50 days. Their position in the mixed layer is assumed to be governed by convection, and is modelled as a purely stochastic process. The survival index *S* is the average body mass (μg, dry) of all 500 individuals multiplied by the corresponding survival probability (see Equation (10)). The error bars represent 1 SD.

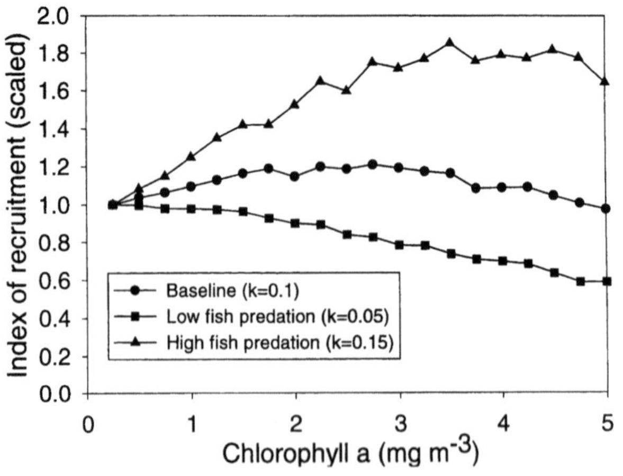

Figure 8. The relative recruitment success (scaled to its lowest value) at increasing phytoplankton densities and three scenarios of fish predation. The simulation is similar to Figure 7, with 7.5 mg prey dw m^{-3} and a mixing depth of 20 m (no wind).

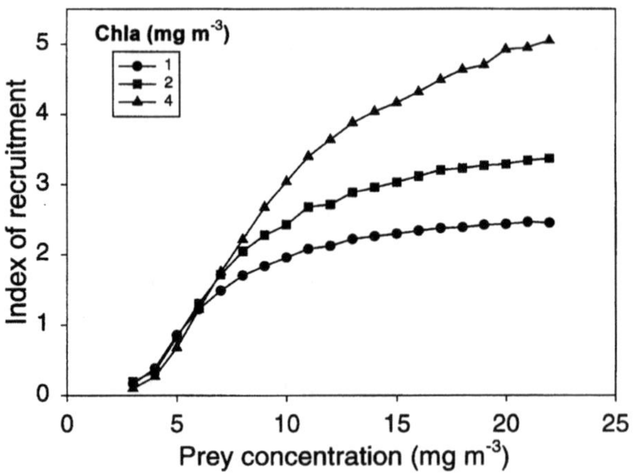

Figure 9. Recruitment success scheduled as a function of prey and chlorophyll concentration. All other parameters are as in Figures 7 and 8.

ors. The copepods can then stay in deeper and darker waters and still maintain high growth rates, but at lower risk of predation from fish. Also the predation risk estimation is sensitive to behaviour, as small pelagic fish and macroplankton are of the sizes that would most benefit by remaining in the turbid layer (Fig. 6). In a study of optimal vertical migration in mesopelagic fish, Rosland & Giske (1994) found that the fish would benefit by remaining in the turbid surface water for prolonged feeding during dusk and dawn.

Our model does not include behaviour, but we outline general mechanisms of how turbidity will modify growth and survival in larval fish. Although flexible behaviour can modify our conclusions, the direction of the predictions should persist even if behaviour were included. The role of behavioural flexibility in both prey and predators is obviously a topic that deserves further attention in aquatic ecology (Verity & Smetacek, 1996).

Mechanisms other than those included in the model may operate on the success of larvae at high phytoplankton concentrations. Some fish larvae can feed on phytoplankton directly (van der Meeren, 1991), but this is not evaluated in the model presented here.

Turbidity in laboratory and field studies

A range of experiments have been conducted to assess the role of turbidity on foraging of fish (e.g., Vinyard & O'Brien, 1976; Boehlert & Morgan, 1985; Chesney, 1989; Miner & Stein, 1993; Utne, 1997; Utne-Palm, 1999). These studies have shown various relations between turbidity and foraging success. Surprisingly, fish seems to have higher feeding rates at intermediate levels of turbidity (Boehlert & Morgan, 1985; Bristow & Summerfelt, 1994; Bristow et al., 1996; Utne, 1997; Utne-Palm, 1999), at least at high ambient light conditions. At low light, the turbidity reduces ingestion success (Miner & Stein, 1993). One mechanism put forward to explain these observations is that the contrast between the background and the prey is increased in water with more particles (Hinshaw, 1985; Utne-Palm, 1999). At low light levels, the detection of prey may instead be limited by absolute light levels rather than contrast. Miner & Stein (1993) found that more turbidity caused larval bluegill to select smaller prey (at high light levels), and Vinyard & O'Brien (1976) and Reid et al. (1999) observed that the reactive distance of large prey was severely reduced as turbidity increased. Utne (1997) and Utne-Palm (1999) also found that gobies had longer reactive distance under slightly turbid conditions, but at higher turbidity the reactive distance dropped. These results conform well to predictions from the model of visual range used here (Aksnes & Giske, 1993; Aksnes & Utne, 1997), where larger prey will be less frequently encountered in turbid water relative to smaller prey (Equation (2)).

A fairly small improvement of contrast with turbidity is required to reverse the negative effect of turbidity at a given light intensity (Fig. 5). However, from the empirical studies listed above, it does not seem likely that the effect of turbidity on prey contrast will persist at high turbidity, as assumed in the simulations presented in Figure 5. It is likely that the benefit of contrast levels off at increasing turbidity (Utne-Palm, 1999), and that the reduced visibility eventually becomes the dominant factor.

In the cultivation of fish larvae, it has been shown that larva walleye in turbid tanks feed and grow at rates superior to larvae in clear water (Bristow & Summerfelt, 1994; Bristow et al., 1996). If the water is clear, their larvae showed a tendency to cling to the walls (where light is reflected) of the tank and have low feeding success. The more diffuse light in turbid tanks may have prevented light reflection from the walls and thus the phototactic behaviour of clinging to the walls, and at the same time provided improved contrast to the food particles (Bristow & Summerfelt, 1994; Bristow et al., 1996). Bristow et al. (1996) compared the effect of a dye (Aquashade, absorbing light at wavelengths required for photosynthesis) with clear and turbid (from adding clay) water on growth and survival of larval walleye (*Stizostedion vitreum*). It turned out that the dye reduced performance of the larvae, while clay was beneficial. The cones of perch (*Perca fluviatilis*) absorb light at wavelengths similar to those of Aquashade, and therefore may block out visual stimuli (Bristow et al., 1996). It is difficult to speculate on whether different types of algae may produce similar results. Given the potential importance of algae in generating turbidity, there are surprisingly few studies reported on their effect on the reaction distance, growth and survival of fish. These mechanisms are not included in the model, but then they may also be 'tank-effects' that do not apply to natural settings.

The role of estuaries, river plumes and fronts in the recruitment success of fish has been examined at several locations (e.g., Fox et al., 1999; Grimes & Finucane, 1991; Grimes & Kingsford, 1996; Sirois & Dodson, 2000). Such areas are characterised by high productivity and are often more turbid and turbulent than other oceanic sites. Fox et al. (1999) found that tidally induced turbidity probably reduced the fraction of the water column suitable for feeding of herring larvae in the Blackwater Estuary (Essex, England). However, no reduction in feeding incidence was recorded over the tidal cycle, and feeding seemed more correlated to surface irradiance. Similarly, Sirois & Dodson (2000) did not find significant differences in feeding rate among larval rainbow smelt (*Osmerus mordax*) from a turbid and a less turbid region of the St. Lawrence estuary. Laboratory experiments confirmed this, and showed that larvae in turbid water grew better although feeding rates were similar. Larvae in turbid tanks probably had a lower activity level and allocated a higher proportion of ingested energy to somatic growth (Sirois & Dodson, 2000). Grimes & Finucane (1991) found higher chlorophyll and zooplankton concentrations and growth rates of ichthyoplankton in frontal areas compared to adjacent waters. Grimes & Kingsford (1996) reviewed the literature of the effects of river discharge plumes on larval fish, and their conclusion was unequivocal. Apparently, some species seem to be more adapted to take advantage of enhanced food production than others, therefore various species may respond differently to fronts.

In relation to our model, these observations emphasise the potential of individual behaviour in modifying the consequences of environmental factors such as turbidity. In clear water, the risk of being detected by predators increases, and the behavioural response may be to reduce the level of activity or to descend in the water column. At intermediate turbidity, the larvae will be less susceptible to visual predators, and thus less vigilant and more efficient in searching for food or in transforming ingested food into growth. At some point, however, turbidity will certainly reduce both feeding and growth rates.

Nutrient enrichment and recruitment to fish stocks

Hjort (1914) suggested that the variations in year class strength of major fish stocks could be caused by lack of food during the first feeding period. Fossum (1996) found clear evidence that the survival of first feeding Norwegian spring-spawning herring larvae were correlated to the temporal overlap with prey, and Ellertsen et al. (1989) found similar results for Arcto-Norwegian cod. Additionally, the model developed by Fiksen & Folkvord (1999) supported the 'critical period' concept, since first feeding larvae have short reactive distance, low prey capture success and limited foraging abilities. Although there is a general consensus that fish larvae are food-limited (Cushing, 1996), the stage at which food is most limiting is not clear. Cushing (1996) reviewed this issue, and concluded that the late larval stage was the best candidate to predict recruitment. This would imply that as the larva grows, its food demand would increase more than its ability to find and capture prey. Naturally, density-dependent survival at older stages would impede the effects of enhanced survival during the early larval phase.

In conclusion, the model predicts higher phytoplankton concentrations to be favourable to recruitment. The most prominent effect of higher algal biomass is obtained when both predation risk from visual predators and the food availability (zooplankton) is high. If there is linear food-chain dynamics, so that more nutrients are converted into phytoplankton and zooplankton biomass, then nutrient enrichment is likely to augment the recruitment success of fish. Whether the food-web response is likely to be linear is, however, not obvious (e.g., Hulot et al., 2000). Several studies have looked at how egg production of *Calanus finmarchicus* relates to primary production, and there seems to be a positive correlation when corrected for the effect of lipid reserves (Richardson et al., 1999). Therefore, higher primary production and growth or egg production of zooplankton in the first feeding areas of larval fish probably enhance the survival of larvae and recruitment to fish stocks.

Acknowledgements

We thank Anne Christine Utne-Palm for valuable discussions and the Research Council of Norway for a grant to ØF (133562/122).

References

Aksnes, D. L. & J. Giske, 1993. A theoretical model of aquatic visual feeding. Ecol. Model. 67: 233–250.

Aksnes, D. L. & A. C. W. Utne, 1997. A revised model of the visual range in fish. Sarsia 82: 137–147.

Bailey, K. M. & E. D. Houde, 1989. Predation on eggs and larvae of marine fishes and the recruitment problem. Adv. mar. Biol. 25: 1–83.

Boehlert, G. W. & J. B. Morgan, 1985. Turbidity enhances feeding abilities of larval Pacific herring, *Clupea harengus pallasi*. Hydrobiologia 123: 161–170.

Bristow, B. T. & R. C. Summerfelt, 1994. Performance of larval walleye cultured intensively in clear and turbid water. J. World Aquacult. Soc. 25: 454–464.

Bristow, B. T., Summerfelt R. C. & R. D. Clayton, 1996. Comparative performance of intensively cultured larval walleye in clear, turbid, and colored water. Prog. Fish Cult. 58: 1–10.

Chesney, E. J., 1989. Estimating the food requirements of striped bass larvae *Morone saxatilis*: effects of light, turbidity and turbulence. Mar. Ecol. Prog. Ser. 5: 191–200.

Cushing, D. H., 1996. Towards a science of recruitment in fish populations. In: Kinne, O. (ed.), Excellence in Ecology, Vol. 7. Ecology Institute, Oldendorf/Luhe.

Ellertsen, B., P. Fossum, P. Solemdal & S. Sundby, 1989. Relation between temperature and survival of eggs and first-feeding larvae of northeast Arctic cod (*Gadus morhua* L.). Rapp. PV. Reun. Cons. Int. Exp. Mer. 191: 209–219.

Fiksen, Ø. & A. Folkvord, 1999. Modelling growth and ingestion processes in herring *Clupea harengus* larvae. Mar. Ecol. Prog. Ser. 184: 273–289.

Fiksen, Ø. & J. Giske. 1995. Vertical distribution and population dynamics of copepods by dynamic optimization. ICES J. mar Sci. 52: 483–503.

Fiksen, Ø., A. C. W. Utne, D. L. Aksnes, K. Eiane, J. V. Helvik & S. Sundby, 1998. Modelling the influence of light, turbulence and ontogeny on ingestion rates in larval cod and herring. Fish. Oceanogr. 7: 355–363.

Folkvord, A. & J. R. Hunter, 1986. Size-specific vulnerability of northern anchovy, *Engraulis mordax*, larvae to predation by fishes. Fish. Bull. 84: 859–869.

Fossum, P., 1996. A study of first-feeding herring (*Clupea harengus* L.) larvae during the period 1985–1993. ICES J. mar. Sci. 53: 51–59

Fox, C. J., R. Harrop & A Wimpenny, 1999. Feeding ecology of herring (*Clupea harengus*) larvae in the turbid Blackwater estuary. Mar. Biol. 134: 353–365.

Giske, J. & A. G. V. Salvanes, 1999. A model of enhancement potentials in open ecosystems. In Howell, B. R., E. Moksnes & T. Svåsand (eds), Stock Enhancement and Sea Ranching. Fishing News Books, Blackwell: 22–36.

Giske, J., D. L. Aksnes & Ø. Fiksen, 1994. Visual predators, environmental variables and zooplankton mortality risk. Vie Milieu 44: 1–9

Grimes, C. B. & J. H. Finucane, 1991. Spatial distribution and abundance of larval and juvenile fish, chlorophyll and macrozooplankton around the Mississippi River discharge plume, and the role of the plume in fish recruitment. Mar. Ecol. Prog. Ser. 75: 109–119.

Grimes, C. B., M. J. Kingsford, 1996. How do riverine plumes of different sizes influence fish larvae: do they enhance recruitment? Mar. freshwat. Res. 47: 191–208.

Heath, M. R., 1993. The role of escape reactions in determining the size distribution of prey captured by herring larvae. Environ. Biol. Fish. 38: 331–344

Hinshaw, J. M., 1985. Effects of illumination and prey contrast on survival and growth of larval yellow perch *Perca flavescens*. Trans. am. Fish. Soc. 114: 540–545.

Hjort, J., 1914. Fluctuations in the great fisheries of Northern Europe viewed in the light of biological research. Rapp. PV. Reun. Cons. Int. Exp. Mer. 20: 1–228

Hulot, F. D., G. Lacroix, F. Lescher-Moutoué & M. Loreau, 2000. Functional diversity governs ecosystem response to nutrient enrichment. Nature 405: 340–344.

Kirk, J. T. O. 1980. Relationships between nephelometric turbidity and scattering coefficients in certain Australian waters. Austr. J. mar. Freshwat. Res. 31: 1–12.

Kiørboe, T., P. Munk & K. Richardson, 1987. Respiration and growth of larval herring *Clupea harengus*: relation between specific dynamic action and growth efficiency. Mar. Ecol. Prog. Ser. 40: 1–10.

Leggett, W. C. & E. DeBlois, 1994. Recruitment in marine fishes: Is it regulated by starvation and predation in the egg and larval stages? Neth. J. Sea. Res. 32: 119–134.

MacKenzie, B. R., T. J. Miller, S. Cyr & W. C. Leggett, 1994. Evidence for a dome-shaped relationship between turbulence and larval fish ingestion rate. Limnol. Oceanogr. 39: 1790–1799

McGurk, M. D., 1986. Natural mortality of marine pelagic fish eggs and larvae: role of spatial patchiness. Mar. Ecol. Prog. Ser. 34: 227–242.

Miner J. G. & R. A. Stein, 1993. Interactive influence of turbidity and light on larval bluegill (*Lepomis macrochirus*) foraging. Can. J. Fish. aquat. Sci. 50: 781–788.

Peterson, I. & J. S. Wroblewski, 1984. Mortality rate of fishes in the pelagic system. Can. J. Fish. aquat. Sci. 41: 1117–1120.

Reid, S. M., M. G. Fox & T. H. Whillans, 1999. Influence of turbidity on piscivory in largemouth bass (*Micropterus salmoides*). Can. J. Fish. aquat. Sci. 56: 1362–1369.

Richardson, K., S. H. Jónasdottir, S. J. Hay & A. Christoffersen, 1999. *Calanus finmarchicus* egg production and food availability in the Faroe-Shetland Channel and northern North Sea: October–March. Fish. Oceanogr. 8 (Suppl. 1): 153–162.

Riley, G. A., 1956. Oceanography of the Long Island Sound 1952–1954. II Physical oceanography. Bull. Bing. Oceanogr. Coll. 15: 15–46.

Rosland, R. & J. Giske, 1994. A dynamic optimization model of the diel vertical distribution of a pelagic planktivorous fish. Prog. Oceanogr. 34: 1–43.

Sirois, P. & J. J. Dodson, 2000. Influence of turbidity, food density and parasites on the ingestion and growth of larval rainbow smelt *Osmerus mordax* in an estuarine turbidity maximum. Mar. Ecol. Prog. Ser. 193: 167–179.

Svåsand, T., T. S. Kristiansen, T. Pedersen, A. G. V. Salvanes, R. Engelsen, G. Nævdal & M. Nødtvedt, 2000. The biological and economical basis of cod stock enhancement. Fish Fish. 1: 173–205.

Utne, A. C. W., 1997. The effect of turbidity and illumination on the reaction distance and search time of the marine planktivore *Gobiusculus flavescens*. J. Fish Biol. 50: 926–938.

Utne-Palm A. C. 1999. The effect of prey mobility, prey contrast, turbidity and spectral composition on the reaction distance of *Gobiusculus flavescens* to its planktonic prey. J. Fish Biol. 54: 1244–1258.

van der Meeren, T., 1991. Algae as first food for cod larvae, *Gadus morhua* L.: filter feeding or ingestion by accident? J. Fish Biol. 39: 225–237.

Verity, P. G. & V. Smetacek. 1996. Organism life cycles, predation and the structure of marine pelagic ecosystems. Mar. Ecol. Prog. Ser. 130: 277–293.

Vinyard, G. L. & W. J. O'Brien, 1976. Effects of light and turbidity on the reactive distance of bluegill (*Lepomis macrochirus*). J. Fish. Res. Bd Can. 33: 2845–2849.

Hydrobiologia **484**: 61–73, 2002.
O. Vadstein & Y. Olsen (eds), Sustainable Increase of Marine Harvesting: Fundamental Mechanisms and New Concepts.

Grid-generated turbulence in a mesocosm experiment

Signild Nerheim[1], Jan Erik Stiansen[2] & Harald Svendsen[3]
[1]*Department of Oceanography, Göteborg University, Box 460, SE 405 30 Göteborg, Sweden*
E-mail: sine@oce.gu.se
[2] *Institute of Marine Research, P.O. Box 1870, Nordnes, N-5817 Bergen, Norway*
E-mail: jan.erik.stiansen@imr.no
[3]*Geophysical Institute, University of Bergen, N-5007 Bergen, Norway*
E-mail: harald.svendsen@gfi.uib.no

Key words: turbulence, mesocosm, energy dissipation rate, entrainment, mixing

Abstract

In two mesocosm experiments designed to study the effect of nutrients and turbulence on trophic levels below fish larvae in the ecosystem, turbulence was created by oscillating grids. Utilizing two different frequencies, two levels of turbulence ('high' and 'low') were generated in a total of eight mesocosms in both experiments. In each mesocosm a two-layer density structure was created. We succeeded in generating turbulence that mimics turbulence found in natural ecosystems, with turbulent eddies on larger scales than those relevant for organism interactions. Energy dissipation rates (ϵ) were calculated from the frequency spectrums of the measured turbulent velocities. In the second experiment the average energy dissipation rates in the upper layer were 1.9×10^{-7} W kg^{-1} and 5.5×10^{-8} W kg^{-1} in the 'high' and 'low' enclosures, respectively. Finding 'background' turbulence levels between 5.3×10^{-9} and 3×10^{-8} W kg^{-1}, we suggest that future experiments consider turbulent measurements as a standard variable.

Introduction

Entrapping a volume of the natural ecosystem in flexible wall mesocosms gives a partly controllable experimental platform on a larger scale than that which is possible in the laboratory, allowing a study of the interactions between organisms within a limited ecosystem. Compared to studies in the field, mesocosms have to follow the same fluctuations as the surrounding water masses in environmental physical parameters like temperature and light. However, the effect of other parameters such as nutrients, turbulence and advection can be controlled and investigated.

Turbulence is almost always present in the marine environment. In his review article, Sanford (1997) states that: 'Enclosed experimental ecosystems, if they are to mimic the function of natural ecosystems, also must mimic natural turbulence and its effects'.

Turbulence in nature is typically generated on large scales. The energy is transferred to smaller and smaller scales, through a process of cascading instabilities, until it reaches scales where molecular forces transform the momentum to heat (e.g., Richardson, 1922; Batchelor, 1967). The largest turbulence scale possible in a mesocosm is the dimension (the smallest of the width and the depth) of the mesocosm itself. It is therefore important that the size of the experimental system is much larger than the turbulent scales relevant to the species of interest (e.g., Osborn, 1995; Osborn & Scotti, 1996). It was pointed out by Osborn & Scotti (1996) that in laboratory tanks of limited size the range between the energy-containing eddies and the dissipation scale is not large enough to achieve the $\kappa^{-5/3}$-spectral relation found in natural isotropic turbulence (Kolmogorov, 1941) for the wave number (κ).

The importance of turbulence and mixing on the biological ecosystem has received increasing attention the last decade. Turbulence affects the ecosystem by changing processes such as the encounter rate (e.g., Rothschild & Osborn, 1988), particle capture (e.g., Shimeta & Jumars, 1991), phyto-plankton diffusion layer (e.g., Mann & Lazier, 1996), aggregation and dis-aggregation (e.g., Kiørboe, 1993) and small-scale patchiness (e.g., Moore et al., 1992).

Generating and controlling turbulence and mixing in mesocosms is a challenge. Mismatched eddy length scales are responsible for difficulties in mimicking natural turbulence mixing in experimental enclosures (Sanford, 1997). Large systems demand innovative turbulence generating designs. According to Sanford (1997), grids, together with paddles, are the best available techniques for water column mixing in ecosystem experiments.

Experiments with grid-generated turbulence in stratified fluid were first conducted by Rouse & Dodu (1955). Extensive discussions on the subject can be found in the book by Turner (1973), and in more recent reviews by Hopfinger (1987) and Fernando (1991). A turbulent upper layer will entrain fluid from a non-turbulent lower layer, resulting in an increase of the upper layer thickness. The entrainment velocity, defined as the change of layer thickness with time, is found to decrease with distance from the generating source and with the density difference between the layers.

Though some experiments on the effect of turbulence and mixing on the ecosystem have been conducted in mesocosms, few have thoroughly measured and quantified the turbulence intensity. Howarth et al. (1993) investigated the effect of grid-generated turbulence on the rate of nitrogen fixation by cyanobacteria in 3-m^3 homogeneous mesocosms. Harada et al. (1996) conducted ecosystem studies with mixing in large stratified mesosms (5 m in diameter, 18 m deep), and showed a high ability to reproduce the physical and biochemical conditions of the water column. Prins et al. (1999) conducted an experiment in land based mesocosms, investigating the effect of nutrient loading on phytoplankton, zooplankton and macrozoobenthos. Using a two-layer density system, turbulence was generated in both layers by stirring. With continuous flow to both layers, the interface was maintained for 7 months. However, they did not measure the turbulence.

In 1996, a multidisciplinary project, Nutrients and Pelagic Production (NAPP), was launched to address the effect of nutrients and turbulence on the ecosystem on trophic levels below fish larvae. The experiments involved the use of large nutrient enriched enclosures with grid-generated turbulence and a two layer density profile. The project was a joint venture with groups working with microbiology, sedimentation, phytoplankton, zooplankton and physical oceanography. This article addresses only the physical environment in the enclosures, with special emphasis on turbulence and mixing. Further results from NAPP can be found in Nejstgaard et al. (2001a,b) and Svensen et al. (2001).

The aim of the physical oceanographic group of NAPP was to generate turbulence that mimics turbulence found in natural ecosystems, both in scales and intensities. The main issue was to generate scales large enough for a well-developed inertial subrange to develop, and relevant for the interactions between the organisms in the mesocosms.

a)

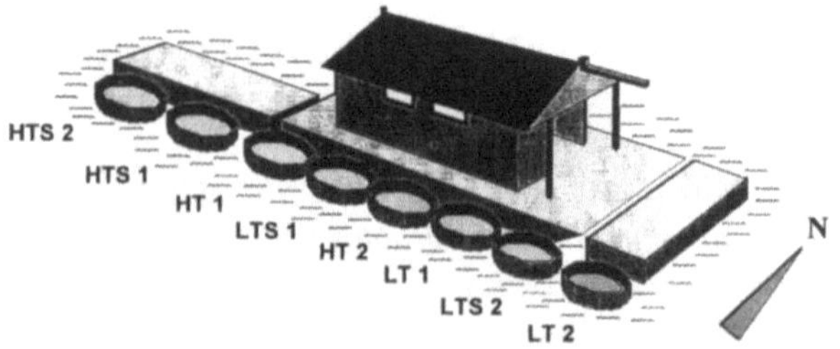

b)

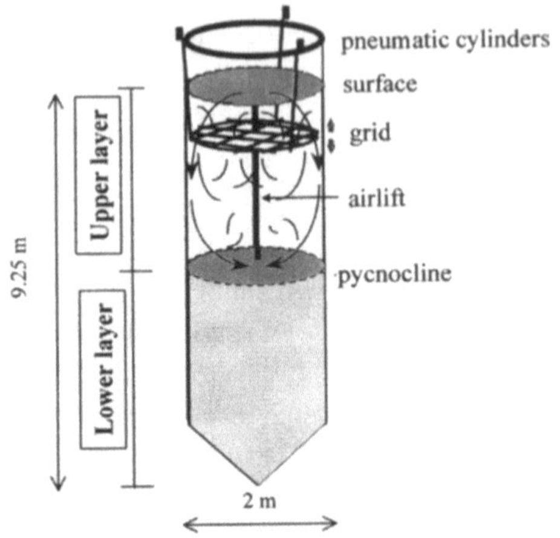

Figure 1. Eight mesocosms were moored to floating frames on the south side of the three rafts (a). See text for details. (b) A mesocosm. In the first experiment, stratification was created by adding fresher water to the upper layer, whereas in the second experiment, salt was added to the lower layer. The grids oscillated with a stroke of 40 cm in the upper part of the enclosures, and an airlift pump created a large scale, slow circulation in the upper 4 m.

Materials and methods

The experiments were conducted at the Marine Biological Field Station in Raunefjorden outside Bergen, Norway, during 3 weeks (3–26 September) in 1996 and 3 weeks (4–29 April) in 1997. Eight 27-m^3 polyethylene enclosures, 9.25 m deep (the lower 1.5 m were conical) with a 2-m diameter were moored to floating frames at the south side of three connected rafts, situated in the middle of a bay sheltered by islands. Figure 1a,b shows principle sketches of the rafts and the enclosures.

Each enclosure was named with regard to the treatment of nutrients and turbulence, LT (low turbulence) and HT (high turbulence) denotes the turbulence treatment, while S denotes silica enrichment. Two enclosures were given the same treatment, indicated by numbers 1 and 2. The main raft was more stable than the side rafts, but observations indicate a small oscillating period of 1–3 s for all rafts, even in calm conditions, due to incoming swells.

September 1996

Setup

The enclosures were filled separately with water from 4 m depth. Stratification was created by adding fresh water at the surface of the enclosures at the end of the filling. To allow the system to settle, no treatment was given the first 24 h, except for the airlift (see below). During this time two of the enclosures started to surface, due to the mean density within the enclosure being lower than the surrounding water. This resulted in the water column to be totally mixed in LT 2, and HT 2 becoming more saline due to saltier seawater flowing into the mesocosm (this is evident in Fig. 3). To compensate for lower mean density in the enclosures compared to the surrounding water, a weight was attached to the bottom of all enclosures. Despite the weights one of the enclosures (HTS 1) floated to the surface on 10 September. Small holes in the wall of HT 2, suffered when it surfaced, resulted in a slow leakage during the experiment. On 17 September this enclosure completely overturned, and was terminated.

An airlift system (a 4-meter long vertical tube) provided homogeneity in the upper layer. A sketch of the enclosures with the mean circulation pattern and airlift system is shown in Figure 1b. The flow through the airlift was 40 l/min.

Due to water loss caused by evaporation and sampling, 0.1–0.2 m^3 water (0.8–1.5% of the upper layer volume) was added to the upper layer on four occasions.

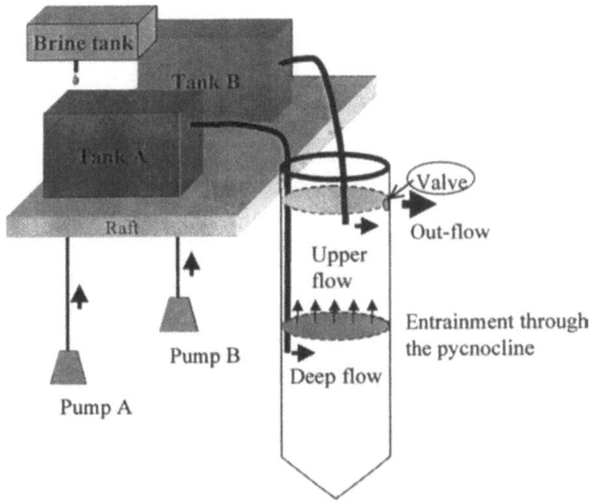

Figure 2. Schematical representation of the flow to the upper and lower layers in the second experiment. Salt-enriched seawater was added below the pycnocline to balance entrainment (salt saturated seawater from a brine tank was mixed with the seawater in tank A before water was apportioned to the lower layer). Unfiltered seawater was added to the upper layer. The out-flow went through a valve at the water surface. Pumps with level switches maintained an even water level in the two tanks, and thereby a constant flow to the two layers. One tank supplied flow for the specific layer in all mesocosms.

Turbulence generation

Turbulence was generated in the upper layer by large oscillating Plexiglas grids attached to a circular steel frame, 1.5 m diameter (Fig. 1b). Grid bars were 0.5 cm thick, 5 cm wide, and the mesh size was 10 cm. The stroke was 40 cm, centered at 1.10 m depth. The grids were driven by three pneumatic cylinders. Two turbulence levels ('high' and 'low') were used. At the beginning of the experiment both had an oscillation period of 40 s. The grid in the 'high turbulence' enclosures moved down during 7 s where it paused for 13 s before rising during 7 s and paused for 13 s again. The grid in the 'low turbulence' enclosures moved down during 16 s and paused for 4 s. On 15 September the periods were changed in order to increase the difference in intensity between the two turbulence levels. In the high turbulence enclosures the oscillating period was set to 14 s (moving 5 s and pausing 2 s each direction), while the grids in the low turbulence were terminated with only the 'background' turbulence (discussed later) as a generating source.

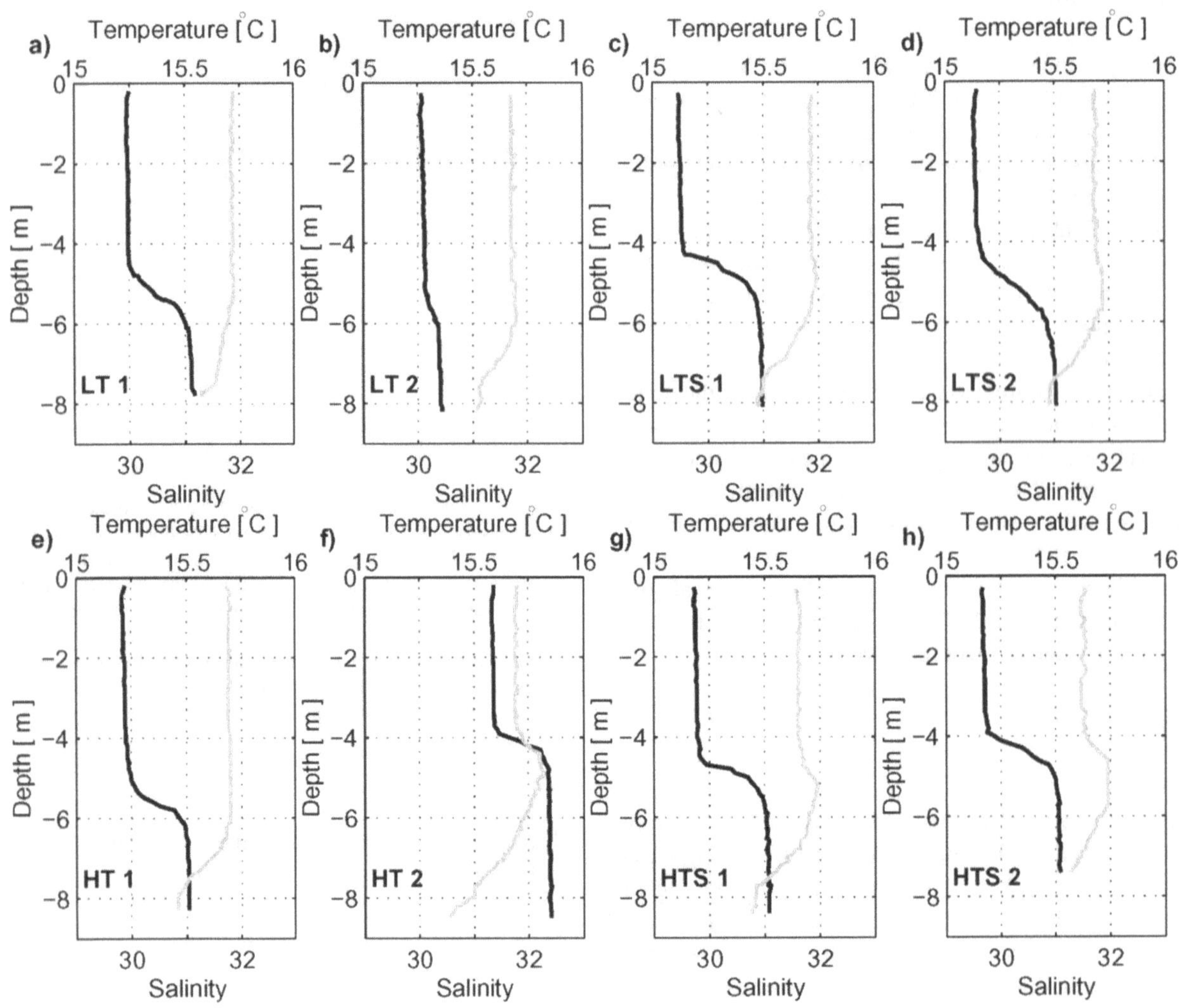

Figure 3. Initial salinity (black line) and temperature (gray line) profiles for the first experiment, 8 September 1996.

Measurements

Every second day the turbulence systems were stopped for 3 h so that CTD profiles and biological samples (not presented) could be taken. Turbulence was measured with an Acoustic Doppler Velocimeter (ADV, see Lohrmann et al. (1996) for details). Ten-minute velocity time series were taken in all enclosures at several different depths, some repeatedly. Measurements were also made with grids at rest to investigate the 'background' turbulence.

April 1997

In general, the experimental setup in the second experiment (April 1997) was the same as in the first experiment (September 1996). However, some modifications were done, which are explained in the following.

Setup

The enclosures were filled simultaneously by pumping seawater from 5 m depth into a large dispersion tank on the raft, and further into the enclosures through eight separate outlets. This procedure was chosen in order to prevent variability in the biological initial condition of the enclosures due to patchiness in the sea.

Stratification was created by using salinity-enriched water when filling the lower layer of the enclosures. Seawater saturated with NaCl (a salinity of approximately 250) was added to the seawater in the dispersion tank during the filling of the lower layer.

The upper layer with natural seawater was then poured gently on top. The enclosures were negatively buoyant compared to the surrounding water.

In an attempt to minimize the effect of waves, an oil boom was arranged in a circle around the raft. The boom penetrated about 1 m down into the sea, and was anchored to the bottom. This worked well for smaller wind-waves, but had little or no effect on larger waves (e.g., incoming swells).

Flow to upper and lower layers

A major issue in this experiment was to keep the density interface stationary in order to achieve stationary and equal turbulent conditions in the LT and HT mesocosms. Since turbulence in the upper layer was stronger than in the lower layer a net amount of water was entrained to the upper layer from the lower. To compensate for this effect, a constant flow of water with equal density as the water in the lower layer had to be added to the lower layer to replace the entrained fluid. Seawater was pumped into a tank (tank A in Fig. 2) from 5 m depth. Saturated salt water was added to the tank at a flow rate set to ensure an increase in salinity of 1.5–2 to correspond to the salinity in the lower layer. The tank was continually stirred to ensure homogeneous conditions, and the salinity-enriched seawater was added continuously to the enclosures well below the pycnocline.

Flow to the upper layer was added for three reasons. First, a certain exchange of water in the upper layer was necessary to prohibit increase of pH. Second, seawater flow to the upper layer would counteract an otherwise steady salinity increase in the upper layer due to the entrainment of the more saline water from the lower layer. Third, this kept a steady 'overpressure' within the enclosures, thereby maintaining the shape of the mesocosms throughout the experiment. The upper layer received unfiltered seawater from tank B, pumped from 5 m depth. A valve mounted at the surface of each enclosure permitted water to seep out of the enclosures. Figure 2 illustrates the main principles of the flow systems.

Initially, the volume to the lower layer was set to counter the entrainment observed in the first experiment. Approximately 0.8 l min^{-1} was added to the HT enclosures and 0.25 l min^{-1} to the LT enclosures, corresponding to 40 and 10 cm day^{-1}, respectively. As the total water volume added to the enclosures exceeded 10% of the upper layer, which was too high for biological considerations, the flow was therefore adjusted to 0.28 l min^{-1} on 10 April to the lower layer in all mesocosms. The immediate consequence was a lowering of the pycnocline in HT enclosures to a new equilibrium level. From 10 April the daily exchange was 5% of the surface waters in the HT enclosures, and 7% of the LT enclosures.

Grid system

Two changes were made to the grid system. A new attachment system raised the center of oscillation 10 cm to a depth of 1 m (stroke area 0.8–1.2 m depth). The oscillating period for the high turbulence was set to 14 s for the HT enclosures, and 56 s for the LT enclosures. The high turbulence sequence was equal to the final setting used in the first experiment. In the LT enclosures grids were moving for 21 s and pausing 7 s in each direction.

Due to malfunctioning the grids stopped intermittently between the start and 10 April, with some minor malfunctions until 12 April and again on 20 April. The grid in HTS 2 never performed well and was permanently stopped on 13 April. HTS 2 is thus referred to as a No-turbulence enclosure, where only the air lift pump created motion.

Measurements

A Seacat CTD was used for hydrographical measurements. Non-uniform lowering speed caused spikes in the conductivity data below the upper homogeneous layer. Additional salinity measurements were made with a handheld salinotherm. The vertical resolution was 25 cm, the salinity resolution 0.1. In the upper layer, a comparison between salinotherm and CTD values showed good agreement, and in CTD profiles without spikes, agreement was good through the entire water column.

The flow to the upper and lower layer was measured every day.

Calculation of energy dissipation rates

Calculations of energy dissipation rates follow the method of linear regression described by Stiansen & Sundby (2001). The mean velocity and spikes were removed, and 5-min interval time series were transformed to the frequency domain with a Fast Fourier Transform. Noise was removed in the frequency domain. The white noise was higher by an order of magnitude in the horizontal components than the vertical component, a result of probe configuration (Gytre et al., 1996). The other dominant noise feature was an

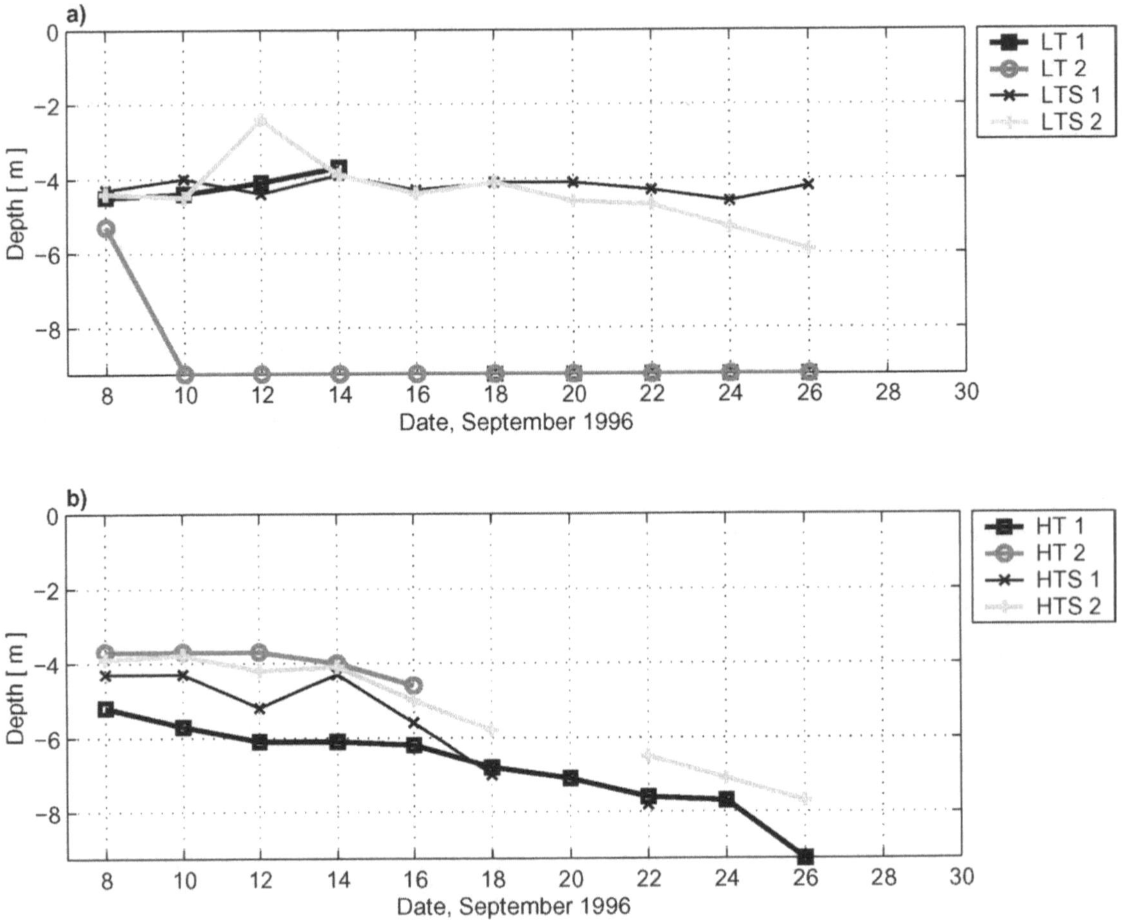

Figure 4. The depth of the pycnocline during the first experiment, September 1996. (a) LT mesocosms; (b) HT mesocosms. Note that HT 2 was terminated after 16 September, as the whole mesocosm floated to the surface creating a complete turnover.

energy peak around 0.5 Hz, corresponding well to the observed raft movement period.

Assuming homogeneous, isotropic, stationary turbulence the energy dissipation rate ϵ was found by using the Eulerian frequency spectrum suggested by Tennekes (1975).

$$S(\omega) = B\epsilon^{2/3}{u_{\mathrm{rms}}}^{2/3}\omega^{-5/3}.$$

Here $S(\omega)$ is the energy spectral density function, ω the frequency, u_{rms} the root-mean-square velocity (after subtracting the mean velocity), ϵ the turbulent dissipation rate and B a constant of order 1 (Tennekes & Lumley, 1972). In our calculations B=1 was used.

The energy dissipation rate ϵ was calculated for all three cartesian velocity components by fitting a regression line to the $\omega^{-5/3}$ relationship in the inertial-advective subrange.

In some spectra, a clear subrange could not be identified, and these components were removed before further analysis. The energy dissipation rate was calculated as a component average. A logarithmic average was chosen in favor of a sample mean as the logarithmic average is more robust to overestimating dissipation rates if one component is much larger than the others. Series where two out of three components had no distinct subranges were discarded.

Results

Hydrographic data, September 1996

The initial salinity and temperature profiles, measured on 8 September, are shown in Figure 3. The temperature was either homogeneous or had a weak stable stratification. The salinity in the upper layer was homogeneous, and the halocline was weakly stably stratified. Both the thickness of the halocline and the

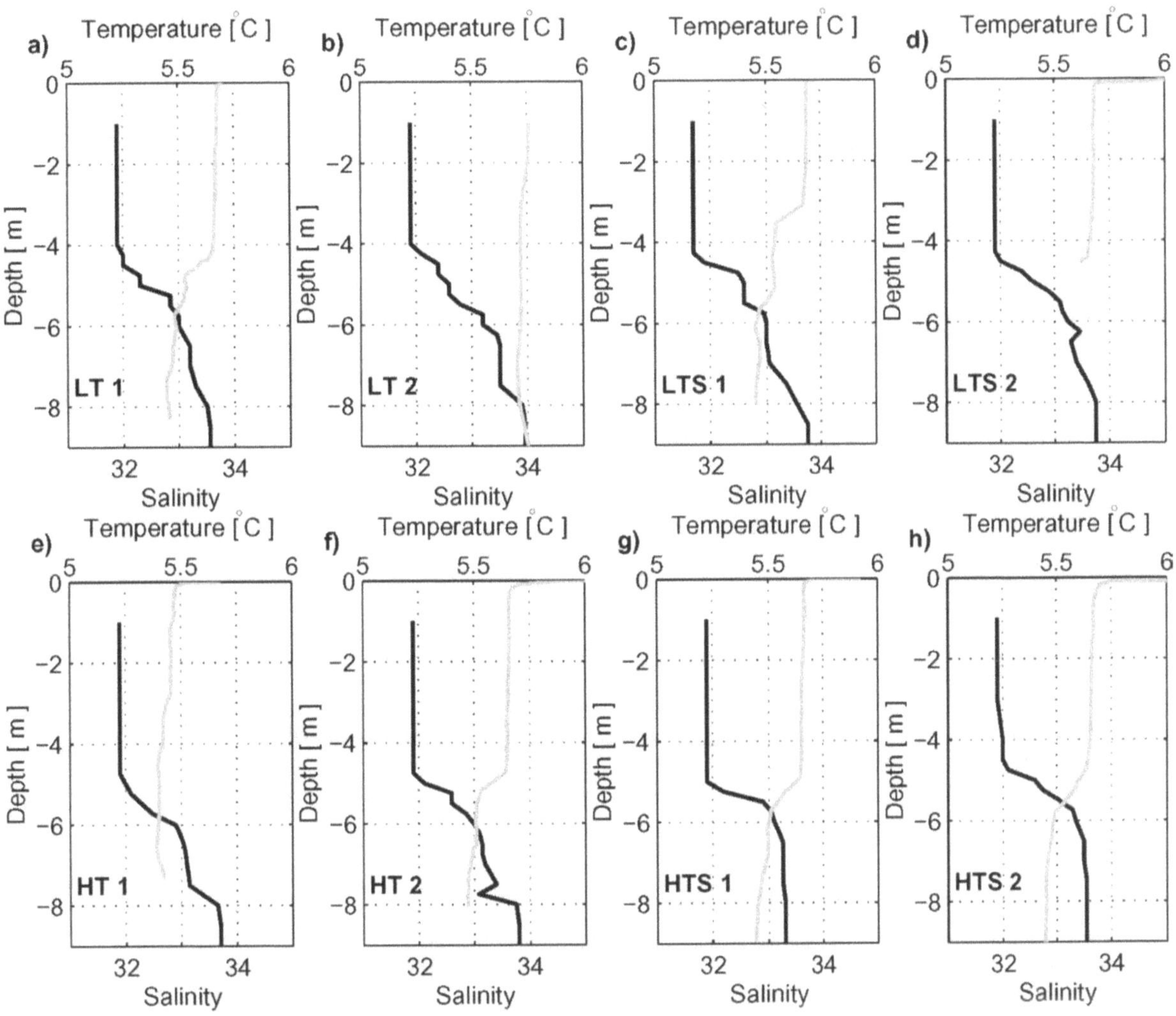

Figure 5. Initial salinity (black line) and temperature (gray line) profiles for the second experiment, 10 April 1997.

difference in salinity between the upper and lower layer varied. Since the density was mostly affected by salinity in our experiment, the halocline is referred to as the pycnocline in the following. The mesocosms that surfaced, LT 2 and HT 2, differed from the other mesocosms. The first was almost homogeneous (Fig. 3b), the other had a higher salinity (Fig. 3f).

The pycnocline was determined from the salinity profiles, while Chl-*a* values supported the observed pycnocline depths. The pycnocline is shown in Figure 4. In the HT enclosures the deepening of the upper layer was quite rapid (Fig. 4b), whereas in the LT enclosures (Fig. 4a) the erosion was relatively slow. Exceptions were LT 2, HT 2 and HTS 1. In LT 2 the initial density gradient was small, but constant from 10 September. HT 2 overturned completely on 17 September. HTS 1, which surfaced on 10 September, maintained some stratification. However, the density difference and the pycnocline decline rapidly after the incident.

The density difference between the upper and lower layer initially ranged between 0.3 kg m^{-3} (LT 2) to 1.1 kg m^{-3}, and decayed rapidly in all mesocosms because of entrainment from the lower saltier water into the fresh upper layer. The most rapid deepening of the upper layer was found in the mesocosms with low density difference. HT 1 had the highest density difference of the HT mesocosms, and the entrainment rate was slower than in HTS 1 and HTS 2. The entrainment velocity in HT 1 was approximately 25 cm day^{-1} initially. In HTS 1 and HTS 2, the entrainment increased after the adjustment of the grid frequency,

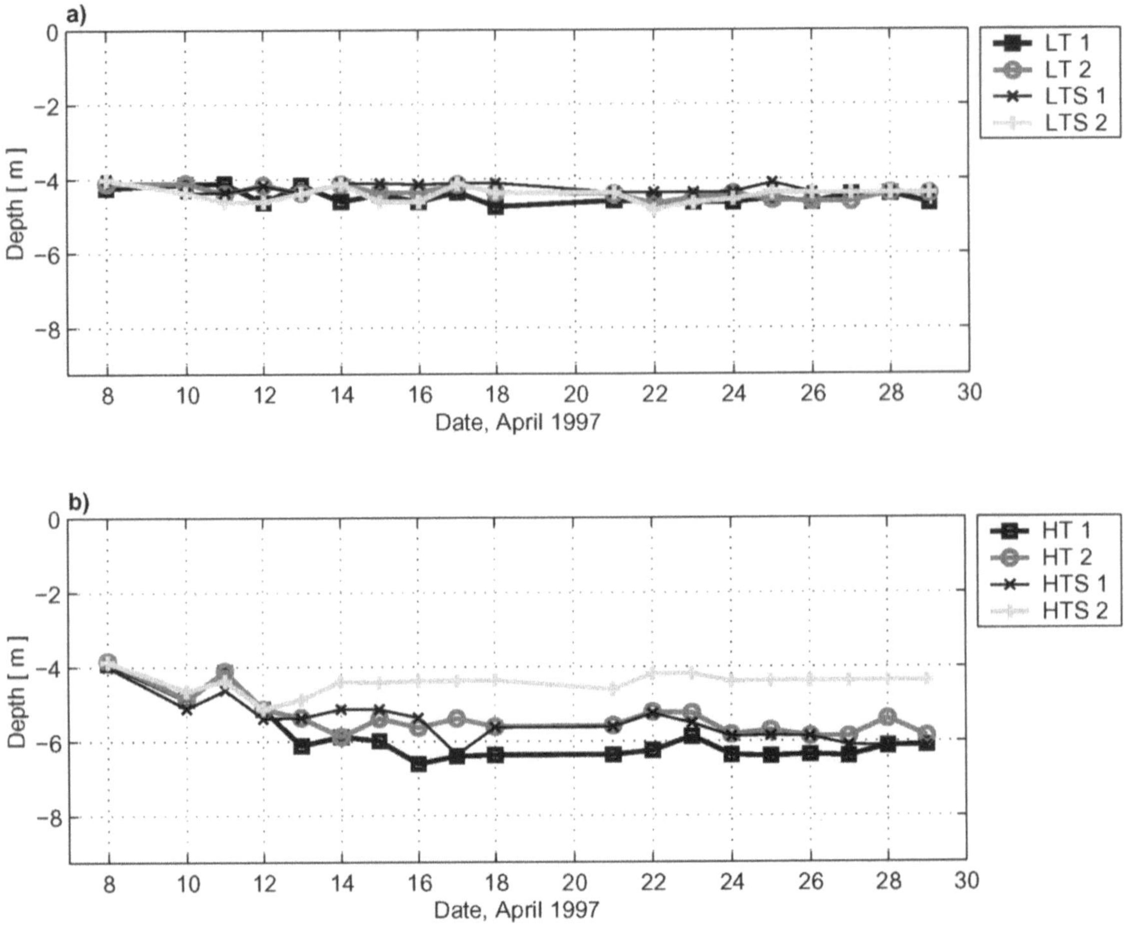

Figure 6. The depth of the pycnocline during the second experiment, April 1997. Note that the pycnocline in HTS 2 returned to the low turbulence levels on 13 April, after the grid was terminated. The temporary rise of the HT pycnoclines between 10–11 April was caused by malfunctioning of the grid system.

reaching velocities of 40 cm day^{-1}. In the LT enclosures, the entrainment velocity ranged from 0 to 20 cm day^{-1}.

The vertical temperature profiles in the mesocosms were approximately homogeneous during the experiment, decreasing slowly from 15.7 to 12.6 °C.

Hydrographic data, April 1997

Initial values (9 April) of temperature and salinity are shown in Figure 5. The presence of a well-mixed upper layer with a thickness of 4–4.5 m is clearly seen in the salinity profiles. The average salinity was higher than in the first experiment, and the average temperature lower. The difference in salinity between the upper and lower layers was larger than in the first experiment, indicating a stronger stratification. As the CTD profiles were disregarded because of spikes, the pycnocline depth was determined from salinotherm profiles. CTD profiles without spikes support the pycnocline depth found from the salinotherm.

The depth of the pycnocline is shown in Figure 6. The pycnocline remained fairly stable in all enclosures. In the LT mesocosms (Fig. 6a) the pycnocline was around 4 m throughout the experiment. In the HT mesocosms (Fig. 6b) the pycnocline fell to around 5.5–6 m after the flow was reduced on 10 April, where it stayed for the duration of the experiment. The pycnocline in HTS 2 returned to the low-turbulence level after the grid stopped (Fig. 6b, 13 April). The large variation seen in HTS 1 on 17 April is most likely a clerical error.

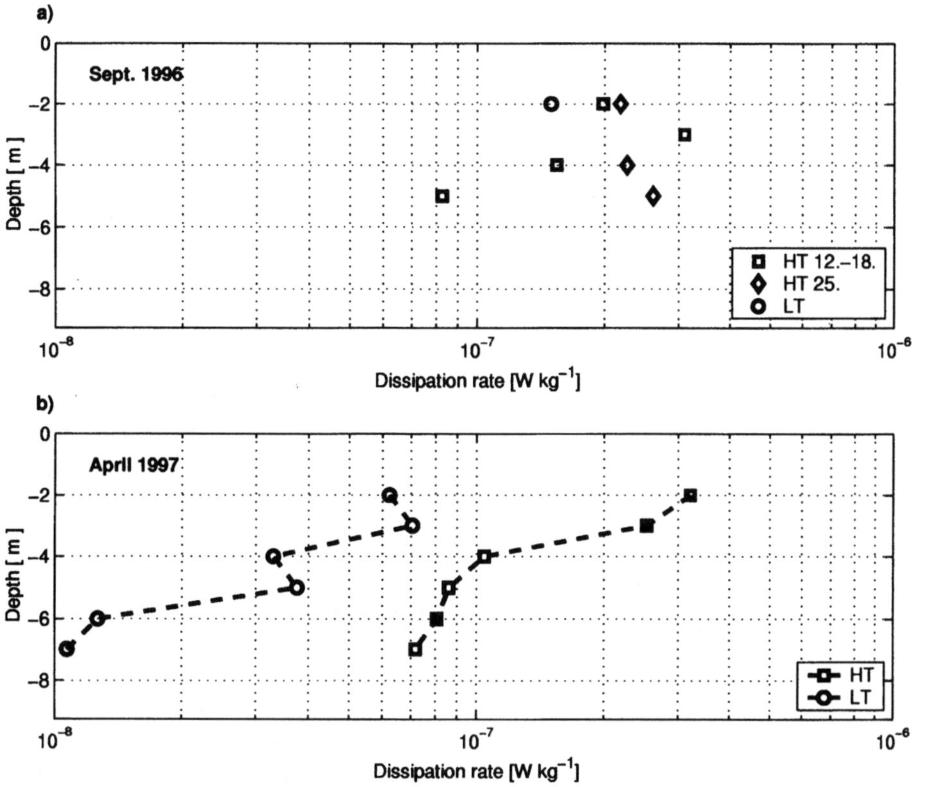

Figure 7. The distribution of the mean energy dissipation rate. (a) The mean dissipation rates at each meter for the first experiment. For LT, there is only one value at 2 m depth. The dissipation rates in HT enclosures have been bulked (12–18 September, squares; 25 September, circles) because of the time development of the pycnocline. (b) The mean dissipation rates for the second experiment.

The temperature was initially 5.5 °C, and increased to 7.3°C during the experiment. The largest gradient between mesocosm and sea was 0.1 °C.

The measured flow was relatively constant over time, 0.28 and 0.34 l min^{-1} to to the lower layer and the upper layer, respectively. The volume flowing into the lower layer is equal to 403 l day^{-1}, and corresponds to a height of 13 cm in the mesocosms. With a stationary pycnocline, this is equal to the entrainment velocity. Details on the flow can be found in Nerheim (1999).

Energy dissipation rates

After removing bad spectra, a total of 52 dissipation values from the first experiment, and 78 values from the second experiment at different depths were used for further analysis. In the second experiment, with a stable pycnocline, dissipation rates can be treated as bulk values, independent of time. In the first experiment, the likelihood of the pycnocline-development having influenced the distribution of the turbulent energy warrants some caution in our treatment of the turbulent energy.

Because of the deepening of the pycnocline in the first experiment, dissipation values have been sorted with respect to the state of the pycnocline. Figure 7a shows the mean dissipation rate for each meter for HT 1 for the time period 12–18 September, when the pycnocline depth was approximately 6 m, and the values measured on 25 September, when the whole mesocosm was well-mixed. The grid frequency was changed on 15 September, but this had no visible impact on the turbulence levels. For this paricular grid geometry we have found that the velocity of the grid is the main variable, not the frequency of the cycle. Note how the turbulence was stronger at 5 m depth in HT 1 on 25 September when the mesocosm was mixed throughout the water column. For comparison the mean value for the LT mesocosms at 2 m is also shown. Measurements below 2 m were disregarded because of bad spectra.

Figure 7b shows the mean dissipation rate per depth from the second experiment. There is a strong

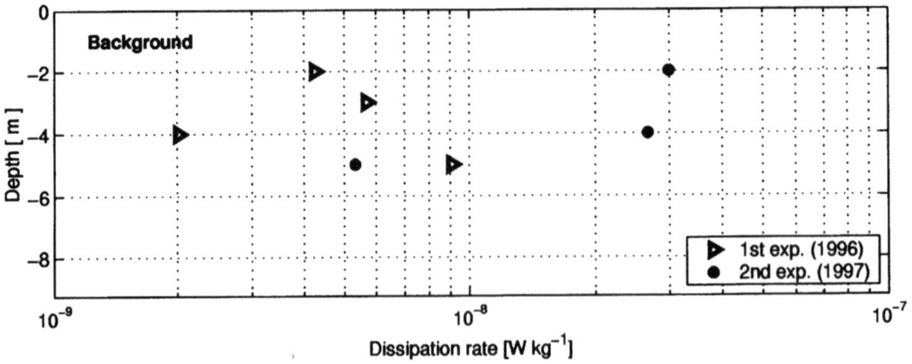

Figure 8. Mean energy dissipation when grids are at rest. Note the difference in axis scale from Figure 7.

Table 1. Mean energy dissipation rates, April 1997: units: W kg^{-1}

	HT	LT	HTS 2
Upper layer	1.9×10^{-7}	5.5×10^{-8}	2.8×10^{-8}
Lower layer	7.6×10^{-8}	2.0×10^{-8}	5.3×10^{-9}

decay of turbulence between 2 and 3 m in the HT enclosures. Overall, turbulence was higher in the HT mesocosms than in LT, and the decay with distance from the grid is clearly seen. The presence of the pycnocline, 4 m in LT mesocosms and 6 m in HT mesocosms, is not evident in the dissipation rate profiles.

Measurements taken when the grids are at rest are presented in Figure 8. The values from the first experiment (September 1996) are taken from LTS 1 and LTS 2 where the pycnocline was stable. The values in the second experiment (April 1997) are from HTS 2, where the grid was at rest from 10 April. The latter values are generally higher than the values from the first experiment.

The layer-average dissipation rates for the second experiment (April 1997) are presented in Table 1, divided into the three intensity regimes; 'high', 'low' and HTS 2 (no grid motion).

Discussion

The first experiment served in many ways as a pilot experiment. Because of adjustments during the experiment, the results are not conclusive. However, the results from the first study were very useful for the planning of the second experiment. The main difference between the two experiments was the introduction of a compensation flow to the lower layer in the latter, in order to keep a steady interface between the two density layers. In the first experiment, the pycnocline depth increased during the experimental period while a stationary state was reached early in the second.

Biological considerations put some limitations on the design. In the second experiment, the decreased flow in the HT enclosures led to a deeper upper layer than in LT enclosures. As a consequence, the volume in which the turbulent energy was distributed was larger in HT than LT enclosures. The effect on the immobile organisms is that they will be mixed lower down in the HT enclosures than in the LT enclosures. In general, the turbulence generation mechanism should aspire to do as little damage as possible to the organisms in the study, which is the case with slowly moving grids.

In both experiments, the interface was not as sharp as desired. The advantage of a sharp density interface is that it prevents energy loss through radiation by internal waves on a continuous interface (Linden, 1975). Another aspect is that the stratification should be equal in all enclosures to give the same buoyancy fluxes and the same decay of turbulence in the parallel enclosures. However, there is always a trade-off between which variables to be held constant in a multidisciplinary experiment. In this case a constant flux of water was chosen on behalf of equal depth of the pycnocline.

The airlift system was used to ensure that the organisms were well distributed. It generated a mean circulation pattern in the upper layer in the enclosures. According to Egge & Aksnes (1992) a particle passes through the airlift in average 4 times a day at a flow of 40 l min^{-1}. The airlift system can generate turbulence in two ways. First, inside the tube the environment will be highly turbulent, with water squirting out of the upper end. Second, the flow from the airlift was in average slow, but wall-layer friction cannot be ruled out. However, the slow flow in the enclosure and the small volume (compared to the volume of the upper layer) of the air-tube indicates that this is a minor turbulence source, compared to the grid.

In order to generate natural turbulence, the grid mechanism needs to be able to create turbulent motion on a scale large enough to form an energy cascade with a well-developed inertial subrange, which only occurs in a high Reynolds number (Re) regime. The Reynolds number is defined as Re = DU/ν, where D is a typical scale, U a typical velocity and ν the kinematic viscosity. In the HT mesocosms the grid geometry gives a grid Reynolds number of 32 000 in the HT enclosures (D is a stroke of 40 cm, U the grid velocity 40 cm/5 s), or 11 400 in the LT enclosures (U = 40 cm/14 s). Another reasonable choice of D and U is the enclosure diameter, 2 m, and the mean velocity in the upper layer, 0.5 m s^{-1}. This choice gives a Reynolds number of 10 000.

The largest possible turbulence scale is determined by the size of the system, while the smallest possible scale is determined by the Kolmogorov microscale. Accordingly, the largest turbulent scale is of the order of 1 m. Dissipation rates of 10^{-7} and 10^{-5} W kg^{-1} corresponds to a Kolmogorov microscale of 0.18 and 0.32 cm, respectively. Hence, the system should be more than large enough for a well developed inertial subrange to be present, which also is supported by the form of the turbulent spectra.

The form of our turbulence spectra support the analysis of Tennekes (1975), that the Eularian frequency spectra should follow a $\omega^{-5/3}$slope. This relation has not been explained on a theoretical level, but experiments favor Tennekes (Nelkin, 1992). Carruthers and Hunt (1986) use a slope of $\omega^{-5/3}$, whereas in Yeung (2001), numerical simulations show stronger support for the Lagrangian ω^{-2} spectrum. In Stiansen & Sundby (2001), the linear regression method used in this paper has been compared with other well-known methods such as the energy dissipation law (e.g., Frisch, 1995) and Taylor's 'frozen turbulence' hypothesis (e.g., Tennekes & Lumley, 1972), and proved to be an equally well suited method for calculating the energy dissipation rate.

Approximately 90% of the turbulent kinetic energy dissipates. The remainder is used to increase the potential energy of the turbulent layer by entraining fluid from below. If both layers are turbulent, entrainment from both sides will happen and the position of the interface is the sum of the two entrainment rates. Studies of turbulence in homogeneous and stratified fluid have showed that the presence of a pycnocline or an interface will alter the decay of turbulence (e.g., Hopfinger & Toly, 1976; Gargett et al., 1984; Hopfinger, 1987). When the vertical motion is suppressed, three-dimensional turbulence is no longer present.

In both experiments, entrainment was taking place. In the first experiment, the entrainment velocity varied due to different initial conditions and adjustments of the grid frequency. In the second experiment, a stable interface corresponded to a 13 cm day^{-1} entrainment velocity, calculated from the flow needed to keep the interface stationary. The reason why the entrainment velocity was the same in both HT and LT was the deeper pycnocline in HT enclosures.

The energy dissipation rate, shown in Figure 7, was higher in the HT mesocosms than the LT mesocosms, and even lower when the grids were at rest (Fig. 8). A comparison between the upper layer in the HT mesocosms in the first and second experiment show that the dissipation rate in the two lies within the same order of magnitude. Further, Figure 7 indicates that the presence of a strong pycnocline prevents penetration of the energy to the lower layer, while no such 'obstacle' appears in a homogeneous enclosure (ref. measurements at 5 m on 25 September, Figure 7a).

As one objective in the study was to create two different levels of turbulence, the relatively high level of energy in the LT enclosures gave a small difference (a factor of 0.5) between 'high' and 'low' levels. In the work by Stiansen & Sundby (2001) the same frequencies used in a laboratory tank gave a difference of an order of magnitude. The reasons why the grid frequencies were not adjusted more from the first to the second experiments to get a larger difference are as follows: (1) the background turbulence limited the turbulence intensity from below; (2) the pneumatic cylinders performed unevenly below a certain speed; (3) from above, a higher turbulence level would cause a higher entrainment of the pycnocline, which could not be counteracted by an increase in flow to the lower layer, since the flow should be constant.

According to Yamazaki & Osborn (1988), energy dissipation rates in the ocean lie in the range 10^{-10}–10^{-2} W kg^{-1}including coastal waters and estuaries. The latter value is an extreme situation of strong tidal current, and can be considered a rare event. Veth (1983) assessed the interval range 10^{-8}–10^{-6} W kg^{-1}

as typical for the upper mixed layers. Our results, ranging from 10^{-8} to 10^{-7} W kg^{-1} (see Table 1, Figures 7 and 8), are within the typical range in the mixed layers. Peters & Redondo (1997) pointed out that most turbulence-plankton studies have been performed with turbulence intensities well above the normal dissipation rates (between 10^{-6} and 10^{-2}W kg^{-1}), so in this respect, our values represent a more realistic case. Mann & Lazier (1996), following Purcell (1977), derived that an energy dissipation rate of 10^{-6} W kg^{-1} gives approximately 5% higher diffusive flux to a perfectly absorbing sphere of 100 μm. Smaller cells need higher turbulence levels in order to have any effect, i.e. the diffusive flux is a function of both the turbulence intensity and the size of the cell. The choice of natural levels versus desired effect is an important consideration.

The dissipation rate when the grids were at rest, shown in Figure 8, varied between the first and second experiments. Dissipation rates in the HTS 2-enclosure from the second experiment are an order of magnitude higher than the values measured in the first, despite the protection of the oil boom. In contrast, measurements taken in an adjacent mesocosm (6 m deep and 2 m wide, with an airlift, but without a grid, running parallel to our second experiment) gave a mean dissipation rate of 8.3×10^{-8} W kg^{-1} (B. Hansen, pers. comm). The reason for this higher 'background' turbulence might be a higher airlift flow in the mesocosms with higher background energy, as well as a variation in the surrounding waters. If the airlift does produce 'background' turbulence that varies with an order of magnitude, efforts should be taken to investigate this further. The weather is another factor. Even though measurements were taken with little wind and waves, the prevailing weather conditions were better in the first than the second experiment, and may have been important. As all our measurements were done in the best possible conditions, it is likely that our estimates of background turbulence are low.

The 'background' turbulence can be related to a number of sources. Steele et al. (1977) suggested wind effects (waves, shear flow instabilities), convection, movements of bag wall and thermal forcing as sources for turbulent motion present in mesocosms. The diffusivities found by Steele et al. (1977) was of the order 0.1 cm^2s^{-1} which are smaller than normal turbulent diffusion by an order of magnitude (Sanford, 1997). Gust (1977) found that enclosures narrower than the prevailing wavelength were generally better mixed than wider enclosures.The levels of background turbulence found in our experiments suggest that some turbulence always will be present in mesocosms. In the future, we suggest that turbulence becomes a standard parameter to measure when doing ecosystem experiments.

Concluding remarks

We succeeded in creating a grid-generating turbulence system appropriate for generating turbulence in flexible-wall mesocosms. The two experiments show good reproducibility, with turbulent energies within the same order of magnitude. The system produced energy dissipation rates ranging from 1.9×10^{-7} to 5.5×10^{-8} W kg^{-1} (upper layer mean, second experiment), which is within the normal levels in the mixed layer. The dimensions of the system ensured large eddies sufficient to give an inertial subrange in the frequency domain. Measurements of turbulent velocities, when the grids were not moving, indicate that 'background' turbulence is present (2.8×10^{-8}W kg^{-1} , second experiment). As all measurements were made under calm weather conditions, it is likely that the background values can become even higher.We suggest that turbulence measurements in flexible-wall mesocosms, even without turbulence-generation systems, are performed in future ecosystem experiments in order to evaluate the background levels.

Acknowledgements

This study was a part of the MARICULT project network and was supported by funding from the project Nutrients and Marine Pelagic Production (NAPP). We would like to thank our co-workers on NAPP. Dr Svein Sundby contributed with valuable discussions regarding the design of the grid-system and interpretation of the turbulence data. Comments and criticism from Björn Sjöberg were greatly appreciated.

References

Batchelor, G. K., 1967. The Theory of Homogeneous Turbulence. 5th edn. Cambridge University Press, Cambridge.

Carruthers, D. J. & J. C. R. Hunt, 1986. Velocity fluctuations near an interface between a turbulent region and a stably stratified layer. J. Fluid Mech. 165: 475–501.

Egge, J. K. & D. E. Aksnes, 1992. Silicate as regulating nutrient in phytoplankton competition. Mar. Ecol. Prog. Ser. 83: 281–289.

Fernando, H. J. S., 1991. Turbulent mixing in stratified fluids. Ann. Rev. Fluid Mech. 23: 455–493.

Frisch, U., 1995. Turbulence: The Legacy of A.N. Kolmogorov. Cambridge, Cambridge University Press. 296 pp.

Gargett, A. E., T. R. Osborn & P. W. Nasmyth, 1984. Local isotropy and the decay of turbulence in a stratified fluid. J. Fluid Mech. 144: 231–280.

Gust, G., 1977. Turbulence and waves inside flexible-wall systems designed for biological studies. Mar. Biol. 42: 47–53.

Gytre, T., J. E. Ø. Nilsen, J. E. Stiansen & S. Sundby, 1996. Resolving small scale turbulence with acoustic Doppler and acoustic travel time difference current meters from an underwater tower. IEEE: 442–450.

Harada, S, M. Watanabe, K. Kohata, T. Ioriya, M. Kunugi, T. Kimura, S. Fujimori, H. Koshikawa & K. Sato, 1996. Analyses of planktonic ecosystem structure in coastal seas using a large-scale stratified mesocosm: A new approach to understanding the effects of physical, biochemical and ecological factors on phytoplankton species succession. Water Science and Technology 34 (78): 219–226.

Hopfinger, E. J., 1987. Turbulence in stratified fluids, a review. J. Geophys. Res. 92 C5: 5287–5303.

Hopfinger, E. J. & J.-A. Toly, 1976. Spatially decaying turbulence and its relation to mixing across density interfaces. J. Fluid Mech. 78: 155–175.

Howarth, R. W., T. Butler, K. Lunde, D. Swaney & C. R. Chu, 1993. Turbulence and planktonic nitrogen fixation: a mesocosm experiment. Limnol. Oceanogr. 38: 1696–1711.

Kiørboe, T., 1993. Turbulence, phytoplankton cell size, and the structure of pelagic food webs. Adv. Mar Biol. 29: 1–72.

Kolmogorov, A. N., 1941. The local structure of turbulence in incompressible viscous fluid for very large Reynolds number. Dokl. Akad. Nauk SSSR, 30: 9–13. (Reprinted in English in Proc. r. Soc. London A 434: 9–13, 1991.)

Linden, P. F., 1975. The deepening of a mixed layer in a stratified fluid. J. Fluid Mech. 71: 385–405.

Lohrmann, A., R. Cabrera, G. Gelfenbaum & J. Haines, 1996. Direct measurements of Reynolds stress with an acoustic Doppler velocimeter. Proceedings of IEEE Fifth Working Conference on Current Measurement: 205.

Mann, K. H. & J. R. N. Lazier, 1996. Dynamics of Marine Ecosystems: Biological-Physical Interactions in the Oceans. 2nd edn, Blackwell Science, Oxford: 394 pp.

Moore, P. A., R. K. Zimmer-Faust, S. L. Bement, M. J. Weissburg, J. M. Parish & G. A. Gerhardt, 1992. Measurement of microscale patchiness in a turbulent aquatic odor plume using a semi-conductor-based microprobe. Biol. Bull. (Woods Hole) 183: 138–142.

Nejstgaard, J. C, L. J. Naustvoll & A. Sazhin, 2001a. Correcting for underestimation of microzooplankton grazing in bottle incubation experiments with mesozooplankton. Mar. Ecol. Prog. Ser. 221: 59–75.

Nejstgaard, J.C., B. H. Hygum, L. J. Naustvoll & U. Båmstedt, 2001b. Zooplankton growth, diet and reproductive success compared in simultaneous diatom and flagellate-microzooplankton-dominated plankton blooms. Mar. Ecol. Prog. Ser. 221: 77–91.

Nelkin, M., 1992. In what sense is turbulence an unsolved problem? Science 255 (5044): 566–570.

Nerheim, S., 1999. Gittergenerert turbulens i et mesoskala eksperiment. Cand. Scient. thesis, University of Bergen, Norway: 84 pp (in Norwegian).

Osborn, T., 1995. Copepod behavior and turbulence: an evolutionary response? In Skjoldal H. R., C. Hopkins, K. E. Erikstad & H. P. Leinaas (eds), Ecology of Fjords and Coastal Waters. Elsevier, Amsterdam: pp 27–32.

Osborn, T. & A. Scotti, 1996. Effect of turbulence on predator-prey contact rates: where do we go from here? Mar. Ecol. Prog. Ser. 139: 302–304.

Peters, F. & J. M. Redondo, 1997. Turbulence generation and measurement: application to studies on plankton. Sci. Mar. 61: 205–228.

Prins, T. C., V. Escaravaga, L. P. M. J. Wetsteyn, J. C. H. Peeters & A. C. Smaal, 1999. Effects of different N- and P-loading on primary and secondary production in an experimental marine ecosystem. Aquat. Ecol. 33 (1): 65–81.

Purcell, E. M., 1977. Life at low Reynolds number. Am. J. Physics 45: 3–11.

Richardson, L. F., 1922. Weather Prediction by Numerical Process. Cambridge University Press, Cambridge.

Rothschild, B. J., T. R. Osborn, 1988. Small-scale turbulence and plankton contact rates. J. Plankton Res. 10 (3): 465–474.

Rouse, H. & J. Dodu, 1955. Turbulent diffusion across a density discontinuity. La Houille Blanche 10: 522–532.

Sanford, L. P., 1997. Turbulent mixing in experimental ecosystem studies. Mar. Ecol. Prog. Ser. 161: 265–293.

Shimeta, J. & P. A. Jumars, 1991. Physical mechanisms and rates of particle capture by suspension-feeders. Oceanogr. Mar. Biol. Ann. Rev. 29: 191–257.

Steele, J. H., D. H. Farmer & E. W. Henderson, 1977. Circulation and temperature structure in large marine enclosures. J. Fish. Res. Bd Can. 34: 1095–1104.

Stiansen, J. E. & S. Sundby, 2001. Improved methods for generating and estimating turbulence in tanks suitable for fish larvae experiments. Sci. Mar. 65 (2): 151–167.

Svensen, C., J. K. Egge & J. E. Stiansen, 2001. Can silicate and turbulence regulate the vertical flux of biogenic matter? A mesocosm study. Mar. Ecol. Prog. Series. 217: 67–80.

Tennekes, H., 1975. Eulerian and lagrangian time microscales in isotropic turbulence, J. Fluid Mech. 67: 561–567.

Tennekes, H. & J. L. Lumley, 1972. Cambridge, MA, MIT Press: 300 pp.

Turner, J. S., 1973. Buoyancy Effects in Fluids. Cambridge University Press, London.

Veth, C., 1983. Turbulence measurements in the stratified central North Sea with laser-doppler velocimeter system. In Sündermann, J. & W. Lenz (eds), North Sea Dynamics. Spinger, Berlin: 412–428.

Yamazaki, H. & T. R. Osborn,1988. Review of oceanic turbulence: implications for biodynamics. In Rothschild, B. J. (ed.), Toward a Theory on Biological–Physical Interactions in the World Ocean. Kluwer Academic Publishers, Dordrecht: 215–234.

Yeung, P. K., 2001. Lagrangian characteristics of turbulence and scalar transport in direct numerical simulations. J. Fluid. Mech. 427: 241–274.

Hydrobiologia **484:** 75–87, 2002.
O. Vadstein & Y. Olsen (eds), Sustainable Increase of Marine Harvesting: Fundamental Mechanisms and New Concepts.

Micro- and mesozooplankton response to enhanced nutrient input – a mesocosm study

Ingrid Gismervik[1], Yngvar Olsen[2] & Olav Vadstein[2]
[1]*Section of Marine Zoology and Chemistry, University of Oslo, P.O. Box 1064, Blindern, N-0316 Oslo, Norway*
[2]*Trondhjem Biological Station, Norwegian University of Science and Technology, N-7491 Trondheim, Norway*

Key words: nutrients, ciliates, mesozooplankton, growth rates, mortality rates, egg production

Abstract

An 18-day mesocosm study was performed in central Norway to assess the effect of enhanced nutrient input to the marine plankton community. This paper reports the responses of micro- and mesozooplankton to increased food supply due to elevated nutrient input. Seven mesocosms (M1–M7) were added variable doses of N, ranging from 0 to 2.19 μm N l^{-1} d^{-1}. Phosphorus and silicate was added in Redfield ratios. The ciliate community responded rapidly to the treatment, and reached maximum biomass of 88 μg C l^{-1} within a week in the most fertilised mesocosm (M7). *Tontonia* sp. and a small *Strombidium* sp. dominated biomass and numbers, respectively. Ciliate biomass declined rapidly after the peak, returning to initial values by the end of the experiment (Day 18). Mesozooplankton biomass increased from the second week, due to recruitment of *Acartia* spp., *Centropages* spp. and *Oithona* sp. Numbers of *Temora longicornis* and *Pseudocalanus* sp. remained low. Highest biomass of mesozooplankton (116 μg C l^{-1}) was recorded in M6 by Day 18. Egg production rates for *Acartia* spp. peaked in M3 at Day 11, while calculated mortality rates for juvenile copepods was highest in M1 and M7. Estimated community net growth rates were highest in the most fertilised mesocosms for both copepods and ciliates. It is concluded that enhanced nutrient input affected both biomass and the relative species composition of the zooplankton community.

Introduction

Enhanced nutrient input to coastal waters increases primary production and algal biomass (Rosenberg et al., 1990; Smayda, 1990). Although the response of the algal community is generally well studied, the subsequent response of the zooplankton community to high and sustained nutrient loads is poorly known. This concerns both lower trophic levels like single-celled ciliates, as well as more studied taxa like copepods. However, as eutrophication has become a concern for marine waters during the last decades, some effort has been made to assess the effect of increased nutrient supply on the zooplankton community. Monitoring data indicate that zooplankton biomass has increased in some areas (Ciszewski, 1985; Fransz et al., 1992; Leppäkoski & Mihnea, 1996), although smaller crustacean species seemed to replace larger species. In a more heavily influenced area, the mesozooplankton biomass has been reduced, whereas the protozooplankton has become increasingly important (Schiewer, 1998). Studies along geographical gradients in the Baltic Sea further suggest that protozoans respond positively to eutrophication (Arndt, 1991). On the contrary, such relationships were not found for copepods in the northern Baltic (Viitasalo, 1992). Hence, the relationship between eutrophication and zooplankton is still vague, but the indications are that changes are taking place both in terms of biomass and species composition. While the initial phase of the eutrophication process may lead to increased biomass of zooplankton, abrupt negative effects may result at a longer time perspective. Negative effects, such as altered species composition, may come as a result of reduced species diversity of food algae and increased likelihood for nuisance alga (Revelante & Gilmartin, 1979; Smayda, 1990).

There are but a few marine mesocosm studies were metazoans have been given appropriate time to respond numerically to elevated nutrient levels.

These have shown increased mesozooplankton biomass compared to ambient water (Harris et al., 1982; Nejstgaard et al., 1997). Different nutrient regimes favoured different phytoplankton communities, and this was reflected in the zooplankton response. The results from the experiments mentioned above are in accordance with field studies showing increased biomass of zooplankters following short-term perturbations supplying new nutrients to the euphotic zone (Cowles et al., 1987; Nielsen & Kiørboe, 1991). Long-term field studies also suggest that there is a coupling between algal biomass and copepod production (Kiørboe & Nielsen, 1994). Literature compiled by Klein Breteler & Schogt (1994) suggests that copepods are food limited, at least for periods. However, food limitation does not necessarily apply equally to all ontogenetic stages, as the early stages of copepods seem to be less food limited than egg production by females (Kiørboe, 1998). Furthermore, copepod population growth may rely more on mortality rates of early stages than on numbers of eggs produced (Kiørboe, 1998). Thus, if these early stages are not subject to food limitation, increased algal biomass due to enhanced nutrient input may not change the copepod population growth rate, although egg production may be enhanced.

Compared to copepods, ciliates show very rapid numerical responses to increased food availability in laboratory experiments, as well as in mesocosm experiments (Olsson et al., 1992; Kivi et al., 1993; Kivi et al., 1996; Nejstgaard et al., 1997). The response is usually short-lived, and the ciliate populations tend to disappear as fast as they appeared, most likely because of enhanced predation pressure from a growing copepod population. However, the response may also be short-lived in systems were predators have been excluded (Christaki & Van Wambeke, 1995), implying that ciliates decimate their own food source, or reach some saturation in their ability to reproduce asexually at such high speeds. There is some evidence from field studies that ciliate populations are frequently predator-controlled (Kivi et al., 1993; Nielsen & Kiørboe, 1994), but may form blooms when copepods are scarce, like during the spring bloom (Nielsen & Kiørboe, 1994). Hence, increased protozoan biomass in eutrophicated areas (Arndt, 1991; Schiewer, 1998) may be a response to increased food levels, and/or to an alteration of the predator community. In this paper we investigate these relationships using data from an 18-day mesocosm study, where seven mesocosms received different nutrient loading rates. We also examine the numerical response of the mesozooplankton community to fertilisation, and look into responses on the species level.

Material and methods

The mesocosms (38 m^3, 12 m deep) were suspended in a coastal lagoon, Hopavågen, just south of the outlet of the Trondheimsfjord in Central Norway. The experiment was run from 19 August to 4 September 1997, with seven mesocosms (in short notation M1 to M7), of which six were fertilised with exponentially increasing doses of inorganic N, Si and P in atomic ratios of 16:16:1. N was added as ammonium nitrate in 1:1 atomic ratios. Nitrogen doses were 0, 0.15, 0.26, 0.44, 0.75, 1.27 and 2.16 μmol N l^{-1} d^{-1}. Salinity and temperature profiles were determined by CTD casts.

Integrated mixed samples (0–10 m) were taken every second day using a 2 m long tube sampler (volume 4.2 l) and transferred to a 25 l light-protected can. Samples of algae and ciliates were taken from the mixed sample and fixed in acid Lugol's iodine at final concentrations of 1 and 2%, respectively. For further counting, samples were settled in 10 or 50 ml Utermöhl chambers. Ciliates were counted in M1, M3, M5 and M7, whereas phytoplankton was counted in all mesocosms. Cells were counted and sized at appropriate magnification and cell volumes calculated using simple geometrical formulas. Volumes were converted to units of carbon by a factor of 0.19 pg C μm^{-3} for ciliates (Putt & Stoecker, 1989) and according to Strathman (1967) for the phytoplankton. Selected samples were protargol stained according to Montagnes (1993) in order to identify dominant ciliate species. Samples for metazooplankton (0–10 m, totally 21 l, see above) were concentrated on a 35 μm net, transferred to a 250 ml plastic vial, and fixed with acid Lugols iodine at final concentration of 1%. All samples from M1, M3, M5 and M7 were enumerated, whereas M2, M4 and M6 were enumerated from Day 11 onwards. Copepodites were measured in a Leica M10 stereomicroscope at 32× magnification, and assigned to one of 5–6 size categories. Numbers of copepodites in stage C1 and C2 (for mortality analysis) were determined from size categories.

Copepods (*Acartia, Temora, Pseudocalanus* and *Centropages*) for length-carbon estimates were taken from ambient water, anaesthetised with carbonated water, picked individually and quickly rinsed in distilled water before they were put into tin capsules for CN analysis. A joint length–weight regressions

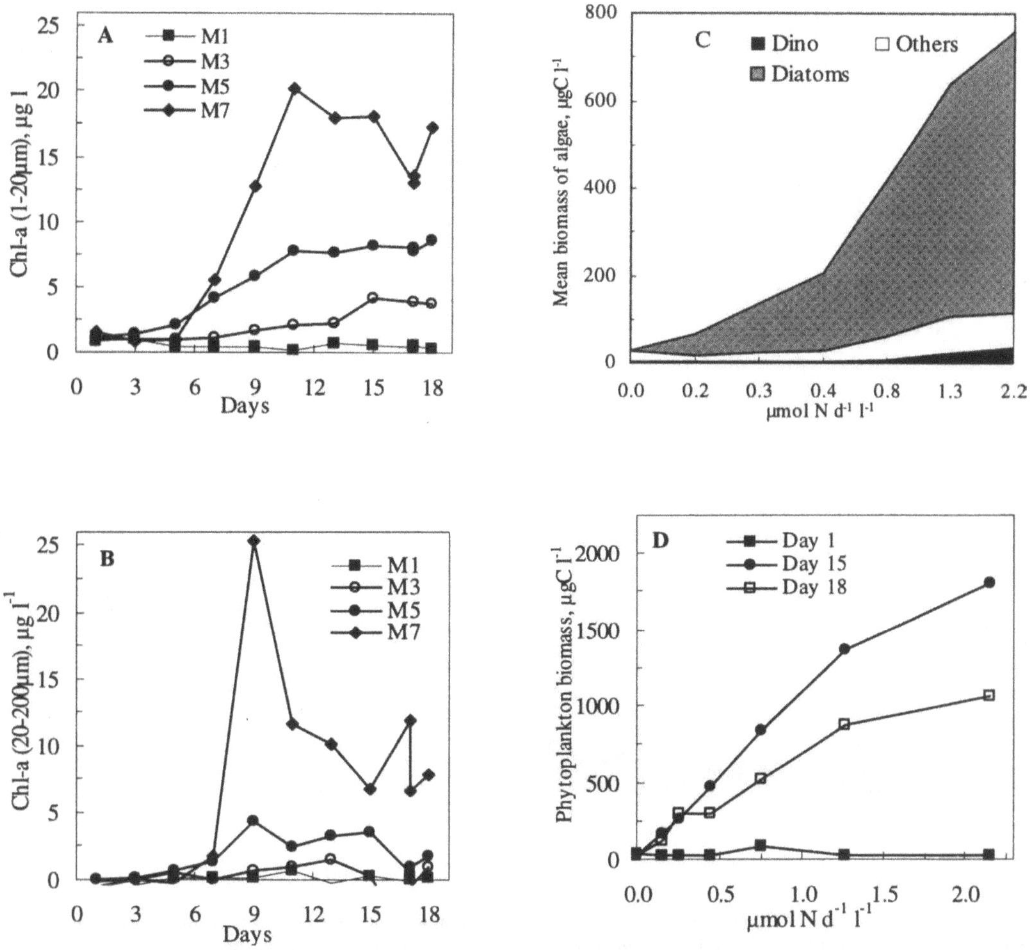

Figure 1. Phytoplankton development in selected mesocosms added variable doses of nutrients. (A) Concentration of chlorophyll *a* in 1–20 μm particles as a function of time. (B) Concentration of chlorophyll *a* in 20–200 μm particles as a function of time. (C) Average biomass distribution of the main groups of phytoplankton as a function of nutrient dose. (D) Biomass of phytoplankton at Day 1 (initial value), Day 15 (maximum value), and Day 18 (final value) as a function of nutrient dose.

($W = a^*L^b$) was established by the statistical package JMP. The following constants gave the best fit; $a = 11.8$ (95% CL of 9.8 and 13.7), and $b = 3.1$ (95% CL of 2.1 and 4.1), df=9. Other length–weight relationships were taken from Sabatini & Kiørboe (1994) (cyclopoids and harpacticoids), Berggreen et al. (1988) (nauplii), Nakamura et al. (1997) (appendicularia), Kiørboe & Sabatini (1995) (egg), Kiørboe & Sabatini (1994) (egg sacs, assuming 11 eggs in a sac (Nielsen & Sabatini, 1996)). Small medusas were given a carbon weight of 4.4 μg C based on measurements of small *Sarsia* sp. ($L = 1.6$ mm, $n = 3$), whereas carbon weights of cladocera were taken from Gismervik (1997).

Egg production experiments with *Acartia* spp. were run in water from Mesocosms 1, 3, 5 and 7 from Day 2 to 5 (M1 was excluded at day 5 due to loss of females), and from Day 7 to 13. Egg production experiments with *Temora longicornis* were run in water from Mesoscosm 5 from Day 8 to 11. Copepods were collected from ambient water with a 200 μm net with closed cod end. Eight females were picked individually and placed in 1 l glass bottles, filled with 200 μm pre screened mesocosm water. Bottles were incubated in a running water bath at *in situ* temperatures, and inverted every 8 h. After 24 h, eggs and females were gently poured into a filter cup, submerged in water. Eggs produced were counted, whereas the females were transferred to new bottles with fresh water from the mesocosms. On some occasions eggs were transferred to hatching chambers. These were plastic cups containing 200 ml filtered seawater. Air was supplied

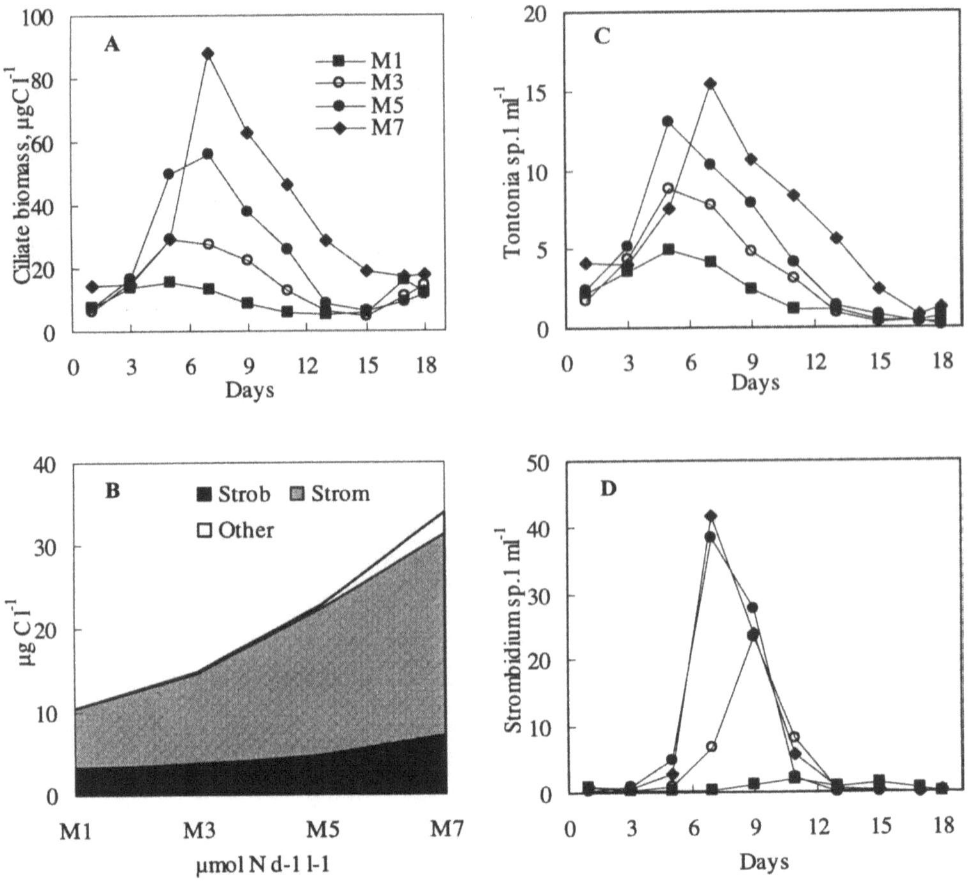

Figure 2. (A) Ciliate biomass (μg C l^{-1}) in selected mesocosms as a function of time. Average biomass distribution of the main groups of ciliates (Strom. = Strombidids and Strob. = Strobilids, Other includes all other heterotrophic ciliates) in the different mesocosms. (C and D) Abundance of *Tontonia* sp. 1 and *Strombidium* sp.1, respectively.

through a pasteur pipette, to ensure that the eggs were in suspension. Hatching percentages of eggs were recorded after 48 h. Specific egg production rate was calculated assuming a constant egg weight of 0.0365 μg C (Kiørboe & Sabatini, 1995), and a measured female body weight of 3.67 μg C.

The net growth rate of the ciliate community (μ) was calculated by least square fitting of the equation $N_t = N_o{}^*e^{\mu t}$ to experimental data, were N_o = number of ciliates day 1, and N_t = number of ciliates day 3 to 7 (non-linear fit option of the statistical package JMP). Copepod community net growth rate (g) was calculated likewise, by least square fitting of the equation $W_t = W_0{}^*e^{gt}$ to experimental data, were W_0 = biomass (μg C l^{-1}) of copepods day 11, and W_t = biomass (μg C l^{-1}) of copepods day from 13 to 18.

Mortality rates r (d^{-1}) of juvenile copepods, i.e., from nauplii to copepodite 1 and 2, were calculated as $r = (\ln(N_i) - \ln(C_{i=6})) / (t_i - t_{i+6})$ assuming a developmental time of 6 days (Kiørboe & Sabatini, 1995). N_i is the numbers of nauplii, and $C_{i=6}$ is the numbers of copepodite 1+2 found six days later. The nauplies were not determined as to stage, thus some individuals could still be in the nauplii stage after 6 days, while others could have entered the C3 stage, however the bulk of the nauplii would be found in the C1 and C2 stage. For Mesocosms 1, 3, 5 and 7 mortality rates were calculated for all odd day numbers from Day 1 to Day 11 (i = 1–11), while for Mesocosms 2, 4, and 6, numbers were only available for Day 11 (i = 11).

Results

Hydrography

Temperature was relatively constant and developed equally with time in all mesocosms. The overall av-

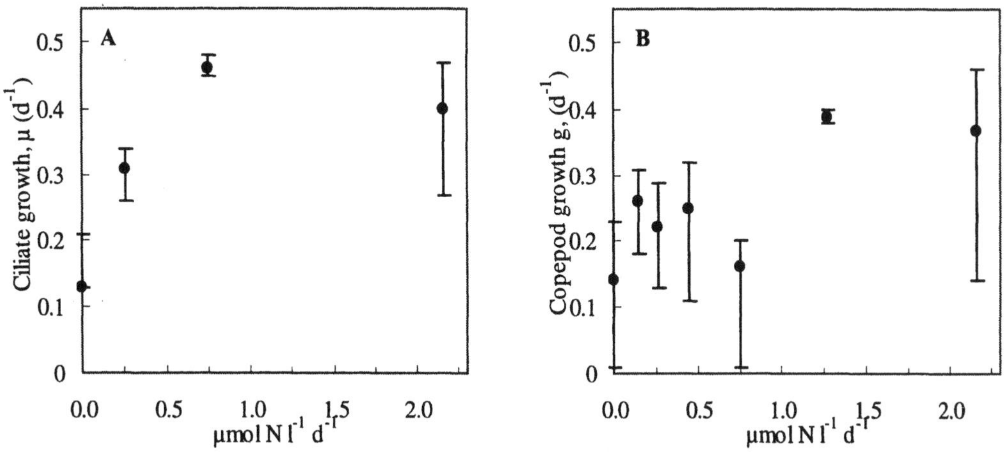

Figure 3. (A) Estimated ciliate community net growth rate, μ (d^{-1}), from day 1 to day 7. Non-linear fit to $N_t = N_o{*}e^{\mu t}$. Upper and lower confidence limits given (95%) for μ. (B) Estimated copepod community net growth rate, g (d^{-1}), from day 11 to day 18. Non-linear fit to $W_t = W_0{*}e^{gt}$. Upper and lower confidence limits given (95%) for g. See text for further explanation.

erage for mesocosm and time was 15.7 °C. Salinity showed some more variation, with mesocosm averages in the range of 30.4–31.6 o/oo and overall average of 30.8 o/oo.

Phytoplankton

The chlorophyll *a* responded relatively quickly after day 5–7, and the response was positively related to the nutrient dose added (Fig. 1A, B). The first peak of chlorophyll in the 20–200 μm fraction culminated at day 9, whereas the 1–20 μm fraction levelled off after day 11. It was notable that chlorophyll *a* containing particles in the size range 20–200 μm were primarily found in significant concentrations for the higher range of nutrient doses.

Diatoms, dinoflagellates, and small flagellates (termed others) all responded to nutrient additions, and the response of the diatoms in particular was strong (Fig. 1C). The diatom biomass was very low at the start, but responded immediately. The dominant blooming species were *Rhizosolenia fragilissima, Skeletonema costatum* and unidentified centric diatoms. The biomass ratio of the species was approximately 6:3:1 at the end in M7. Diatoms were strongly dominant in M5–7 in the final part of the experiment (Fig. 1C). The group made up 25 and 91% of the average phytoplankton biomass and the biomass on day 15 (maximum) in M7, respectively (Fig. 1D).

Only a few of the high number of dinoflagellate species present reacted to nutrient addition. The dinoflagellates that responded most pronouncedly, were *Diplopsalis* sp. and a group of unidentified thecate dinoflagellates. Other groups that responded were small nano-flagellates in the size range of 2–7 μm diameter.

Ciliates

The ciliate community responded rapidly to the enrichment, reaching maximum values within a week (Fig. 2). Net growth rates during the first week ranged from 0.13 d^{-1} in M1 to 0.46 d^{-1} in the enriched M5 (Fig. 3A). The response was, however, short-lived, and the biomass was reduced to the initial levels by Day 18 (Fig. 4A).

Aloricate oligotrichous ciliates of the family Strombidiidae dominated the ciliate community (Fig. 2B), both in biomass and in diversity, with 20 different morphotypes. A roundly-shaped-species of *Tontonia* was most important in terms of biomass, reaching maximum values of 31 μg C l^{-1}, corresponding to 15.4 individuals ml^{-1} (Figure 2C). This species resembled *Tontonia poopsia* (Montagnes & Lynn, 1988), because of its very deep oral groove, and coiled girdle, but it was somewhat smaller; mean length 30 μm. Another less numerous *Tontonia* species resembled *Tontonia simplicidens* (Lynn & Gilron, 1993) (conical form, mean length 29 μm). Ciliate numbers were dominated by a very small (mean length of 14 μm) species of *Strombidium* sp. (denoted *Strombidium* sp. 1 in Fig. 2D). Strobilidiids on average comprised 20–32% of the ciliate biomass (Fig. 2B), and seven morphotypes were identified. The slight increase of

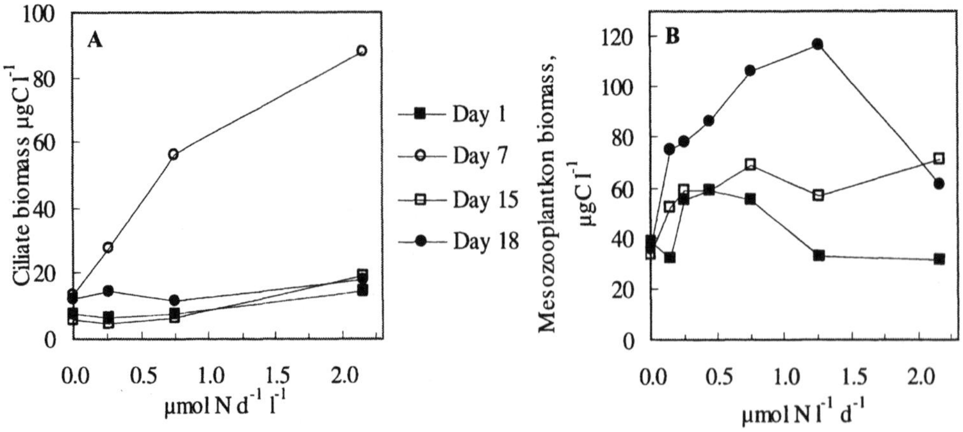

Figure 4. Ciliate and copepod biomass as a function of nutrient loading rate for selected days.

Strobilidiids by the end of the experiment was due to an increase of the large *Strobilidium strobila.* Some prostomes and haptorids (mainly *Balanion* spp.) were found (Fig. 2B), while scuticociliates were barely present (included in 'other' in Fig. 2B).

Metazooplankton

After 18 days, the total mesozooplankton biomass was elevated in mesocosms that were enriched with nutrients (Fig. 4B), but the response levelled off at the highest addition. This was in contrast to the high and almost linear response of the phytoplankton community (Fig. 1D).

Calanoid copepods dominated the zooplankton biomass throughout the experiment, while cyclopoid copepods dominated in numbers at the midway stage (Day 7 to Day 13) (Fig. 5 A–D). Adult specimens dominated at the start, nauplii numbers peaked at day 11 and copepodite numbers increased thereafter (Fig. 5). The positive response in calanoid numbers in the fertilised mesocosms was due to *Acartia* spp. (*A. clausi* and *A. longiremis*) and to *Centropages* spp. (*C. hamatus* and *C. typicus*). The numbers of *Temora longicornis* remained low, despite the high number of females present during the first days of the experiment. There was a slight increase in numbers of *Pseudocalanus* sp. and *Paracalanus parvus* in M3 and M5 (Fig. 5B,C), numbers of *Paracalanus parvus* in this group were negligible. Numbers of *Oithona* sp. increased in all fertilised mesocosms compared to M1, but there was no clear effect of the loading rate (Fig. 5).

The appendicularian *Oikopleura dioica* was abundant at the start of the experiment, disappearing by day 3, but reappearing at the end of the experiment (Fig. 6A). However, these specimens were small, and negligible in terms of biomass. Numbers of hydromedusae (*Sarsia* sp., *Podocoryne carnea*, and a few *Obelia* sp.) increased by the end of the experiment (Fig. 6B). The apparently arbitrary response to nutrient enrichment rate might be an artefact of inadequate sampling.

Egg production, growth and mortality rates of copepods

Copepod females used for egg production measurements were collected in ambient water, and incubated in mesocosm water for three (Exp. 1) or six (Exp. 2) subsequent days. The copepods spent some time acclimating to the new food regime, showing an increasing egg production for *Acartia* the first 3 to 4 days in the laboratory (Fig. 7). After 4 days (Exp. 2), the production became reduced, most likely due to aged females or lack of males. A maximum egg production rate of 19±9 eggs per female *Acartia* was found in treatments with M7 water in the first experiment, while in the second experiment a maximum of 24±9 eggs per female was found for treatments with M3 water (Fig. 7). This corresponds to specific egg production rates of 0.19 and 0.23 d^{-1}, respectively. There was no correlation between egg production rate and chlorophyll levels, or between egg production rate and ciliate biomass. The hatching success for calanoid eggs was high throughout the experiment in all mesocosms (84 – 98%). In contrast to *Acartia* spp., *Temora* females failed to produce eggs during the experiment (Exp. 2, Figure 7). However, this is in accordance with the lack of response of this species in all mesocosms (Fig. 5), and may be explained by the fact that the ex-

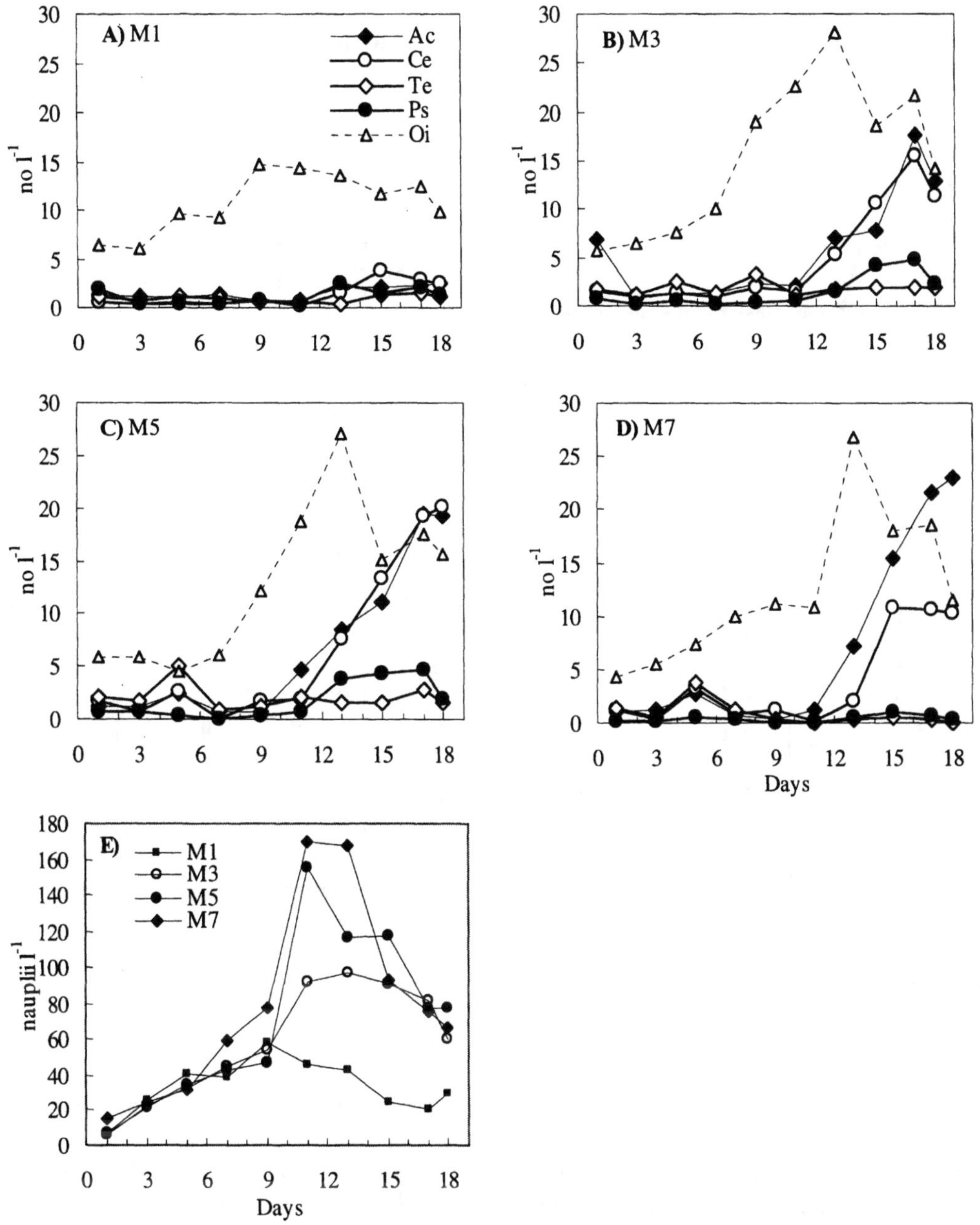

Figure 5. (A–D) Development of different copepod species over time. (E) Total numbers of nauplii as a function of time.

periment was run by the end of this species' seasonal distribution.

Egg sacs belonging to Oithonids were recorded in the mesocosms throughout the study. Numbers of egg sacs per female *Oithona* sp. did not differ between treatments, and changed little during the course of the experiment (Fig. 8).

Calculated copepod community net growth rates from day 11 to day 18, ranged from 0.14 d^{-1} in M1 to 0.39 d^{-1} in M6 (Fig. 3B). By the end of the experiment, the biomass was dominated by juveniles (C1–C3), which indicated potential for further biomass increase.

Calculated mortality rates of juvenile copepods showed some trends (Fig. 9). The mortality rates were lowest in M3, accompanied by M5 in middle of the experiment. The rates were high throughout in M1 and M7. The last period, from day 11 to day 17, mortality

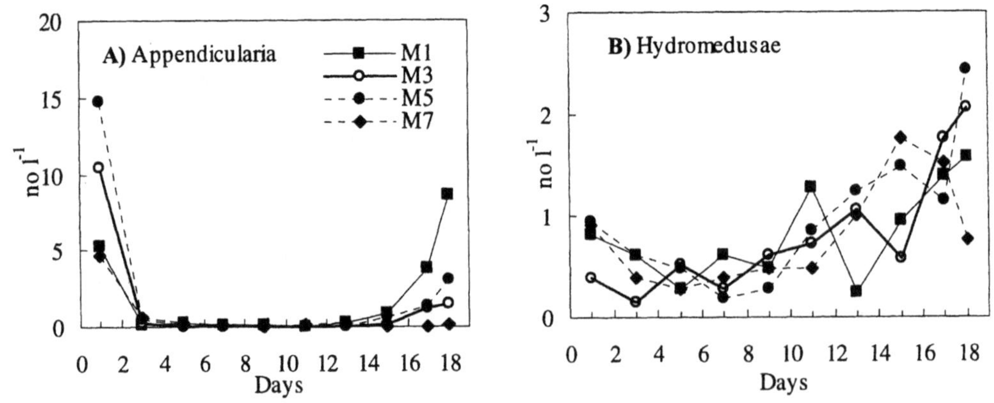

Figure 6. Abundance of Appendicularia and Hydromedusae in the mesocosms as a function of time.

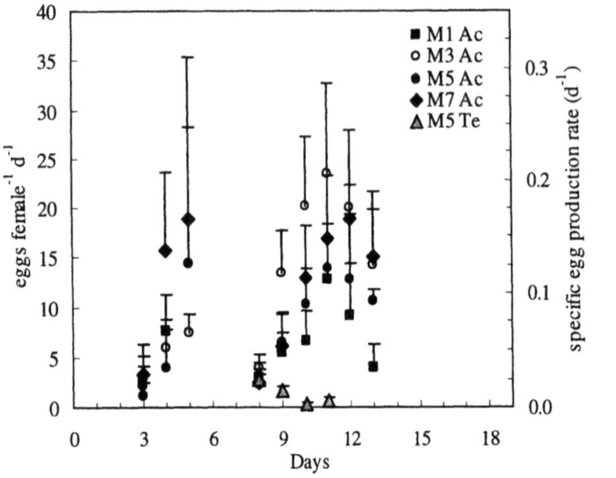

Figure 7. Egg production by *Acartia* spp. in M1, M3, M5 and M7 in exp.1; Day 3 to Day 5, and in exp. 2; Day 8 to Day 13, and by *Temora longicornis* in exp. 2; Day 8 to Day 11. Note that the rates given for the first two days in each experiment are not representative for the mesocosms, as the animals are acclimating to a new food regime. Mean + SD.

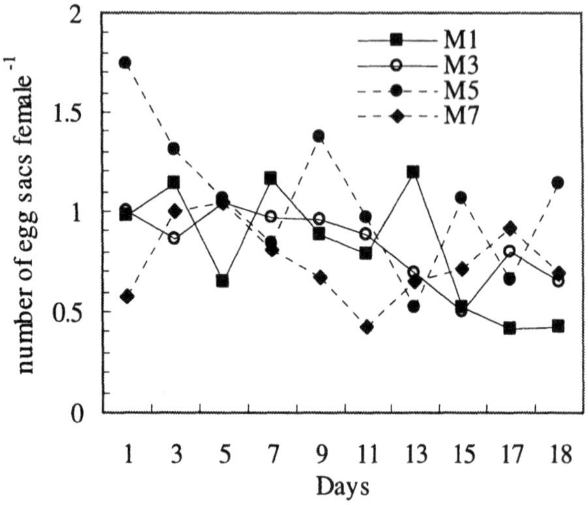

Figure 8. Number of egg sacs recorded per female *Oithona* sp. in the mesocosms.

rates increased in all mesocosms, except M1. For three of the mesocosms, M2, M4, M6, rates where only obtained for the last period. All mesocosms showed rates that corresponded to their 'neighbour' mesocosm (M1=M2, M3=M4 and M5=M6).

Discussion

Ciliates

Rather unexpectedly, *Tontonia* spp. dominated the biomass of ciliates during the experiment. This was easily seen with a dissecting microscope during the ciliate bloom, as *Tontonia* spp. is a conspicuous species when observed live, due to its long tail. In the field, *Tontonia* spp. normally occur in low numbers, and seldom become the dominating taxa (Dale, 1987; Revelante & Gilmartin, 1987; Montagnes et al., 1988; Stoecker et al., 1989; Edwards & Burkill, 1995; Sime-Ngando et al., 1995; Vaqué et al., 1998; Montagnes et al., 1999). This taxon has not been reported to occur in high numbers in other mesocosm experiments. Little is known about their feeding habits, but some species are recognised as mixotrophs (Stoecker et al., 1989). The fact that they dominated the biomass during this experiment may be due to their high initial numbers in the mesocosms (between 1.7 and 4.2 ml^{-1}), while other species occurred in low numbers (<1 ml^{-1}). However, a small strombidid soon dominated ciliate numbers, reaching net growth rates of 1.4 d^{-1} during the first week. Small oligotrichs frequently domin-

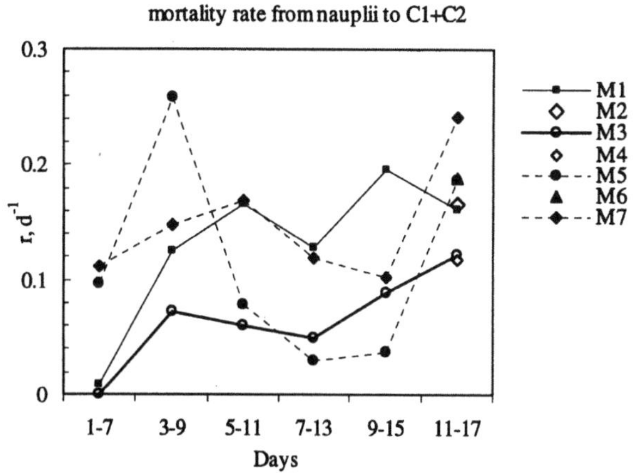

Figure 9. Mortality rates (d^{-1}) from nauplii to copepodite 1 and 2 as a function of time. See text for further explanation. One negative value (M3) assumed to be zero.

ate numbers in enclosed systems (Turner & Graneli, 1992; Christaki & Van Wambeke, 1995; Nejstgaard et al., 1997), although small prostomatids may also be numerous (*Urotricha*; (Olsson et al., 1992)). The stimulation of ciliates in enclosed systems may be a result of reduced predation pressure following reductions in larger zooplankton (during the filling of the enclosures) and increased food supply. The appendicularian *Oikopleura dioica* died after filling of the mesocosms, and there seemed to be some death among the crustaceans, which was also related to the initiation of the experiment. However, the crustaceans recovered quickly, and after 11 days, the nauplies may have exerted a massive predation pressure on the ciliates. The ciliate population collapsed before this peak, and calculated predation pressure (not shown) from crustaceans was not high enough to explain the swift decline in ciliate biomass during this period. Another explanation may be that the ciliates grazed down their food supply. Although numbers of edible nano-flagellates continued to increase at this point, the biomass build up may not have been high enough to support the huge ciliate community.

Metazooplankton

Crustaceans dominated the metazooplankton, and two copepod taxa distinguished themselves with a positive response to fertilisation; *Acartia* spp. and *Centropages* spp. On the contrary, the numbers of *Pseudocalanus* sp. and *Temora longicornis* remained low in all mesocosms, although numbers increased slightly from the first to the second week. The lack of response of *Temora longicornis* was probably due to the timing of the experiment, as *Temora* was approaching the end of its seasonal distribution in the lagoon (Gismervik, unpublished). This may also explain why *T. longicornis* failed to produce eggs during the experiments. Field studies in the eutrophicated Marsdiep area (Dutch coast) have shown that *Temora longicornis* is able to respond positively to enhanced food abundance with increased egg production rates, and that it can prolong its seasonal distribution (Fransz et al., 1992).

Pseudocalanus sp. formed large populations in the lagoon during autumn, thus their lack of response must be otherwise explained. As this species is a sac spawner, which have a lower egg production rate than broadcast spawners (Kiørboe & Sabatini, 1995) and references therein), it may require a longer experimental period to respond to fertilisation and subsequent enhanced food supply. However, few egg sacs belonging to *Pseudocalanus* sp. were found throughout the study. This, together with the fact that the slight response seen in M3 and M5 levelled off in the last couple of days and the continual low numbers in M7, suggest that this species is not responding to fertilisation like *Acartia* spp. and *Centropages* spp.

The cyclopoid *Oithona* sp. is also a sac spawner, known to have lower metabolic rates than calanoid copepods (Lampitt & Gamble, 1982; Kiørboe & Sabatini, 1994). However, this species responded quickly both to being enclosed by doubling its numbers (M1) and to fertilisation (see also Harris et al., 1982)). Numbers were equally high in M3 and in M7, indicating that the food supply in M3 was sufficient to sustain maximum net rates. Furthermore, numbers of egg sacs found per female *Oithona* sp. did not increase during the course of the experiment, nor did numbers increase due to fertilisation. Sabatini & Kiørboe (1994) found that clutch size was not affected by food concentration, while spawning frequency increased with increasing food concentration up to a level of about 200 μg C l^{-1} d^{-1}. Using the formula given by Sabatini & Kiørboe (1994) for population-specific egg production rates (SEP) and temperature-dependent hatching rate, while assuming a constant clutch size of 11 eggs (Nielsen & Sabatini, 1996), an overall mean rate of 0.10±0.03 d^{-1} was found. This rate was in the upper range recorded for *Oithona* sp., suggesting that the egg production by *Oithona* sp. during our experiment was close to this species' potential (Sabatini & Kiørboe, 1994). It has previously been suggested that the Oithonids are not able to take advantage of elevated food concentrations (Paffenhöfer, 1993), and that egg production rates by this species is far more constant than rates recorded for calanoids in the field. This seems to be the case in our study, as indicated both by numbers of copepodites in the mesocosms and the calculated egg production.

Egg production by calanoid copepods in the field is believed to be food limited e.g., Kiørboe (1998), and it was expected that the egg production would increase up to a saturation level with increased fertilisation rate . In the first experiment, egg production by *Acartia* was highest in the fertilised mesocosms after acclimatisation (Day 5). In the second experiment, production was highest in M3, followed by M7 and M5. The variation was high for all treatments, a problem frequently encountered during egg production experiments (Tester & Turner, 1990 and referenced therein). However, the consistency of the results suggested that egg production was food limited in M1 in both experiments and in M3 during the first experiment (Fig. 7). For these experiments food levels were low, between 38 and 55 μg C l^{-1} (including auto- and heterotrophic food in the 2–200 μm size range). Females were not separated from their eggs during the experiments and cannibalism may have underestimated egg production. This should, however, not affect the comparison of the different treatments. Indeed, if egg cannibalism were not constant between the different treatments, it would be expected to be highest in the low or non-fertilised mesocosms, as food availability was lower and clearance rates therefore higher in these bottles. The egg production rates after acclimatisation were in the upper range recorded for *Acartia* sp. in the field (Peterson et al., 1991; Tiselius et al., 1991; Jónasdóttir et al., 1995), but lower than rates recorded for *Acartia clausi* exposed to excess food (Landry, 1978; Ayukai, 1988). The moderate rates in the M5 and M7 mesocosms could be a result of egg cannibalism as mentioned above, poor food quality or incubation of low or non-producing females. However, the initial production rates that were recorded (Day 3 and Day 8), which should indicate the level of egg production taking place in the lagoon, were equal for the experiments and treatments. The assumption of randomly picked females and similar handling of all containers thus seemed to be fulfilled. This indicates that the food regime, both in terms of composition as well as food concentration, was most favourable in M3. On the contrary, the 'unhandy' diatoms in M5 and M7 may have reduced the fitness of copepods. The high hatching success in all mesocosms show that the diatoms, although 'unhandy', was not inhibitory for egg development.

Nauplii numbers were stimulated in all mesocosms, but levelled of in M1. The high nauplii numbers recorded on day 11 must be due to enhanced egg production on day 9 (mean hatching time at 15 °C; 1.5 d^{-1} (Kiørboe & Sabatini, 1995)), and the food converted to these eggs were probably eaten at day 7 (confer Tester & Turner (1990) and references therein). At this point of the experiment the ciliate biomass was still high (Fig. 4), while the algal biomass was moderate (Fig. 1). The ciliate bloom, despite its ephemeral nature, may therefore have stimulated copepod production. Indeed, ciliates are known to be preferred food for copepods, and to have a positive effect on egg production (Stoecker & Egloff, 1987; Wiadnyana & Rassoulzadegan, 1989).

Rough calculations of mortality rates from nauplii to early copepodites (pooled over all species) ranged from 0.01 to 0.25 d^{-1} (if we disregard the negative rate in M3 by the start of the experiment). The rates were in agreement with the data collected from the field (data compiled by Kiørboe, 1998), generally higher than rates reported from mesocosms without carnivores (Hygum et al., 2000), and lower than rates reported from mesocosms with high numbers of carnivores (Harris et al., 1982). Ohman & Wood (1995) found that mortality rates for *Pseudocalanus newmani* depended both on developmental stage and season. According to their study, the C1 stage experiences the highest mortality rate, indicating that our calculated rates may be higher than the rate experienced by the copepod community as a whole. The highest mortality rates were found in the non-fertilised and the heavily fertilised mesocosms, indicating food limitation in M1 and heavy predation pressure or inadequate phytoplankton composition in M6 and M7.

Zooplankton biomass response to increased nutrient and food levels

Both the micro- and the mesozooplankton responded positively to increased nutrient loading rate during our experiment. While the response of ciliates was rapid and short-lived, the mesozooplankton doubled in biomass by the end of the experiment. The response to enhanced nutrient loads were not linear and net growth rates for both ciliates and copepods levelled of at the higher loads. The share of large and 'unhandy' diatoms was considerably higher in M6 and M7 than in the less fertilised mesocosms. This may explain why copepod biomass and corresponding growth estimates levelled of at a higher nutrient loading rate. The growth rates estimated for the copepod community included new recruits and their growth during a week, as well as the somatic growth of the initial copepod community. Thus, our estimated growth rates were higher than rates reported for distinct copepod cohorts (Kiørboe & Sabatini, 1995; Mauchline, 1998). An interesting point was that while there was a clear increase in growth rates with increased nutrient loading, there were no such trends in egg production rates or mortality rates. These rates were highest and lowest, respectively, in the moderately fertilised mesocosms. The high numbers of nauplii in mesocosm 5 to 7 on day 11 and 13 indicated that some increase in recruitment has taken place, and this was reflected in high copepodite numbers by the end of the experiment. The positive biomass response of the copepod community could also be ascribed to enhanced somatic growth rates. Such a relationship between enhanced food levels and copepod growth has recently been described for *Calanus finmarchicus* (Hygum et al., 2000). A cohort of *Calanus finmarchicus*, which was given increased resources compared to a 'control' cohort, did not reduce its development time, but the copepodites increased their weight. These relationships are difficult to uncover when using natural plankton communities.

The ciliate net growth rates were considerably lower than their potential maximum rates (Gismervik et al., 1997), but within the range reported for ciliate *in situ* growth rates (Kiørboe, 1998). However, for some species, rates may have been considerably higher. The small *Strombidium* sp., which dominated numbers after a week in the fertilised mesocosms, obviously had very high growth rates. Although species diversity of ciliates is generally high (38 species or morphotypes were recorded in our mesocosms and about 50 species have been recorded in various field studies (Sime-Ngando et al., 1992; Leakey et al., 1993; Montagnes et al., 1999)), only a few species tend to become successful during bloom events. In our study, only 7 of 38 species reached concentrations above 2 individuals ml^{-1}, and only 2–3 became dominant (>10 individuals ml^{-1}). The same species dominated in all mesocosms. The same was the case for the mesozooplankton, although the number of mesozooplankton species was considerably lower. While *Acartia* spp. and *Centropages* spp. seemed to thrive in this system, *Temora longicornis* and *Pseudocalanus* sp. did not respond to fertilisation and elevated food supply. As yet, there is no obvious explanation to the species-specific responses, but it seems clear that enhanced nutrient levels in natural waters effect the relative species composition of both protozoan and metazoan plankton. We further conclude that moderate fertilisation had a positive effect on copepod community biomass, while ciliates, after their first outburst, were not able to take advantage of the increased food biomass. It is unclear whether the latter is due to predation, food deprivation or other internal effects in the ciliate community.

Acknowledgements

This research was supported by a grant to Ingrid Gismervik from Norsk Hydro AS, and through funding from the European Union via the MAST project COMWEB MAS3-CT96-0052. We thank the staff at Trondhjem Biological Station for valuable help, and two anonymous referees for suggestions which improved the manuscript.

References

Arndt, H., 1991. On the importance of planktonic protozoans in the eutrophication process of the Baltic Sea. Int. Rev. ges. Hydrobiologia 76: 387–396.

Ayukai, T., 1988. Egg production by the planktonic copepod *Acartia omorii* in Onagawa harbor during spring-summer. Bull. Plankton Soc. Japan 35: 127–132.

Berggreen, U., B. Hansen & T. Kiørboe, 1988. Food size spectra, ingestion and growth of the copepod *Acartia tonsa* during development: implications for determination of copepod production. Mar. Biol. 99: 341–352.

Christaki, U. & F. van Wambeke, 1995. Simulated phytoplankton bloom input in top-down manipulated microcosms: comparative effect of zooflagellates, ciliates and copepods. Aqua. Microb. Ecol. 9: 137–147.

Ciszewski, P., 1985. Long-term trends in mesozooplankton biomass development in the southern Baltic. Oceanol. 22: 63–70.

Cowles, T. J., M. R. Roman, A. L. Gauzens & N. J. Copley, 1987. Short-term changes in the biology of a warm core ring: zooplankton biomass and grazing. Limnol. Oceanogr. 32: 653–664.

Dale, T., 1987. Diel vertical distribution of planktonic ciliates in Lindåspollene, Western Norway. Mar. Microb. Food Webs 2: 15–28.

Edwards, E. S. & P. H. Burkill, 1995. Abundance, biomass and distribution of microzooplankton in the Irish Sea. J. Plankton Res. 17: 771–782.

Fransz, H. G., S. R. Gonzalez, G. C. Cadée & F. C. Hansen, 1992. Long-term change of *Temora longicornis* (Copepoda, Calanoida) abundance in a dutch inlet (Marsdiep) in relation to eutrophication. Neth. J. Sea Res. 30: 23–32.

Gismervik, I., 1997. Stoichiometry of some marine planktonic crustaceans. J. Plankton Res. 19: 279–285.

Gismervik, I., T. Andersen & O. Vadstein, 1997. Pelagic food webs and eutrophication of coastal waters: Impact of grazers on algal communities. Mar. Poll. Bull. 33: 22–35.

Harris R. P., M. R. Reeve, G. D. Grice, G. T. Evans, V. R. Gibson, J. R. Beers & B. K Sullivan, 1982. Trophic interactions and production processes in enclosed water columns. In Grice, G. D. & M. R. Reeve (eds), Marine Mesocosms: Biological and Chemical Research in Experimental Ecosystem, Springer-Verlag, New York: 353–387.

Hygum, B. H., C. Rey, B. W. Hansen & K. Tande, 2000. Importance of food quantity to structural growth rate and neutral lipid reserves accumulated in *Calanus finmarchicus*. Mar. Biol. 136: 1057–1073.

Jónasdóttir S. H., D. Fields & S. Pantoja, 1995. Copepod egg production in Long Island Sound, USA, as a function of the chemical composition of seston. Mar. Ecol. Prog. Ser. 119: 87–98.

Kiørboe, T., 1998. Population regulation and role of mesozooplankton in shaping marine pelagic food webs. Hydrobiologia 363: 13–27.

Kiørboe, T. & T. G. Nielsen, 1994. Regulation of zooplankton biomass and production in a temperate, coastal ecosystem. 1. Copepods. Limnol. Oceanogr. 39: 493–507.

Kiørboe, T. & M. Sabatini, 1994. Reproductive and life cycle strategies in egg-carrying cyclopoid and free-spawning calanoid copepods. J. Plankton Res. 16: 1353–1366.

Kiørboe, T. & M. Sabatini, 1995. Scaling of fecundity, growth and development in marine planktonic copepods. Mar. Ecol. Prog. Ser. 120: 285–298.

Kivi, K., S. Kaitala, H. Kuosa, J. Kuparinen, E. Leskinen, R. Lignell, B. Marcussen & T. Tamminen, 1993. Nutrient limitation and grazing control of the Baltic plankton community during annual succession. Limnol. Oceanogr. 38: 893–905.

Kivi, K., H. Kuosa & S. Tanskanen, 1996. An experimental study on the role of crustacean and microprotozoan grazers in the planktonic food web. Mar. Ecol. Prog. Ser. 136: 59–68.

Klein Breteler, W. C. M. & N. Schogt, 1994. Development of *Acartia clausi* (Copepoda, Calanoida) cultured at different conditions of temperature and food. Hydrobiologia 292/293: 469–479.

Lampitt, R. S. & J. C. Gamble, 1982. Diet and respiration of the small planktonic marine copepod *Oithona nana*. Mar. Biol. 66: 185–190.

Landry, M. R., 1978. Population dynamics and production of a marine copepod, *Acartia clausii*, in a small temperate lagoon on San Juan Island, Washington. Int. Revue ges. Hydrobiol. 63: 77–119.

Leakey, R. J. G., P. H. Burkill & M. A. Sleigh, 1993. Planktonic ciliates in Southampton Water: Quantitative taxonomic studies. J. mar. biol. Ass. U.K. 73: 579–594.

Leppäkoski, E. & P. E. Mihnea, 1996. Enclosed seas under maninduced change: A comparison between the Baltic and Black Seas. Ambio 25: 380–389.

Lynn, D. & G. L. Gilron, 1993. Strombidiid ciliates from coastal waters near Kingston Harbour, Jamaica (Ciliophora, Oligotrichia, Strombidiidae). J. mar. biol. Ass. U.K. 73: 47–65.

Mauchline, J., 1998. The Biology of Calanoid Copepods. Academic Press.

Montagnes, D. J. S., 1993. A quantitative protargol stain (QPS) for ciliates and other protists. In Kemp, P. F. et al. (eds), Handbook of Methods in Aquatic Microbial Ecology. Lewis Publishers, Boca Raton: 229–240.

Montagnes, D. J. S. & D. H. Lynn, 1988. A new species of *Tontonia* (Ciliophora: Oligotrichida) from the Isles of Shoals, Maine, U.S.A. Trans. Am. Microsc. Soc. 107: 305–308.

Montagnes, D. J. S., D. H. Lynn, J. C. Roff & W. D. Taylor, 1988. The annual cycle of heterotrophic planktonic ciliates in the waters surrounding the Isles of Shoales, Gulf of Maine: an assessment of their tropical role. Mar. Biol. 99: 21–30.

Montagnes, D. J. S., A. J. Poulton & T. M. Shammon, 1999. Mesoscale, finescale and microscale distribution of micro- and nanoplankton in the Irish Sea, with emphasis on ciliates and their prey. Mar. Biol. 134: 167–179.

Nakamura, Y., K. Suzuki, S.-Y. Suzuki & J. Hiromi, 1997. Production of *Oikopleura dioica (Appendicularia* following a picoplankton 'bloom' in a eutrophic coastal area. J. Plankton Res. 19: 113–124.

Nejstgaard, J. C., I. Gismervik & P. Solberg, 1997. Feeding and reproduction by *Calanus finmarchicus*, and microzooplankton grazing during blooms of diatoms and the coccolithophore *Emiliania huxleyi*. Mar. Ecol. Prog. Ser. 147: 197–217.

Nielsen, T. G. & T. Kiørboe, 1991. Effects of a storm event on the structure of the pelagic food web with special emphasis on planktonic ciliates. J. Plankton Res. 13: 35–51.

Nielsen, T. G. & T. Kiørboe, 1994. Regulation of zooplankton biomass and production in a temperate, coastal ecosystem. 2. Ciliates. Limnol. Oceanogr. 39: 508–519.

Nielsen, T. G. & M. Sabatini, 1996. Role of cyclopoid copepods *Oithona* spp. in North Sea plankton communities. Mar. Ecol. Prog. Ser. 139: 79–93.

Ohman, M. D. & S. N. Wood, 1995. The inevitability of mortality. ICES J. mar. Sci. 52: 517–522.

Olsson, P., E. Graneli, P. Carlsson & P. Abreu, 1992. Structuring of a postspring phytoplankton community by manipulation of trophic interactions. J. Exp. Mar. Biol. Ecol. 158: 249–266.

Paffenhöfer, G. A., 1993. On the ecology of marine cyclopoid copepods (Crustacea, Copepoda). J. Plankton Res. 15: 37–55.

Peterson, W. T., P. Tiselius & T. Kiörboe, 1991. Copepod egg production, moulting and growth rates, and secondary production, in the Skagerak in august 1988. J. Plankton Res. 13: 131–154.

Putt, M. & D. Stoecker, 1989. An experimentally determined carbon:volume ratio for marine "oligotrichous" ciliates from estuarine coastal waters. Limnol. Oceanogr. 34: 1097–1103.

Revelante, N. & M. Gilmartin, 1979. Microplankton diversity indices as indicators of eutrophication in the northern Adriatic Sea. Hydrobiologia 70: 277–286.

Revelante, N. & M. Gilmartin, 1987. Seasonal cycle of the ciliated protozoan and micrometazoan biomass in a Gulf of Maine Estuary. Estuar. Coast. Shelf Sci. 25: 581–598.

Rosenberg, R., R. Elmgren, S. Fleischer, P. Jonsson, G. Persson & H. Dahlin, 1990. Marine eutrophication case studies in Sweden. Ambio 19: 102–108.

Sabatini, M. & T. Kiørboe, 1994. Egg production, growth and development of the cyclopoid copepod *Oithona similis*. J. Plankton Res. 16: 1329–1351.

Schiewer, U., 1998. 30 years' eutrophication in shallow brackish waters – lessons to be learned. Hydrobiologia 363: 73–79.

Sime-Ngando, T., M. Gosselin, S. Roy & J.-P. Chanut, 1995. Significance of planktonic ciliated protozoa in the Lower St. Lawrence Estuary: comparison with bacterial, phytoplankton, and particulate organic carbon. Aquat. Microb. Ecol. 9: 243–258.

Sime-Ngando, T., K. Juniper & A. Vezina, 1992. Ciliated protozoa communities over Cobb Seamount: increase in biomass and spatial patchiness. Mar. Ecol. Prog. Ser. 89: 37–51.

Smayda, T. J., 1990. Novel and nuisance phytoplankton blooms in the sea: Evidence for a global epidemic. In Granéli, E. et al. (eds), Toxic Marine Phytoplankton. Elsevier: 29–40.

Stoecker, D. K. & D. A. Egloff, 1987. Predation by *Acartia tonsa* on planktonic ciliates and rotifers. J. exp. mar. Biol. Ecol. 110: 53–68.

Stoecker, D. K., A. Taniguchi & A. E. Michaels, 1989. Abundance of autotrophic, mixotrophic and heterotrophic planktonic ciliates in shelf and slope waters. Mar. Ecol. Prog. Ser. 50: 241–254.

Strathman, R. R., 1967. Estimating the organic carbon content of phytoplankton from cell volume or plasma volume. Limnol. Oceanogr. 12: 411–418.

Tester, P. A. & J. F. Turner, 1990. How long does it take copepods to make eggs? J. exp. mar. Biol. Ecol. 141: 169–182.

Tiselius, P., T. G. Nielsen, G. Breuel, A. Jaanus, A. Korshenko & Z. Witek, 1991. Copepod egg production in the Skagerrak during SKAGEX, May–June 1990. Mar. Biol. 111: 445–453.

Turner, J. T. & E. Graneli, 1992. Zooplankton feeding ecology – grazing during enclosure studies of phytoplankton blooms from the West Coast of Sweden. J. exp. mar. Biol. Ecol. 157: 19–31.

Vaqué, D., H. Blough & C. M. Duarte, 1998. Dynamics of ciliate abundance, biomass and community composition in an oligotrophic coastal environment (NW Mediterranean). Aquat. Microb. Ecol. 12: 71–83.

Viitasalo, M., 1992. Mesozooplankton of the Gulf of Finland and the northern Baltic proper – A review of monitoring data. Ophelia 35: 147–168.

Wiadnyana, N. N. & F. Rassoulzadegan, 1989. Selective feeding of *Acartia clausi* and *Centropages typicus* on microzooplankton. Mar. Ecol. Prog. Ser. 53: 37–45.

Hydrobiologia **484:** 89–98, 2002.
O. Vadstein & Y. Olsen (eds), Sustainable Increase of Marine Harvesting: Fundamental Mechanisms and New Concepts.

European mussel cultivation along the Atlantic coast: production status, problems and perspectives

A.C. Smaal
RIVO Centre for Shellfish Research, P.O. Box 77, 4400 AB Yerseke, The Netherlands
E-mail: a.c.smaal@rivo.wag-ur.nl

Key words: Mytilus edulis, carrying capacity, culture methods, culture areas

Abstract

Mussel culture in Europe produces about 50% of the annual world-wide harvest of mussels. The main yields of atlantic mussels are from Spain, The Netherlands and Denmark, while the Mediterranean production predominantly comes from Italy. Production in these traditional areas have been stabilised since the 1970s, and the share of Europe in the world production has decreased due to increased production outside Europe. The following factors are important in analysing production status and perspectives. Mussel production is based on an extensive culture and depends entirely on natural resources for food, spat and space. In the main culture areas, production with existing techniques seems to have reached the system's carrying capacity. Spat availability can be an additional limiting factor, particularly in bottom culture. In many traditional mussel culture areas, new functions have developed, such as recreation and nature conservation, and therefore extension of mussel culture is now also space limited. Expansion of mussel culture in Europe takes place in areas like Scottish fjords, Ireland and Greece, and is planned in Norway. Further development of sustainable mussel culture in Europe has different requirements for traditional and for new areas. In new areas the main problems are harmful blooms, local productivity and development of markets. Culture in traditional areas would benefit from optimisation of the system's exploitation capacity, improved spat collection and active participation in coastal zone management. This implies communication with other stakeholders, and exploitation of the mussel stocks within the framework of ecosystem management. Experience with this type of mussel culture is also relevant for development in the new culture areas.

Introduction

The exploitation of mussels as a food source in Europe consists of *Mytilus edulis*, along the Atlantic and North Sea coasts (Fig. 1), and *Mytilus galloprovincialis* in the Mediterranean, Adriatic and Black sea, and in the Rias, Northwest Spain. Culture and harvest in many areas dates back to historic times, such as in Italy, >2000 years ago (Mattei & Pellizzato, 1997), France, >700 years ago (Goulletquer & Héral, 1997), the Netherlands (lease sites first granted in 1870, Dijkema, 1997). In Spain, mussel harvesting was already described in 400 BC, but the actual Galician raft culture emerged only after 1945 (Caceres-Martinez & Figueras, 1997). Culture is now developing in Greece, in the Black Sea and in Norway, although main yields are still produced in the traditional areas (Monfort, 1999).

The annual production within Europe was about 800×10^6 kg fresh weight in 1998, that is almost 50% of world mussel production (Fig. 2). The world largest mussel production by country is from China, where a production increase of various species of mussels was realised from 130×10^6 kg in 1985 to 540×10^6 kg in 1998 (Fig. 3). A rapid expansion of the production of the greenshell mussel *Perna canaliculus* has been shown in New Zealand, where landings increased from 10×10^6 kg in 1985 to 75×10^6 kg in 1998 (Globefish, 1998a,b; FAO, 2000a,b).

In Europe, mussel production has increased dramatically since 1950 (Caceres-Martinez & Figueras, 1997; Edwards, 1997; Goulettquer & Heral, 1997; Kristensen, 1997; Smaal & Lucas, 2000) but in the last decades it has stabilised, and European production decreased relative to total world production (Fig. 3). Apparently, production capacity has reached its

Figure 1. Main mussel culture areas along the Atlantic coast of Europe.

limits in the traditional areas, and expansion into new areas is a long-term process (Haamer, 1997). The main production areas of mussels along the European Atlantic coast are the Rias in NW Spain, The Dutch and German Wadden Sea, the Limfjord in Denmark, and the French coast (Fig. 1). Production increase has been observed in the UK and Ireland, while new developments are being planned in Norway (Anon., 1999).

As mussel farming is an extensive mariculture, it depends entirely on natural resources for feeding, recruitment and space. Analysis of the perspectives of European mussel culture requires knowledge of the exploitation capacity of ecosystems, of factors that regulate natural recruitment, and of coastal zone regulation and management. This paper addresses the state-of-the-art in western European mussel culture by a short introduction per country and an analysis of im-

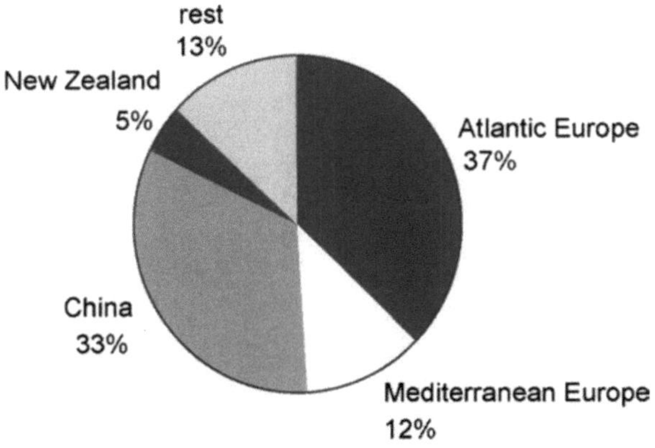

Figure 2. World mussel production in 1998, and contribution per area (after FAO, 2000a,b).

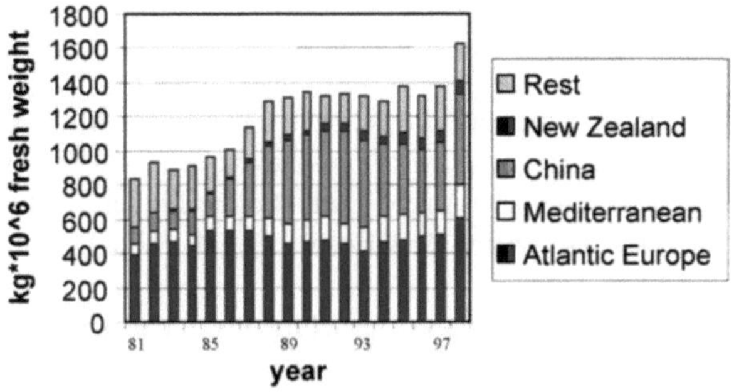

Figure 3. Development of world mussel production in the period 1985–1998 (after FAO, 2000a,b).

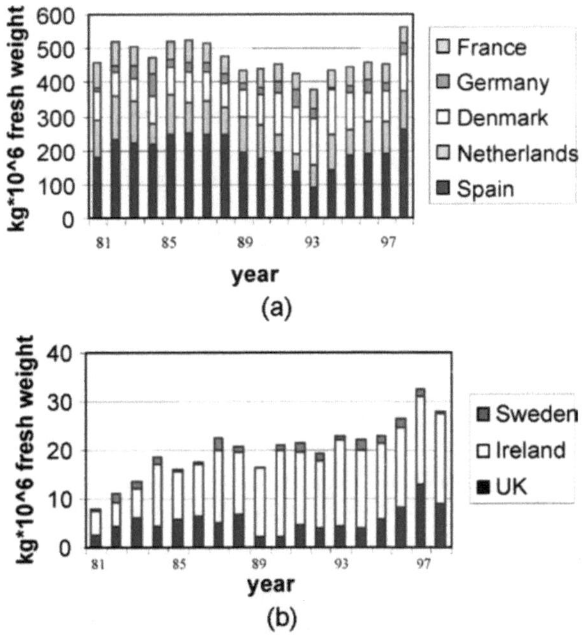

Figure 4. Mussel production in the period 1985–1998 along the European atlantic coast (a) main production areas and (b) areas with increasing production (after FAO, 2000a,b).

portant factors that determine mussel culture. Possible threats and opportunities are evaluated.

Mussel culture along the Atlantic coast of Europe

Spain

The highest mussel production in Europe is realised in the river basins (Rias) of the Galician area, of north-west Spain. Mussels (*Mytilus galloprovincialis*) are cultured on ropes suspended in the water column and attached to rafts. Total production amounts 250×10^6 kg/year, though recent yields have shown lower values, but in 1998 a production of 256×10^6 kg was realised (Fig. 4) (FAO, 2000b). Extension of the number of rafts per area, at present ca. 3500, is not allowed anymore. It was demonstrated that the condition of harvested mussels correlated with the upwelling index and hence with food production (Blanton et al., 1987). Growth of mussels on inshore rafts was less than on rafts in the mouth of the Rias, which illustrates food limitation at inshore sites (Navarro et al., 1991). Mussel seed collection is based on natural spatfall on rocky shores and on special ropes; this gives sufficient supply for the culture, although there is a call for regulation of natural spat collection (Caceres-Martinez & Figueras, 1997). Spatial arrangements of the rafts are based on natural requirements, i.e., food supply and protection against storms and waves, and on other functions of the Rias such as shipping and effluent discharge. Main problems include the regular occurrence of harmful algal blooms, often related to climate conditions, and subsequent closure of harvesting (ICES, 1999). In order to prevent bacterial contamination, mussels are purified in authorised stations before sale (Caceres-Martinez & Figueras, 1997).

France

In France, mussels (*Mytilus edulis*) along the Atlantic coast are grown on poles (bouchots), longlines and on the bottom. There is also harvesting from wild stocks. The main production areas for bouchot mussels are found on the south-west coast up to Normandy, while rope culture is concentrated in some bays in Brittany, like in the Bay of Aiguillon (Fig. 1). Bottom culture exists in Brittany (Brest) and in the English channel (Boulogne); wild stocks are exploited in Normandy. The annual production varies around 60×10^6

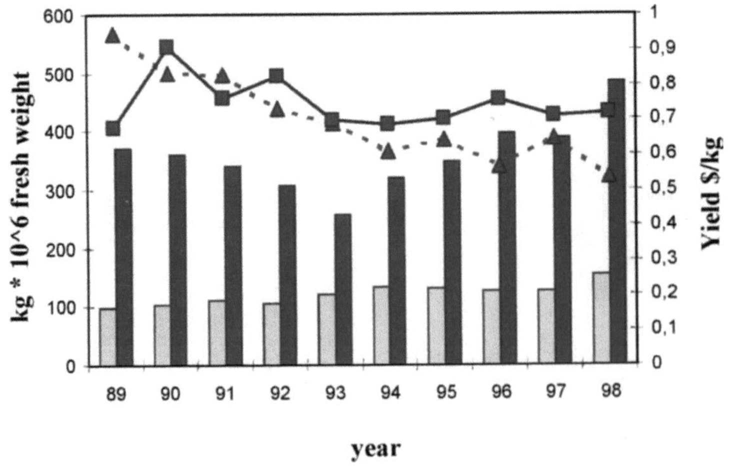

Figure 5. Yield of European production in mass (bars) and value per kg (lines). Light bars and lines with boxes: Mediterranean production, dark bars and dotted line with triangles: Atlantic production (after FAO 2000a,b).

kg, with large fluctuations due to irregular spat availability (Fig. 4). In total there are ca 10^6 bouchots (poles), with a total length of 1600 km, that produce $20–40 \times 10^6$ kg per year. Bottom culture produces 2–3 $\times 10^6$ kg/year, and wild fishery 20–30 $\times 10^6$ kg/year. The major longline production comes from the French Mediterranean coast (Goulletquer & Héral, 1997).

Spat for bouchots and ropes is collected on special collector ropes in the south-western Brittany area, and sold to farmers in Brittany and Normandy, where less spatfall occurs. For bottom culture, spat is collected from wild beds in southern Brittany (Noirmoutier) (Goulletquer & Héral, 1997). Harmful algal blooms are reported for some areas during some months of the year, in which case areas are closed for harvesting (ICES, 1999). In conjunction with biotoxin monitoring there is a programme for water quality and microbiology monitoring (Dosdat & de la Pomelie, 2000).

The Netherlands

Mussel culture in The Netherlands consists almost entirely of bottom culture, carried out on leased sites in the Wadden Sea (3560 ha) and in the Oosterschelde estuary (2040 ha) in the south-west (Fig. 1). Annual production is around 80×10^6 kg, which is the second largest European production of the blue mussel (*M. edulis*) after Denmark (Fig. 4). Owing to limited availability of spat, production was lower in the 1990s and import of half-grown and consumption sized mussels from the German Wadden Sea now occurs (Dijkema, 1997). Particularly in the Oosterschelde it has been shown that the carrying capacity has been fully exploited, as lower growth rates have been recorded in years with lower primary production or higher bivalve stocks (Van Stralen & Dijkema, 1994; Smaal et al., 2001).

Spat are harvested from wild beds in the sublitoral western Wadden Sea, and incidentally from intertidal beds. The latter are now protected within the framework of the fishery and nature conservation policy. All lease sites are presently in use and new sites will not be allowed by the government. Predation by starfish and Eider ducks is considered a problem. Harmful algal blooms causing DSP do sometimes occur in the Wadden Sea area in autumn, but in low frequency. In case of blooms, areas are closed for harvesting. Biotoxin and sanitary quality of mussel culture areas are monitored on a routine basis (Smaal & Lucas, 2000).

Germany

As in The Netherlands, German blue mussel culture is bottom culture, carried out in the Wadden Sea. There are leased sites on 1012 ha in Lower Saxony and 2850 ha in Schleswig-Holstein. Total yield is about 30×10^6 kg (Fig. 4). Spat is fished from intertidal beds in Lower-Saxony and exclusively from subtidal beds in Schleswig-Holstein. Since 1996 it has not been allowed to harvest intertidal mussel spat in this area, owing to nature conservation measures. As large parts of the German Wadden Sea are national parks with a focus on nature conservation, new lease sites are not allowed (Seaman & Ruth, 1997; Schlauch,1999). Harmful blooms causing DSP have been recorded, but they occur at low frequency and are monitored in conjunction with sanitary quality (Seaman & Ruth, 1997).

Denmark

Mussel production in Denmark comes mainly from fishery on wild *M. edulis* beds in the Limfjord and some other fjords, and for a small part from the Danish Wadden Sea and off the coast of Kattegat. Annual production amounts 100×10^6 kg, that is for the larger part processed to canned products due to the low meat content (Fig. 4). There are a few attempts to develop longline culture in various fjords, and experimental relaying has also been tested (Kristensen, 1997). The annual production varies due to variation in mussel mortality, to some extent related to anoxia events in the fjords. Natural recruitment occurs in existing mussel beds and seems sufficient to maintain exploitation of wild beds (Kristensen, 1997).

Harmful blooms and biotoxins are monitored, and occasional DSP has occurred but at low frequency. PSP occurs along the east coast. Water quality and sanitary status is also under routine control and regulation (Kristensen, 1997).

Ireland

In Ireland blue mussels are cultured on bottom leased sites and on ropes, and fishing from wild beds also takes place. Production varies between 10 and 20×10^6 kg annually (Fig. 4b), where 60% is from bottom culture and 40% from rope culture. Production areas are located along the west and south coast, and main production areas for bottom culture are in the southeast and the southwest, and for rope culture the southwest (O'Sullivan, 1998).

Major problems in bottom culture occur with predation by crabs and starfish. Spat is collected from wild stocks.

Harmful blooms causing DSP, PSP and ASP are monitored and are only observed in low frequency. In 1997 a novel toxin, azaspiracid (AZP), was detected in mussels (ICES, 1999). The sanitary status is monitored according to EU directives (EC 91/492), but the focus is on biotoxins (McMahon, 2000).

UK

In England and Wales the major blue mussel culture consists of bottom culture and the main areas are the Wash on the east coast and on the west coast the Menai straits and Morecambe bay (Fig. 1). In many estuaries wild beds occur, but harvesting is prohibited because of poor sanitary quality, or purification is required (Edwards, 1997). In Scotland there is fishery from wild beds in the Tain on the east coast, while on the west coast nowadays suspended culture is under development. In Northern Ireland wild fisheries occur in Belfast and Carlingford loughs (Edwards, 1997). Total production is about 10×10^6 kg/year (Fig. 4b). Limitations are recruitment failure and predation by crabs and starfish (Edwards, 1997) and Eider ducks (Furness, 2000).

Monitoring of harmful algal blooms has shown that in Scotland DSP and PSP occurs annually on the east coast, and on the west coast ASP is now also observed. It appears that these toxins now limit mussel production in Scotland. In England and Wales, DSP and PSP occur regularly on the east coast and along the southwest and west coasts (ICES, 1999). Sanitary status and water quality are routinely monitored (Anon, 2000; Henderson & Davies, 2000).

Sweden

Since the 1970s there has been suspended blue mussel farming on the Swedish west coast (a.o. Ljungskile, Fig. 1); it is a small scale business still under development, and annual yields are 0.5–2 $\times$ 10^6kg (Fig. 4b). Spat are collected by the use of collector strips. Limitations are predation by Eider ducks, fouling with Ascidians, and the occurrence of DSP. Also the limited interest of local consumers hampers development (Haamer, 1997).

Norway

In Norway, blue mussel production started in the seventies and still operates on a small scale. There is bottom culture near Trondheim, and north of Trondheim and near Bergen there are some longline farms. In recent years a rapid expansion has taken place and nowadays there are 60 commercial blue mussel farms in operation. There are plenty of spat, so spat collection is not a problem. Annual yield was not more than 0.5 $\times$ 10^6 kg in 1999, but a rapid increase is expected (www.skjell.com). Main limitations are the occurrence of harmful algal blooms (DSP, PSP) and slow growth in the oligotrophic fjords. In addition, predation by Eider ducks, fouling with Ascidians and other organisms, and a limited local market, are limiting factors (Strand & Vølstad, 1997). New plans are made to dramatically increase mussel production, and to develop large-scale exploitation of the extended coastline (Anon, 1999).

Culture conditions

Carrying capacity

The main mussel cultivation areas along the Atlantic and North Sea coast are located in Spain and in The Netherlands. In these areas, annual production has been relatively stable for more than 15 years (Fig. 4). The question is, if production has reached a plateau, and to what extent systems are now fully exploited.

In Spain the number of rafts has been extended but production has stayed the same (Caceres-Martinez & Figueras, 1997). Blanton et al. (1987) have demonstrated for the Ria de Arosa, that mussel condition was significantly correlated with the upwelling of ocean water rich in nutrients, hence with food supply. Growth was better in years with more upwelling, hence growth is food limited. Production increase would require an increase in food production or a more efficient use of the available food. In The Netherlands, the production target of 100×10^6 kg/year has only been realised twice in the last 15 years. For the Oosterschelde estuary (SW Netherlands), van Stralen & Dijkema (1994) and Smaal et al. (2001) have shown that the carrying capacity is now fully exploited, as mussel growth is food limited and relates to total bivalve stock. For the Wadden Sea, additional factors play a role, such as the availability of spat.

Spatfall

The culture of mussels in Europe depends entirely on natural recruitment. This is a highly variable process, and natural mortality of larvae and first settlement stages is extremely high. A good correlation is found in the Wadden Sea between severe winters and recruitment success in the following season (Beukema, 1992). This is explained by the relatively slow development of predators, like crabs and shrimp, after a severe winter, hence a mismatch of the timing of the occurrence of prey and predator. Benthic predation is lower when spat settles between existing mussels, where protection is given by the byssus threads, as was clearly shown for mussel beds in the Exe estuary (McGrorty et al., 1990). Recruitment success of mussels also seems to be more successful at lower salinity (Limfjord) or in the vicinity of a fresh water source (Wadden Sea), possibly related to lower predator densities or better food quality of decaying freshwater algae.

In bottom culture there is a relatively large need for natural spat, as losses due to predation or climate events are relatively high, compared to suspended culture. The production of 1 kg of consumption-sized mussels from bottom culture in The Netherlands requires on average 0.65 kg of spat (based on gross weight, including tarra), while in rope culture only 0.05–0.3 kg of spat (with low tarra content) would be required. As rope culture often uses spat collectors, it has a better recruitment success than bottom recruitment, and less spat is required to produce a certain yield. It is not surprising that in rope culture, in contrast to bottom culture, spat is normally not a limiting factor.

However, rope culture is often constrained to areas that are sheltered from storms and high currents. These systems typically have good conditions for dinoflagellates, and consequently have a higher risk of harmful algal blooms.

Space: quality of sites and competition with other interests

The selection of sites for mussel culture depends on a number of factors of a generic and site specific nature. In addition to conditions such as hydrodynamic stability and the carrying capacity of the system, culture areas have to meet water quality standards, and are subject to spatial regulation. Water quality at mussel culture sites is determined by the occurrence of harmful algal blooms that cause biotoxins to accumulate in the mussels, the occurrence of pathogenic bacteria and the level of toxic substances.

In many areas there is also spatial competition of shellfish culture with other functions, such as recreation and nature conservation. This has often provoked discussions and conflicts about the environmental impacts of mussel culture.

Harmful blooms

Monitoring of phytoplankton in shellfish culture areas nowadays extends along the whole Atlantic and North Sea coasts, owing to national and EU regulations (ICES, 1999). The monitoring system operates in such a way that areas will be closed to shellfish harvesting when biotoxin levels exceed the standards. However, closure for extended periods can make areas unsuitable for commercial shellfish farming, as was the case in Sweden and Norway in the 1980s and 1990s (Haamer, 1997; Strand & Vølstad, 1997). Measures to prevent harmful algal blooms are presently non-existent, and risks may even increase by anthropogenic influences such as climate change, ballast water discharges and increased transplantation due to the import and export of shellfish. The practice of monitoring, and temporary closure of areas if standards are exceeded, seems feasible for the known toxins.

Sanitary quality and toxic substances

Sanitary quality and level of contaminants is monitored according to national rules, based on EU directives. As implementation can be different in various countries, not all monitoring programmes are similar. However, sanitary quality of culture areas is generally well documented, and graded as A, B or C. Areas with quality A require no purification before sale, B-areas require purification and C-areas are closed for fishing. It should be noticed that in some countries there are four classes (in France: see Dosdat & de la Pomelie, 2000) In general, sources of pathogenic bacteria are well documented and it depends on priorities in water management whether sources will be cleaned up, or whether shellfish culture will be restricted. For toxic substances, monitoring programmes demonstrate possible conflicts with shellfish culture (see Strand & Vølstad, 1997).

Other functions

Traditionally, mussel production has developed in areas characterised by suitable natural conditions and the vicinity of a consumer market. This is particularly relevant for the production for the fresh market, such as in France and in The Netherlands. In these countries other interests in coastal areas have developed, such as nature conservation and recreation. There is nowadays a multifunctional use of water systems and mussel farming has to deal not only with natural conditions but also with other functions. This has provoked limitations on the use of coastal waters for mussel culture and spat collection. Therefore, a major factor now limiting mussel production in traditional areas is the spatial competition with other users. However, mussel culture has the potential to be integrated with functions like nature conservation, as these functions share the requirement of a clean and undisturbed environment. Moreover, sustainable exploitation of mussel populations implies stabilisation of the stock size in an exploited system. Owing to the role of mussels as filter feeders and the subsequent feedbacks in the system, a stable mussel stock may stabilise the ecosystem (Smaal & Prins, 1993; Prins et al., 1998).

Yet, it can be expected that intensified human use of traditional mussel culture areas limits the perspectives of mussel culture in these areas. Expansion is more likely in areas with suitable conditions and less competition for space.

Limitations and perspectives

Market

According to a study by Monfort (1999), mussels produced in Europe are predominantly consumed within Europe, and the major internal trade occurs between neighbouring countries. In a study for the EU, MacAllister et al. (1999) conclude that mussel consumption is considered rather stable and no dramatic changes are foreseen. Various scenarios predict a 1% growth in consumer demand per annum for the next 10 years (MacAllister et al., 1999). In some countries, like The Netherlands, much higher increase in demand has been observed in the last decade, and prices have increased. The average market prices in Europe, however, do not indicate an increased demand in relation to the supply, as they have steadily decreased (Fig. 6). Monfort (1999) expects new opportunities for increases in trade volume and profits from mussel-based value-added products like individual quick frozen (IQF) mussels, ready meals and breaded products. MacAllister et al. (1999) consider environmental measures as major constraints for mussel culture in some areas, particularly for the collection of seed.

Yet, the quality of mussels as a naturally produced human food resource can be considered as a strong potential in the developing market for healthy and safe food. However, the scenario studies mentioned above did not demonstrate clear perspectives for a dramatic market increase.

Regulations

Mussel culture is regulated by EU directives with regard to quality for the consumer in terms of pathogenic bacteria, biotoxins and chemical contaminants. Water quality of shellfish culture areas is under EU regulation (contaminant levels, EU directive 79/923/EEC; biotoxins and pathogenic bacteria, EU directive 91/492/EEC). A recent review (Fernandez et al., 2000) shows that for all European production areas there are EU and national standards for product quality and for the environment where the species is cultured or fished.

Depending on the culture area, a wide variety of national and regional regulations apply. Regulations can deal with aspects of the culture such as the seed stock. In Spain for example wild mussel collection has been regulated since 1969 (Caceres-Martinez & Figueras, 1997). More recently, regulation deals with the multifunctional use of coastal waters. As shellfish culture depends on relatively undisturbed and unpolluted areas, there is a common interest with recreation and nature conservation. However, the regulation of these interests can differ greatly between areas. For the Wadden Sea, for example, directives to regulate mussel culture are different for the Dutch, Danish and German parts; they also differ for the German states Lower Saxony and Sleswig-Holstein (Schlauch, 1999). A similarity is that many areas are closed or very restricted for fishery and culture; seed fishery on intertidal mussel beds is allowed for some areas in Lower Saxony, and in The Netherlands under restrictions. New regulations were implemented in 1993 and 1999 in The Netherlands, and in 1996 in Sleswig-Holstein. In Denmark, mussel fishery in the Wadden Sea is restricted to a limited area, and fishery in the Limfjord is regulated by individual quota and fishery restrictions.

The EU habitat and bird directives are now being implemented in many areas that are used or are relevant for mussel culture. As there is an increase in international trade, mussels are now transplanted on a larger scale. To avoid the introduction of exotic species there is a call for further regulation. It is concluded that regulation has been intensified. Consequently, further development of mussel culture has to take these regulations into account. This requires improved organisation of the mussel farmers, and more focus on environmental management.

Management

Mussel farmers in traditional areas are generally well organised. In Spain, farmers have combined and there is one organisation composed of 80% of the farmers (Caceres-Martinez & Figueras, 1997). In France, shellfish farmers are organised in local committees that are part of regional committees, and the processing industry is organised likewise. They all belong to a national council, and the organisations are partly financed by the government (Goulletquer & Héral, 1997). In The Netherlands mussel farmers are all member of a producers organisation (PO) that has a number of tasks, agreed with the government, and known as co-management. The PO is responsible for mussel seed management, and fishing is organised twice a year, based on extensive stock assessments. Fishery of mussel seed, and also of farmer-owned mussels on cultivation plots is controlled by a black box aboard fishing vessels, registering sailing and fishing activities (Smaal & Lucas, 2000). In the framework of co-management, the PO presents management and fishing plans to the government, as a part of the annual licensing procedures. Recently the shellfish companies have presented a plan for sustainable fishery, based on the triple P approach for sustainable business strategies (ODUS, 2001). Next to Profit, Planet (environment) and People (stakeholders) are considered of equal importance for the Dutch shellfish economy (see: www.schelpdieren.nl).

Management of mussel culture considers the management of their common pool of mussel seed in the first place. Common interests also consider the selection and exploitation of sites. This holds for the different types of culture (bottom lease sites, bouchot sites and areas of suspended culture). A new challenge is to develop a sustainable way of culture, particularly in protected nature conservation areas. This implies that mussel farmers focus not only on mussel seed collection, production and harvest of their product, but also play their roles as managers of mussel stocks, as these stocks are important components of the ecosystems.

The most protected area with mussel culture in Europe is probably the Wadden Sea. In the Dutch part there is experience with co-management (Steins, 1999). Despite good results in terms of seed management and communication with the government, conflicts have arisen with nature conservation organisations. Co-management as such is not sufficient to regulate the common interests of fisheries and nature conservation. A political framework that sets the rules is required.

The perspectives for the European mussel culture depend upon how problems will be managed. This requires a strategy for sustainable exploitation of natural resources. Hence, research and development for improving culture methods are needed, including stock assessments to optimise exploitation and monitoring of environmental parameters and impacts. Communication with stakeholders is also of great importance in many areas of both government and non-governmental organisations and local inhabitants.

As a consequence, mussel farmers are considered responsible not only for the production of mussels but also for the effects on the environment, and its users (nature conservation, recreation). As shown, experience has now been developed in traditional culture areas with mussel culture within the framework of environmental management. In new production areas, primary perspectives depend on the selection of the most suitable sites, in terms of productivity, protection against storms and other threats, such as harmful algal blooms. Long-term perspectives will rely on factors similar to the traditional areas, hence on experience with sustainable culture.

Conclusions

1. Mussel production in the main cultivation areas in Europe has reached a plateau, and further developments depend on new production areas or methods. Relative to the world production there has been a decrease of European mussel yields.
2. The availability of food, spat and space sets limits to production increase. Food production and quality is a function of the level of exploitation and the occurrence of harmful algal blooms. Risks of low spatfall in some years are a threat for bottom culture. Limitation in space for culture and spat collection is apparent in highly regulated nature conservation areas like the Wadden Sea.
3. Perspectives for European mussel culture depend on the possibility to adapt to new regulations, that aim at managing multi-functional use of coastal zones. Managing shellfish stocks in a sustainable way is in the interests of mussel culture as well as nature conservation, both in traditional and new areas.

Acknowledgements

The author is grateful to Per Dolmer, Lars-Ove Loo and Cedric Bacher for providing information for this paper, and to Tammo Bult, Pauline Kamermans, Marnix van Stralen and two anonymous referees for valuable comments on the manuscript. Part of this work was supported by EU ESSENSE contract nr FAIR-CT98-4201

References

Anonymous, 1999. Norges muligheter for verdiskaping innen havbruk. DKNVS/NTVA, Trondheim: 39 (In Nordic).

Anonymous, 2000. Shellfish production in the UK in 1999. Shellfish News 10: 26–30.

Beukema, J. J., 1992. Expected changes in the Wadden Sea benthos in a warmer world: lessons from periods with mild winters. Neth. J. Sea Res. 11: 42–55.

Blanton, J. O., K. R. Tenore, F. Castillejo, L. P. Atkonson, F. B. Schwing & A. Lavin, 1987. The relationship of upwelling to mussel production rias on the western coast of Spain. J. mar. Res. 45: 497–511.

Caceres-Martinez, J. & A. Figueras, 1997. The mussel, oyster, clam and pectinid fisheries of Spain. NOAA Technical Report NMFS 129: 165–190.

Dosdat, A. & C. de la Pomelie, 2000. Regulation and monitoring of marine aquaculture in France. J. Appl. Ichtyol. 16: 157–162.

Dijkema, R., 1997. Molluscan fisheries and culture in The Netherlands. NOAA Technical report NMFS 129: 115–136.

Edwards, E., 1997. Molluscan fisheries in Britain. NOAA Technical report NMFS 129: 85–100.

FAO, 2000a. Yearbook of Fishery Statistics, Capture Production. Rome, Vol. 86/1.

FAO, 2000b. Yearbook of Aquaculture Production. Rome, Vol. 86/2.

Fernandes, T. F., K. Miller & P. A. Read, 2000. Monitoring and regulation of marine aquaculture in Europe. J. Appl. Ichtyol. 16: 138–143.

Furness, B. 2000. How to keep eider ducks off mussel farms. Shellfish News 10: 8–10.

Globefish, 1998a. Commodity update of bivalves. FAO, Rome.

Globefish, 1998b. The world market for mussels. FAO/Globefish Research Programme, Vol. 55. Rome: 123 pp.

Goulletquer, P. & M. Héral, 1997. Marine molluscan production trends in France: from fisheries to aquaculture. NOAA Technical report NMFS 129: 137–164.

Haamer, J. 1997. The mussel industry of Sweden. NOAA Technical report NMFS 129: 1–6.

Henderson, A. R. & I. M. Davies, 2000. Review of aquaculture, its regulation and monitoring in Scotland. J. Appl. Ichtyol. 16: 200–208

ICES, 1999. Report of the ICES-IOC working group on harmful algal bloom dynamics. ICES cm 1999/C:4.

Kristensen, P. S., 1997. Oyster and mussel fisheries in Denmark, NOAA Technical report NMFS 129: 25–38.

MacAlister Elliott & partners Ltd, 1999. Forward study of community aquaculture, summary report. EU DG XIV, Brussels.

McGrorty, S, R. T. Clarke, C. J. Reading & J. Goss-Custard, 1990. Population dynamics of the mussel *Mytilus edulis*: density changes and regulation of the population in the Exe estuary, Devon. Mar. Ecol. Prog. Ser. 67: 157–169

Mattei. N. & M. Pellizzato, 1997. Mollusc fisheries and aquaculture in Italy. NOAA Technical report NMFS 129: 201–216.

McMahon, T., 2000. Regulation and monitoring of marine aquaculture in Ireland. J. Appl. Ichtyol. 16: 177–181.

Monfort, M. C., 1999. The European market for bivalves. FAO/Globefish Research Programme, Vol. 62, Rome, 81 pp.

Navarro, E., J. I. P. Iglesias, A. Perez Camacho, U. Labarta & R. Beiras, 1991. The physiological energetics of mussels from different cultivation rafts in the Ria de Arosa. Aquaculture 94 (2/3): 197–212.

ODUS, 2001. Uit de schulp, visie op duurzame ontwikkeling van de Nederlndse schelpdiervisserij. Yerseke, NL: 51 pp. (in Dutch).

O'Sullivan, G., 1998. Present situation and future prospects for cultivation, commercialisation and industrialisation of mussels in Ireland. Globefish 55: 55–60.

Prins T. C., A. C. Smaal & R. F., Dame, 1998. A review of feedbacks between bivalve grazing and ecosystem processes. Aquat. Ecol. 31: 349–359.

Schlauch, J. 1999. Entwicklung und Struktur der deutschen Molluskenfisherei und -kultur im trilateralen Vergleich mit Danemark und den Niederlanden. Bamberger Wirtschaftsgeografischen Arbeiten, heft 10 (in German).

Seaman, M. N. L. & M. Ruth, 1997. The molluscan fisheries of Germany. NOAA Technical report NMFS 129: 57–84.

Smaal A. C. & L. Lucas, 2000. Regulation and monitoring of marine aquaculture in The Netherlands. J. Appl. Ichtyol. 16: 187–191.

Smaal, A. C. & T. C. Prins, 1993. The uptake of organic matter and the release of inorganic nutrients by bivalve suspension feeder beds. In: Dame, R. F. (ed.), Bivalve Filter Feeders in Estuarine and Coastal Ecosystem Processes. Springer, Berlin: 271–298.

Smaal A., M. van Stralen & E. Schuiling, 2001. The interaction between shellfish culture and ecosystem processes. Can. J. Fish. aquat. Sci. 58: 991–1002.

Steins, N., 1999. All hands on deck – an interactive perspective on complex common-pool resource management based on case studies in the coastal waters of the Isle of Wight (UK), Connemara (Ireland) and the Dutch Wadden Sea. Thesis, Wageningen Universiteit, Wageningen: 225 pp.

Strand, O. & J. H. Vølstad, 1997. The molluscan fisheries and culture in Norway. NOAA Technical report NMFS 129: 7–24.

Van Stralen, M. R. & R. D. Dijkema, 1994. Mussel culture in a changing environment: the effects of a coastal engineering project on mussel culture (*Mytilus edulis* L.) in the Oosterschelde (SW Netherlands) Hydrobiologia 282/283: 359–379.

Hydrobiologia **484:** 99–109, 2002.
O. Vadstein & Y. Olsen (eds), Sustainable Increase of Marine Harvesting: Fundamental Mechanisms and New Concepts.

Effect of nutrient supply on growth of blue mussels (*Mytilus edulis*) in a landlocked bay

Gunvor Øie[1], Kjell Inge Reitan[1], Olav Vadstein[2] & Helge Reinertsen[3]
[1]*NTNU, Department of Botany, Brattøra Research Centre, N-7491 Trondheim, Norway*
[2]*SINTEF Fisheries and Aquaculture, Brattøra Research Centre, N-7491 Trondheim, Norway*
[3]*NTNU, Trondhjem biologicalstation, N-7491 Trondheim, Norway*
Tel: +47 73 59 05 81. Fax: +47 73 59 63 63. E-mail: Gunvor.Oie@sintef.no

Key words: blue mussels, *Mytilus edulis*, growth, nutrients, primary production, environmental conditions, algae

Abstract

The growth of blue mussels (*Mytilus edulis*) was studied in the landlocked bay Hopavågen in central Norway for 3 years, of which in 2 years (1998 and 1999) nutrients were added to increase the primary production. Nutrients (N:Si:P) were added daily from May to October in 1998 (molar ratio 15:5:4.1) and 1999 (molar ratio 16:8:1). The doses of nutrients correspond to 0.4 and 0.8 μg P l^{-1} day^{-1} in 1998 and 1999 respectively. The growth of blue mussels (*Mytilus edulis*) in 1997 was followed at four depths (1, 2, 4 and 7 m). In 1998 and 1999 growth was followed at 2 and 10 m depths at four locations in Hopavågen and at a control station outside on the coast. The nutrient supply in 1998 only slightly increased the algal biomass (chlorophyll *a*), whereas mean daily primary production during the summer remained at the same level as the previous year. The increased nutrient supply in 1999 caused a nearly 50 and 100% increase in mean summer biomass and daily primary production, respectively. The growth of blue mussels in Hopavågen in 1997 and 1998 was within the same size range during the summer. In 1999 the shell length of blue mussels kept at 2 m depth was significantly higher than in the previous year at end of the growth season. The recorded growth was also significant higher than for mussels at 2 m depth at the control station. No difference in shell length was observed on mussels grown at 10 m depth in Hopavågen and in the control stations in 1998 and 1999. A higher tissue content was found in blue mussels grown at 2 m depth in Hopavågen, both in 1998 and 1999 when compared to the control groups. At 10 m depth no differences were recorded.

Introduction

Seasonal growth of blue mussels (*Mytilus edulis*) is influenced by the interaction of environmental variables such as temperature, current velocity, salinity, food supply and food quality. The most important factor in determining growth rate is food supply because this provides the necessary energy to sustain growth. Food availability has often been found to exert a greater influence on growth rate than temperature and salinity (Ceccherelli & Barboni, 1983; Page & Hubbard, 1987). Food supply is generally considered to be the limiting factor in both suspended (Navarro et al., 1991; Perez-Camacho et al., 1991) and seabed culture systems (Smaal & von Stralen, 1990). Page & Hubbard (1987) concluded that the temporal patterns of growth in *Mytilus* were determined mainly by variations in phytoplankton, and that temperature could be virtually eliminated as a growth regulating factor over the 10–18 °C range normally experienced by these mussels.

The primary production in Norwegian coastal areas is relatively low, about 100–230 g C m^{-2} year^{-1}, and the main production takes place in the period between April and September (Eilertsen & Taasen, 1984; Lindahl, 1995). This illustrates that the main growth season of blue mussels is to a great extent restricted to a six month period from spring to early autumn. However, along the 2500 km long Norwegian coastline the growth conditions for blue mussels will vary considerably. This is demonstrated by a variation in growth time from seed to commercial size

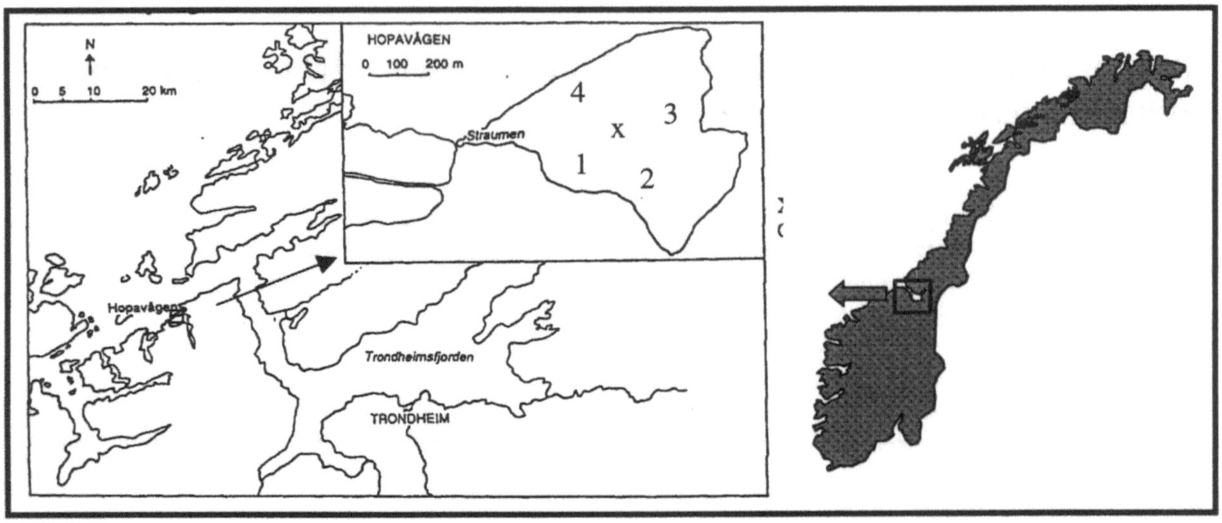

Figure 1. Hopavågen is connected to coastal current in central Norway. The small map of Hopavågen shows the stations (1, 2, 3 and 4) where blue mussels were placed. Water samples for chlorophyll *a* and primary production analysis were collected from station X. In 1997 blue mussels were placed at station 2 (1, 2, 4 and 7 m depths). In 1998 and 1999 blue mussels were placed at all four stations in Hopavågen (2 and 10 m depths) and control stations outside Hopavågen.

from about 1 to 3 years from the south to the north of Norway (Eriksen & Fagerholt, 1998). Parts of this variation may be explained by differences in length of growth season and temperature, but differences in food availability may be more important. Local differences in primary production are caused by differences in natural and anthropogenic impacts of nutrients. It is well documented that primary production, algal biomass and production in general increase with nutrient impact (Boynton, 1982; Mann, 1982; Nixon et al., 1986).

The main aim of this experiment was to test if expected increased food availability caused by enhanced nutrient supply to coastal waters during the summer improved the growth of blue mussels. Effects on the primary producers were monitored, and growth in shell length and mussel weight was used as response parameters for the blue mussels.

Material and methods

The experiment took place in the landlocked bay, Hopavågen (370 000 m^2 and 6.66 10^6 m^3), in central Norway. The bay has a mean depth of 18 m and a maximum depth of 31 m, and is connected to the coast by a narrow and shallow channel (Marion, 1996). The depth of the sill is approximately 1 m at low tide, but varies dependent on the tidal cycle. The water exchanges, mainly caused by the tidal forces, correspond to 0.11±0.04 of the total volume day^{-1}, or 0.19±0.07 of the photic zone (0–10 m) day^{-1}(unpublished results).

No nutrients were added to the bay in 1997, and data (temperature, salinity, chlorophyll *a*, primary production and shell length) were provided for an undisturbed situation. In 1998 and 1999 nutrients were added in rates equal to the natural supply and twice the natural supply, respectively. The nutrients entering Hopavågen in an undisturbed situation was calculated from data of sedimentation rates measured in 1996 and 1997. The dose of nutrients was based on the average sedimentation rate of phosphorus during summer, which corresponded to 0.4 μg P l^{-1} day^{-1}. Addition of silicate and nitrogen was calculated assuming a molar ratio of 16:8:1 for N:Si:P. This corresponds to the optimal N:P ratio of phytoplankton and 50% of the optimal Si:P ratio of diatomes. N was supplied as NH_4NO_3, P as H_3PO_4 and Si as waterglass; $Na_3Si_2O_7$ in aqueous solutions.

No differences between years were observed in dissolved inorganic nitrogen or phosphorus (P = 0.788 and 0.122) in the euphotic zone (Vadstein et al., in prep), which was also the case for the N:P-ratio (P = 0.943). For dissolved inorganic silicate an increase in the summer average concentrations was observed (P = 0.0125), but only 1997 and 1998 were significantly different. Therefore, the N and P nutrients added in 1998 and 1999 were quantitatively consumed.

The nutrients were added daily from early May to mid October to the strong tidal current. Nutrients were automatically added during the early rising tide period with inflowing water to the bay, corresponding to approximately two additions of nutrients daily. The nutrients were supplied through separate dosing channels in order to avoid chemical precipitation and control the dose amount and nutrient composition.

The biomass of algae, measured as a concentration of chlorophyll *a*, and primary production were measured every week from June to September in 1997, 1998 and 1999. Water samples were collected from the central area of the bay as integrated mixed samples from the photic zone (normally 0–10 m, Fig. 1). Samples of the coastal water outside Hopavågen were collected from the surface at the inlet to the bay (sampled at rising tide). This water served as a control. For chlorophyll *a* measurements, water was and filtered through a 200 μm screen to remove mesozooplankton and a modified method of Parson et al. (1984) was performed. This method includes filtering of exact volumes of seawater on GF/F filter (Whatman), extraction in cold acetone and quantification by fluorometry using a Turner Designs fluorometer. The primary production was estimated with the ^{14}C-method. Water was filled in 8 light and 2 dark tissue culture bottles (Nunc, total volume 62 ml), and 3.6 μCi of ^{14}C-bicarbonate was added to each bottle (The international Agency for ^{14}C Determination). The bottles were incubated *in situ* at 0.5, 1, 3, 5, 7, 9, 11 and 13 m depth for 3-4 h. Water was filtered through 0.45 μm membrane filters of mixed cellulose. Inorganic ^{14}C on the filters was removed by HCL fume. Incorporated ^{14}C was determined by liquid scintillation counting (Ultima Gold XR, Packard and Packard model 1900TR Liquid Scintillation counter). Primary production was calculated from net ^{14}Ci incorporation, dissolved inorganic carbon concentration and added ^{14}C. Primary production per day was calculated by multiplying with a light factor: total daily insolation divided by insolation during the incubation period (μmol m^{-2}, 400–700 nm). Area of primary production was estimated by trapeze integration.

In all 3 years, temperature was registered every week at 2 and 10 m depths in the centre of Hopavågen. In 1999, temperature was registered at 2 and 10 m depths every half-hour at four different stations in Hopavågen, and at a control station, using a Tinytag Temperature Logger (Gemini Data Loggers LTD, UK). The water current was registered continuously during a 14-day period (September – October) at 5 and 10 m depths at station 2 in 1997, and at 10 m depth at all stations in Hopavågen and the control station (C1) in 1998. The salinity was determined as practical salinity units (PSU) from conductivity measurements done with a profiling instrument.

Blue mussels were in all years transported to Hopavågen from a commercial farm (Taroskjell in Roan) and placed in net trays at the selected locations. The first year, 1997, 50 mussels (40–44 mm shell length) were put in each tray and placed at station 2 (1, 2, 4 and 7 m depths), as representative for the control situation before addition of nutrients. In 1998 and 1999, groups of 50–90 blue mussels (36.5–40.1 mm shell length) were placed at 2 and 10 m depths at the four stations in Hopavågen and at the control station outside (station C1 in 1998, and stations C1 and C2 in 1999).

The shell length of all individuals was measured every month (from May or June to November or December) using a sliding calliper. The growth rate (μ) was calculated for the period with exponential growth, from the end of June to the end of September, by linear regression of ln$L$ vs. time:

$$\ln L_t = \ln (L_0) + \mu \cdot t, \quad (1)$$

where L_t and L_0 is the shell length at time t and at start. The daily increase in shell length (DLI) is expressed as per cent per day:

$$\mathrm{DLI} = (e^{\mu} - 1)\,100. \quad (2)$$

In 1998 and 1999 the blue mussels were harvested in November, and wet weight, dry weight and ash-free dry weight of the shell tissue was determined. Dry weight was determined after 3 days drying at 60 °C, and the ash-free dry weight was measured after combustion at 600 °C for 12 h.

Results are presented as mean values and standard errors of the mean (SE), both in tables and figures.

Results

Water current, temperature and salinity

Total current velocity was low (1.1–1.8 cm s^{-1}) at all stations in Hopavågen (Table 1). The highest velocity and the lowest number of registrations between 0 and 1 cm s^{-1} (30.7%) were recorded at 10 m depth at station 2. At the control station (C1, Table 1) total current

Table 1. Current direction and current velocity at 5 and 10 m depths in Hopavågen in 1997, and at 10 m depth in stations 1, 2, 3 and 4 in Hopavågen and at Værnes in 1998

	1997, Station 2		1998, 10 m				
	5 m	10 m	St. 1	St. 2	St. 3	St. 4	Control
Main current, degree	198	236	72		91	70	134
Total current velocity (cm s^{-1})	1.5	1.8	1.2		1.1	1.2	1.2

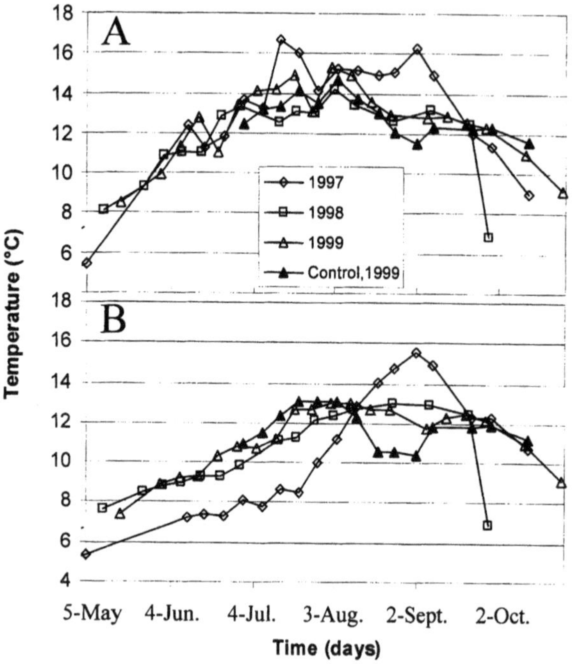

Figure 2. Recorded temperature at 2 m (A) and 10 m (B) depths in the central area of Hopavågen in 1997, 1998 and 1999, and in the control station in 1999.

velocity was in the same range as that measured in Hopavågen.

The recorded temperature at 2 m depth in 1997 increased gradually to mid July (16.6 °C). From the middle of July to the middle of September, temperature was between 14.2 and 16.6 °C, before the temperature decreased rapidly in September. In 1998 the temperature at 2 m depth was between 10.9 and 14.2 °C from June to September. In 1999 the same pattern was registered, except for a slightly higher temperature in July and August. No significant difference in temperature was recorded between Hopavågen and the control station at 2 m depth in 1999. The recorded temperature at 10 m depth in 1997 varied between 5.4 and 9.9 °C from May to the end of July (Fig. 2). From July to the end of August the temperature gradually increased to 15.5 °C, before gradually decreasing. In 1998 the temperature showed a slow increase from 8.5 °C in the middle of May to 12.0–13.0 °C in late July, and no future increase took place during autumn. In 1999, the temperature at 10 m depth varied between 7.5 and 13 °C from June to October, whereas the temperature variation in the same period was in the range of 11–16 °C at 2 m depth (Fig. 2). At 10 m depth temperature in the control station was higher in July (0.2 °C in average), and lower in August and September (1.5 °C in average). Unfortunately, no temperature data from the control are available for 1997 and 1998.

Salinity in Hopavågen varied between 27 and 33¡, but with most values in the 30–32¡ range. There was a tendency towards more brackish water in 1997. Statistically significant differences were observed between 1996 vs. 1997 and 1998. However, with averages of 29.8¡ in 1996 vs. 31.2 and 31.7¡ in the following two years, the biological significance of the difference must be negligible.

Biomass of phytoplankton and primary production

No differences between years were observed in dissolved inorganic nitrogen or phosphorus ($P = 0.788$ and 0.122) in the euphotic zone (Vadstein et al. in prep), which was also the case for the N:P-ratio ($P = 0.943$). For dissolved inorganic silicate an increase in the summer average concentrations was observed ($P = 0.0125$), but only 1997 and 1998 were significantly different. Therefore, the N and P nutrients added in 1998 and 1999 were quantitatively consumed.

In 1997, when no nutrients were added, the biomass measured as chlorophyll *a* was 3.1 $\mu g\ l^{-1}$ on average during the first growth period (June) in Hopavågen, and for rest of the growth season the average was 1.0–1.6 $\mu g\ l^{-1}$(Fig. 3). At the control station chlorophyll *a* was 4.1 $\mu g\ l^{-1}$ during the first growth interval (Table 2, Fig. 3), and for rest of the season chlorophyll *a* content varied between 1.2 and 2.1 $\mu g\ l^{-1}$. Thus, in most cases concentrations in Hopavå-

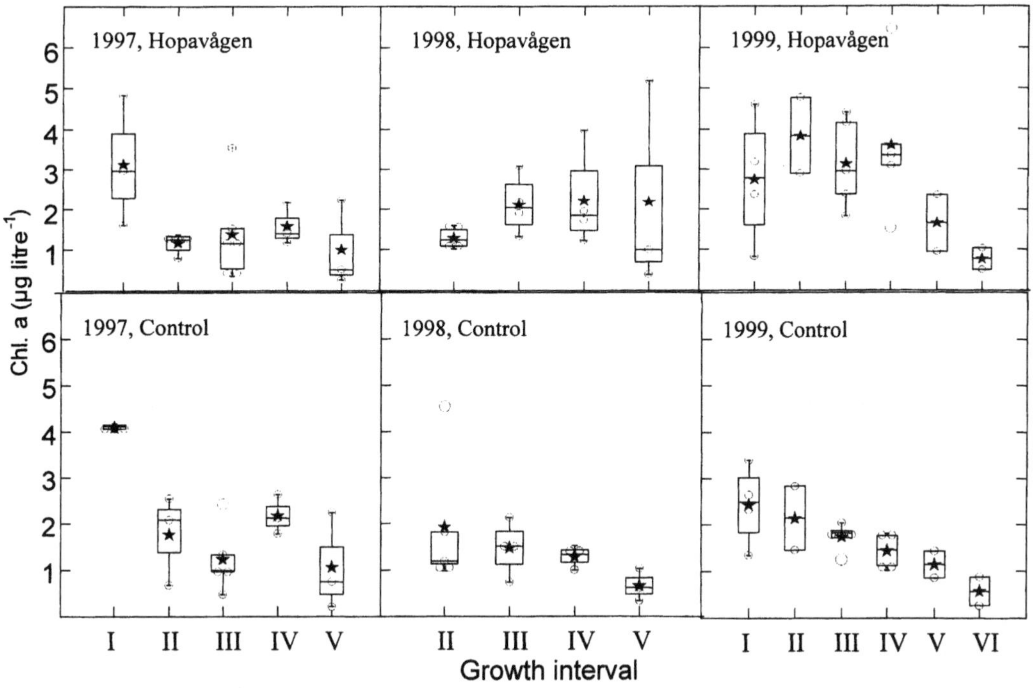

Figure 3. The biomass of phytoplankton (μg chlorophyll *a* l^{-1}, mixed sample, 0–10 m depths) in 1997, 1998 and 1999 in Hopavågen and the control (incoming water to the bay). The growth intervals are described in Table 2. The star shows the average value, and 50% of all observations are inside the square.

Table 2. Abbreviations for the growth interval in 1997, 1998 and 1999. The interval indicates the period between two shell length measurements

	Growth interval					
	I	II	III	IV	V	VI
1997	4 June–25 June	25 June–23 July	23 July–8 Sept	8 Sept–9 Oct	9 Oct–16 Dec	16 Dec–5 Feb
1998		24 June–29 July	29 July–28 Aug	28 Aug–28 Sept	28 Sept–26 Nov	
1999	4 June–30 June	30 June–19 July	19 July–23 Aug	23 Aug–22 Sept	22 Sept–15 Oct	15 Oct–17 Nov

gen and the control were similar this year. In the year with low doses of nutrients added (1998) the mean chlorophyll *a* content was 1.3 μg l^{-1} in July, and average values for rest of the period was in the range of 2.1–2.2 μg l^{-1}. The average chlorophyll *a* content in the control decreased from 2.0 μg l^{-1} to 0.5 μg l^{-1} during the growth season. In 1999, the year with high doses of nutrient added, the chlorophyll *a* content increased from 2.7 μg l^{-1} in June to 3.6 μg l^{-1} at the end of September, and thereafter the chlorophyll *a* content decreased to 1.7 μg litre. In the period October to November the chlorophyll *a* content averaged 0.8 μg l^{-1}. The average chlorophyll *a* content in the control station gradually decreased in 1999 from 2.5 μg l^{-1} to 0.5 μg l^{-1} during the season.

For the growth intervals II–IV the average chlorophyll *a* level in Hopavågen during the three years was 1.47±0.88, 2.07±1.21 and 3.35±1.29, respectively, whereas coastal water had average concentrations of 1.85±1.03 in 1997, 1.57±0.97 in 1998 and 2.18±1.68 in 1999. Addition of nutrients to Hopavågen increased the chlorofyll *a* level in periods II–IV, compared to the control.

The primary production in the period with exponential growth of the blue mussels (growth intervals II, III, IV) showed an average of between 200 and 450 mg C m^{-2} d^{-1} in Hopavågen in 1997 (Fig. 4). In the following year the average primary production was to 300–400 mg C (m^{-2} d^{-1}) for the same intervals (growth intervals II, III, IV). The daily primary

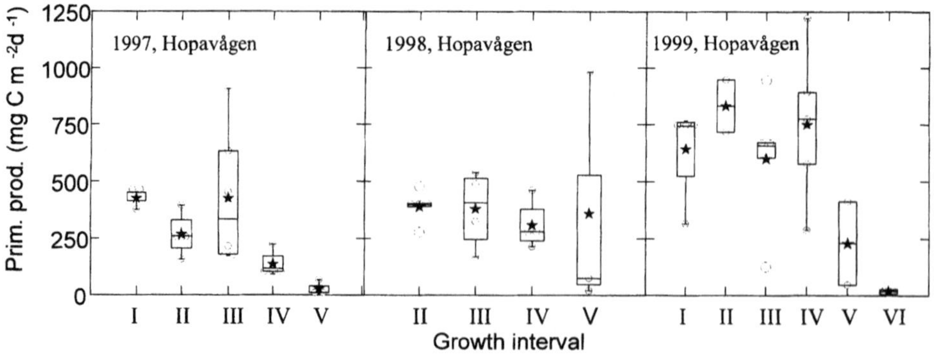

Figure 4. Primary production (mg C m^{-2} day^{-1}) in Hopavågen measured every week from June to September in 1996, 1997 and 1998. The growth interval is described in Table 2. The star shows the average value, and 50% of all observations are inside the square.

Table 3. Daily length increases (%, day^{-1}, ±SE) of blue mussels in the period from the end of June to the end of September at 2 and 10 m depths

	Station 1	Station 2	Station 3	Station 4	Control 1	Control 2
2 m						
1997		0.23±0.02				
1998	0.23±0.02	0.29±0.02	0.24±0.02	0.36±0.02	0.26±0.03	
1999	0.25±0.02	0.24±0.02	0.26±0.02	0.27±0.02	0.21±0.02	0.21±0.02
10 m						
1997						
1998	0.20±0.02	0.13±0.02	0.20±0.02	0.09±0.02	0.22±0.03	
1999	0.20±0.02	0.14±0.02	0.21±0.03	0.14±0.02	0.18±0.02	0.15±0.03

production increased to 600–800 mg C (m^{-2} d^{-1}) from June to the end of September, 1999, when the highest amount of nutrients was added. The primary production increased by 100% from 1998 to 1999. Unfortunately no results for primary production in the control was available.

Growth of blue mussels

The shell length of blue mussels in Hopavågen increased exponentially from 40.9±0.7 – 43.1±0.8 mm in June to 53.5±0.8 – 56.3±0.9 mm in end of September in 1997 (Fig. 5), and no significant differences were recorded at the four depths. In 1998 the size of the mussels at 2 m depth showed no significant differences between the stations in Hopavågen and the control station at the end of the experiment. High growthrate was recorded in 1998 at three of the stations in Hopavågen (2, 3 and 4) from the end of June (36.5±0.5 – 39.1±0.4 mm) to the end of September (49.7±0.4 – 50.6±0.6 mm). At station 1 and the control station the growth terminated at the end of August and the shells reached a mean length of 47.4±0.5 and 48.3±0.7mm, respectively. In 1999 significant differences in sizes of mussels grown in Hopavågen and at the control station were observed at the end of the growth period. The period with exponential growth was from June (39.9±0.5 – 40.1±0.5 mm) to the end of September (51.6–53.6 mm in Hopavågen and 50.3±0.7 mm in the control). At the end of the growth season shell length in Hopavågen was measured to 52.6–53.6 mm, and shell length in control stations was 49.8–51.4 mm.

At 10 m deep no difference in shell length between mussels grown in Hopavågen and at the control station was observed, neither in 1998 nor in 1999. The shell length at the end of June 1998 was 37.1–40.1 mm and at end of November the shell length was 43.5–47.9 mm. In 1999 the shell length increased from 38.8–40.2 mm in June to 46.1–51.6 mm at end of November at 10 m depth.

Daily length increase (DLI) in the blue mussels was calculated for the period with exponential growth, from the end of June to the end of September (Table 3).

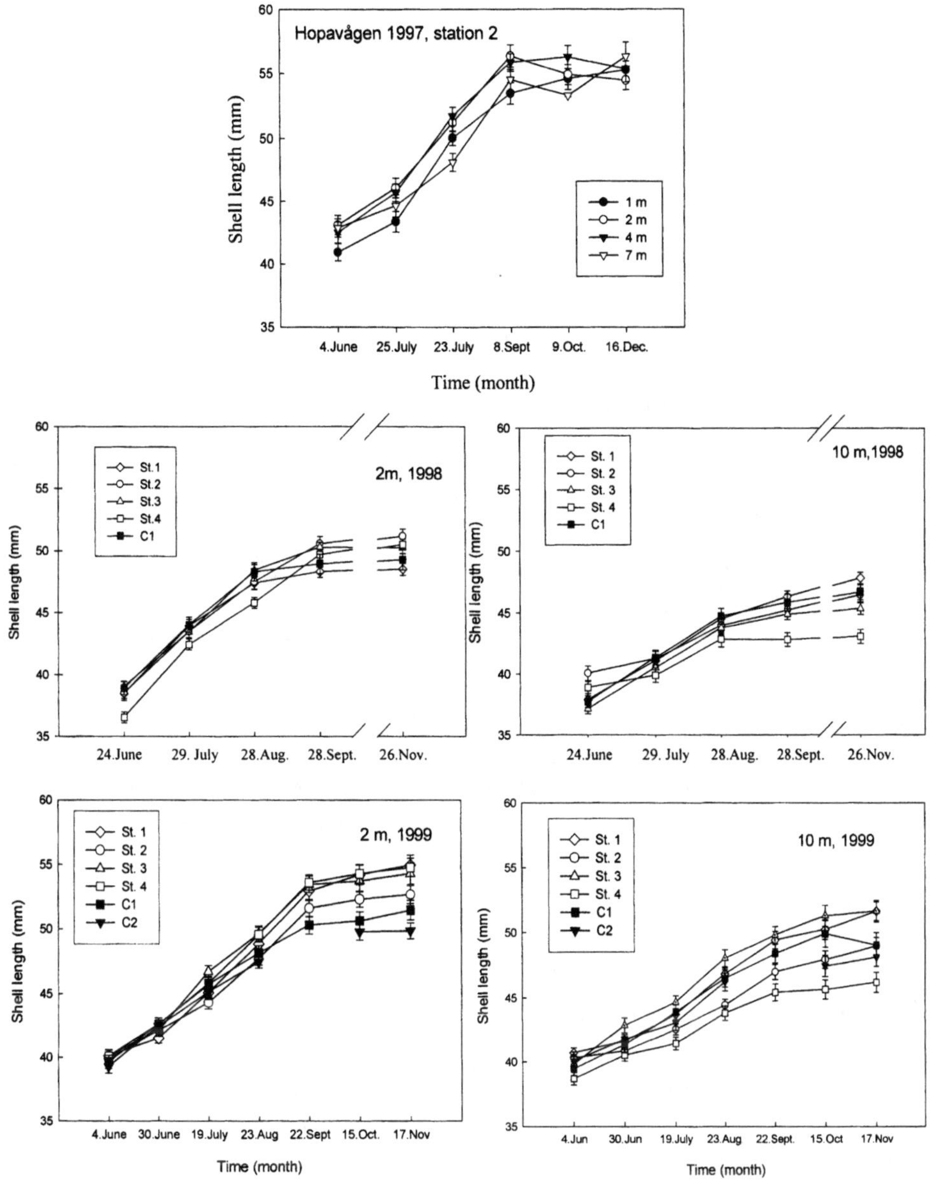

Figure 5. Growth of blue mussels (shell length in mm) in Hopavågen and in control stations on the coast. In 1997 growth was measured at station 2 (1, 2, 4 and 7 m depths). The following years shell length was measured at 4 stations in Hopavågen (2 and 10 m depths), and one control station (2 and 10 m depths) outside the bay. In 1999 two control stations were included (C1 and C2).

In 1997 the daily length increase was 0.23% d^{-1} at 2 m depth. In 1998 the DLI varied between 0.23 and 0.36% d^{-1} for all stations in Hopavågen, and 0.26% d^{-1} at the control station (C1). The daily length increase in 1999 was 0.24–0.27% d^{-1} in all stations in Hopavågen, and 0.21% d^{-1} in both control stations. At 10 m depth DLI was calculated to 0.09–0.20% d^{-1} in Hopavågen and 0.22% d^{-1} in the control group in 1998. In 1999 the DLI was 0.14–0.21% d^{-1} in Hopavågen and 0.15–0.18% d^{-1} in the control group.

The daily length increase was also calculated for each growth period (II, III and IV) and compared with the content of chlorophyll *a*, primary production and temperature in the same periods (Fig. 6). The DLI in the first growth period (II) was similar in all years, independent of increase in chlorophyll *a* or primary production at 2 m. In growth period III, DLI increased with increasing chlorophyll *a* content from 1997 to 1998. In 1999 no increase in DLI was registered although chlorophyll *a* was higher than earlier years.

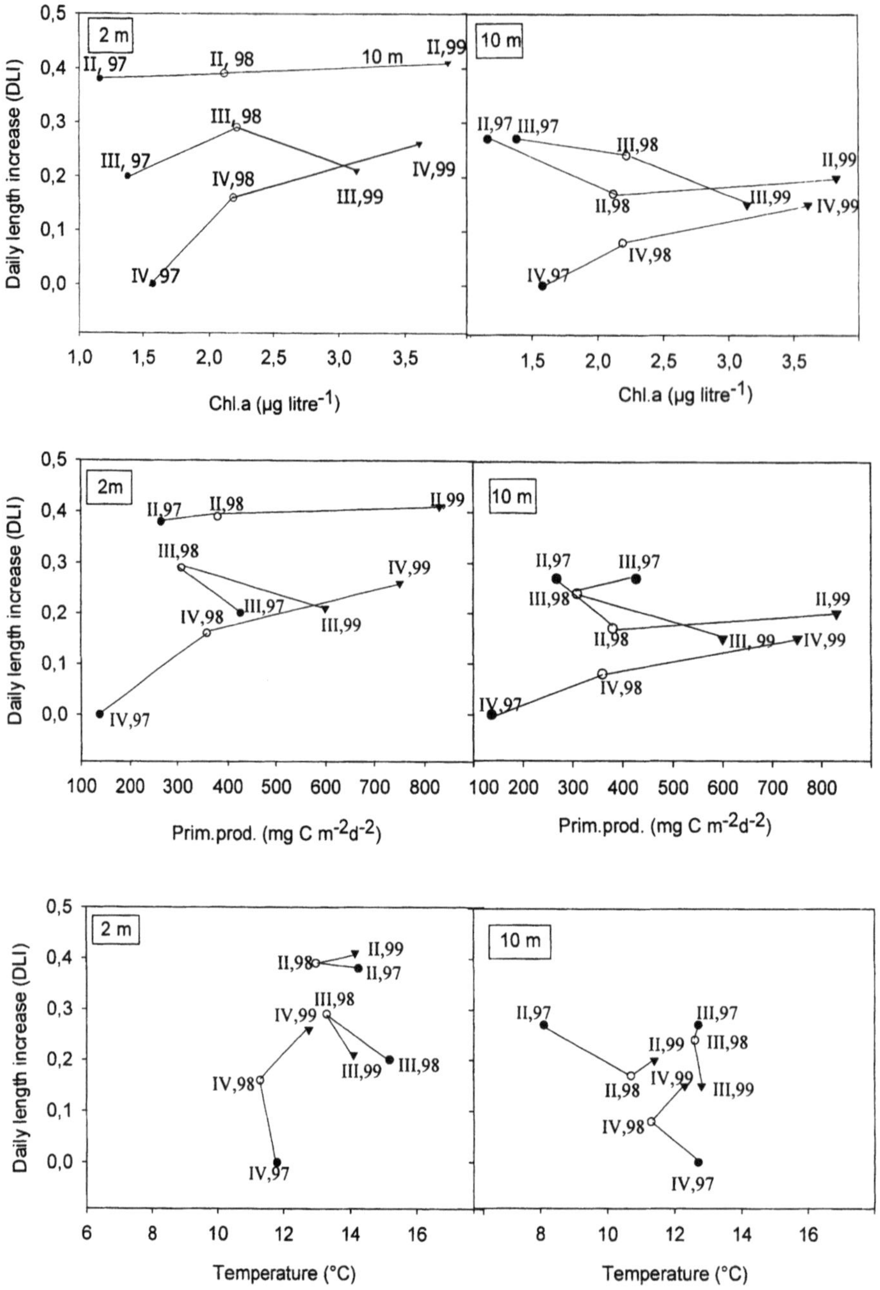

Figure 6. The daily length increases (DLI) in blue mussels at 2 and 10 m as a function of chlorophyll *a* content, primary production and temperature in the same periods. The DLI was calculated for growth periods II, III and IV in 1997 (•), 1998 (o) and 1999 (▼).

The primary production in period III was between 309 and 600 mg C m^{-2} d^{-2} for all 3 years, and no systematic increase in DLI was observed with increasing primary production. In growth period IV, DLI in mussels grown at 2 m increased with increasing content of chlorophyll *a* and increased primary production in 1998 and 1999. No relation between DLI and temperature was observed during the three periods, but there was an tendency to increased DLI in period IV with increased nutrient additions, even though the temperature did not increase at 2 or 10 m depth.

The wet weight content of the tissue at 2 m deep (Fig. 7) was significantly higher ($P < 0.05$) in shell grown in Hopavågen (range: 6.9±0.3-8.1±0.2 g) compared to the control station (4.9±0.3 g) at the end of the experiment in 1998. Dry weight and ash

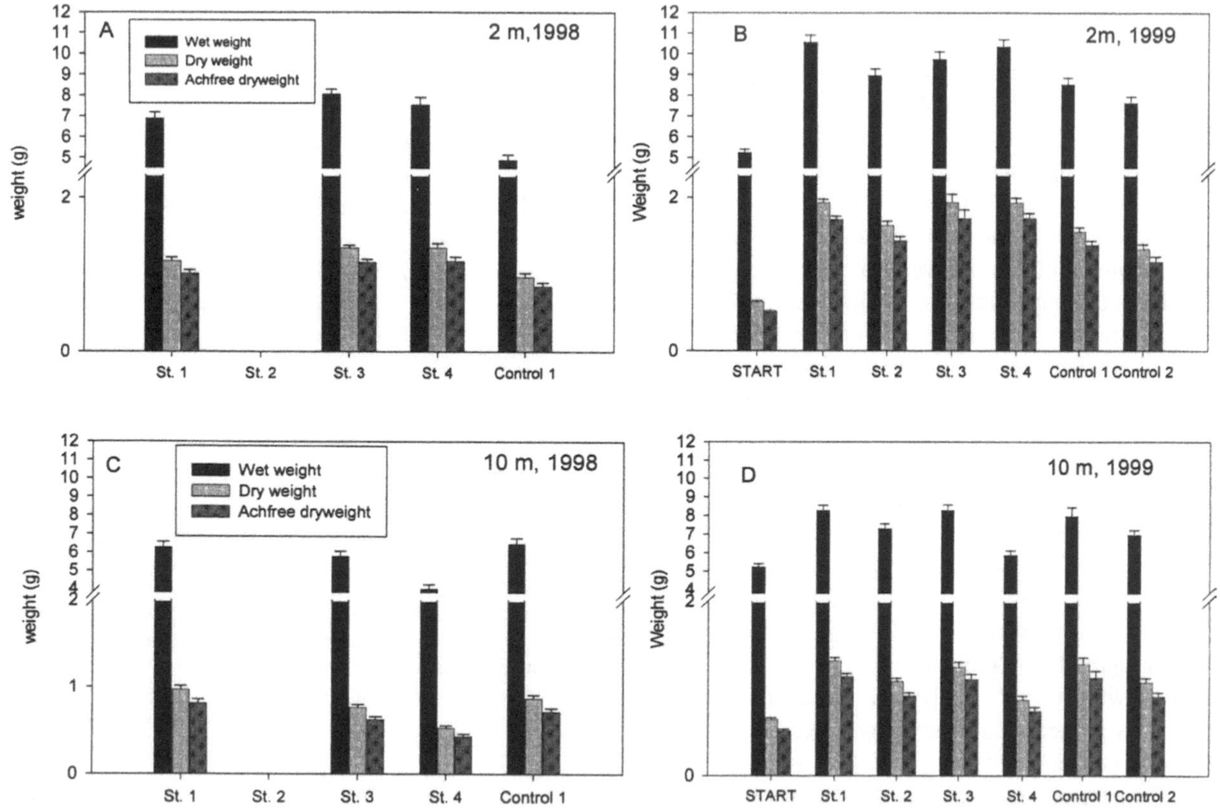

Figure 7. Wet weight, dry weight and ash-free dry weight in blue mussels (2 and 10 m depths) from Hopavågen and the control stations in 1998 and 1999. A and C shows results from 26 November 1998, and B and D shows results from 4 June (start) and 17 November, 1999.

free dry weight content was higher in mussels grown in the Hopavågen compared to the control. The same pattern was found in 1999. Mussels from all stations in Hopavågen had higher wet weight (range: 8.9±0.3–10.6±0.4 g), dry weight and ash-free dry weight than at the control stations (range: 7.6±0.3–8.5±0.3 g wet weight).

At 10 m depth no significant difference in the wet weight was obtained for blue mussels. The only exception was station 4 in Hopavågen, which was significantly lower both in 1998 and 1999.

Discussion

In all years (1997–1999) a high growth rate of blue mussels was recorded from end of June to the end of September in Hopavågen and at the control stations. The termination of growth in September coincided with the recorded decline in primary production and algal biomass. A significant increase in shell length during the experimental period, compared to the control stations, was only recorded at 2 m depth in Hopavågen in the last year of the experiment (1999). The growth of mussels at 2 m depth was both in Hopavågen and at the control station significantly higher than at 10 m depth. The difference in growth at 2 and 10 m can be explained by differences in temperature and food supply, although the food supply was not measured particularly at 2 and 10 m in this work. However, *in situ* fluoresence profiles suggests that the euphotic zone did not always extend down to 10 m in the beginning of the summer, whereas in early autumn 10 m depth was within the euphotic zone by a good margin (unpublished). This indicates that the food supply increased at 10 m depth as the season proceeded. This is supported by the fact that differences in DLI between 2 and 10 m reduced through summer and autumn.

Increased meat content was observed both in 1998 and 1999 for mussels grown at 2 m depth in Hopavågen compared to control, but not for mussels kept at 10 m depth. Bricelj & Shumway (1991) have indicated that somatic growth is more sensitive to environmental conditions than shell growth, and that effects due to improved environmental conditions (e.g., food and temperature) are more evident in somatic growth curves than for shell growth. Increased somatic growth was recorded in 1998 even though the nutrient addition to Hopavågen caused only a minor increase in algal biomass. A doubling of added nutrients the following year caused a clear increase in biomass and primary production during the summer, and both shell length and somatic growth improved that year. The effects of nutrient addition were most pronounced in period IV (Fig. 6). In the spring and early summer (period II) the DLI at 2 m depth was high and independent of nutrient addition. At 10 m depth the variations in period II and III may have been caused by low temperatures in the spring. The mussel tray, at 10 m, was below the upper stratification layer in early summer.

The highest temperature during the growth season was recorded in 1997. However, in all years the temperature at 2 m was above 10 °C (during the periods I to V), which is considered as the lower limit for optimal growth. A relationship between temperature and growth rates is clearly demonstrated when shell length is plotted against age day degree and temperature (Seed & Suchanek, 1991). However, growth rates expressed in these terms are not always consistent, which suggest that factors others than temperature (e.g. food supply) are involved (Wilson, 1977; Thomson, 1984). As the highest temperature was recorded in 1997, temperature can be excluded as the factor responsible for the improved growth recorded in 1998 and 1999. Other registered physical conditions did not reveal significant variations from year to year in Hopavågen. The recorded mean current velocity was low at all stations (1.1–1.8 cm s^{-1}), and reduced current is one of the main reasons why mussels have lower growth at the bottom than mussels grown on ropes and lines (Rodhouse et al., 1984). Current measurements reported from mussel culture sites are typically in the range of 2–10 cm s^{-1} (Hickman, 1992). Hence, in our case the lower current may have contributed to the observed food limitation.

We conclude that the growth of blue mussels in the study area is food limited. Increased primary production and algal biomass due to enhanced supply of nutrients first of all affected the somatic growth (5–65% increase compared to the control) of blue mussels. An increase in shell length was only recorded when the nutrient supply was tripled compared to the estimated natural supply of nutrients (estimated as phosphorus supply). This illustrates the importance of including mussel biomass production as a criterion for selection of sites for blue mussel farming.

Acknowledgements

We thank Katja Hohne Reitan, Jørund Larsen, Svenn Ove Linde and Lasse M. Olsen for technical assistance, and Knut Arve Johansen, Kjersti Rennan, Kjerti Andresen for doing chemical analysis. The work was supported financially by Norsk Hydro ASA and the MARICULT research programme.

References

Bricelj, V. M. & S. Shumway, 1991. Physiology: energy aquisition and utilization. In Shumway, S. (ed.), Scallops: Biology, Ecology and Aquaculture. Elsevier Science Publishers: 333–337.

Boynton, W. R., W. M. Kem & C. W. Keefe, 1982. A comparative analysis of nutrients and other factors influencing estuarine phytoplankton production. In Kennedy, V. S. (ed.), Estuarine Comparison. Academic Press, New York: 69–90.

Ceccherelli, V. U. & A. Barboni, 1983. Growth survival and yield of Mytilus galloprovencialis Lmk. On fixed suspended culture in a bay of the Po River delta. Aquaculture 34: 101–114.

Eilertsen, H. C. & J. P. Taasen, 1984. Investigation on the plankton community of Balsfjorden, northern Norway. The phytoplankton 1976–1978. Environmental factors, dynamics of growth and primary production. Sarsia, 69: 1–15.

Eriksen, J. & A. Fagerholt, 1998. SNDs strategi for skjellnæringen. SND-rapport (in Norwegian).

Hickman, R. W. Mussel cultivation. In Gosling, E. (ed.), The Mussel *Mytilus*: Ecology, Physiology, Genetics and Culture. Elsevier Science Publishers: 465–510.

Kemp, P. F., S. Y. Newell & C. Krambeck, 1990. Effects of filter-feeding by the ribbed mussel *Geukensia demissa* on the water column microbiota of the *Spartina alterniflora* saltmarsh. Mar. Ecol. Prog. Ser. 59: 119–131.

Lindahl, O., 1995 Long-term studies of primary phytoplankton production in the Gullmar fjord, Sweden. In Skjoldal, H. R., C. Hopkins, K. E. Erikstad & H. P. Leinaas (eds), Ecology of Fjords and Coastal Waters: 105–112.

Mann, K. H., 1982. Ecology of Coastal Waters: A Systems Approach. Studies in Ecology Vol. 8. Blackwell, Oxford.

Marion, P. van, 1996. Ecological studies in Hopavågen, a land-locked bay at Agdenes, Sør Trøndelag, Norway. Gunneria 71: 39 pp.

Navarro, E., J. I. P. Iglesias, A. Perez-Camacho, V. Labarta & R. Beiras, 1991. The physiological energetics of mussels (*Mytilus galloprovencialis* Lmk.) from different cultivation rafts in the Ria de Arosa (Galicia, N.W. Spain). Aquaculture, 94 (2/3): 197–212.

Newell. C. R., S. E. Shumway, T. L. Cucci, & R. Selvin, 1989. The effects of natural seston particle size and type on feeding rates, feeding selectivity and food resource availability for the mussel *Mytilus edulis* Linnaeus, 1758 at bottom culture sites in Marine. J. Shellfish Res. 8: 187–196.

Nixon, S. W., C. A. Oviatt, J. Frithsen & B. Sullivan, 1986. Nutrients and the productivity of estuarine and coastal ecosystems. J. Limnol. Soc. sth. Afr. 12: 47–71.

Page, H. M. & D. M. Hubbard, 1987. Temporal and spatial patterns of growth in mussels *Mytillus edulis* on an offshore platform: relationships to water temperature and food availability. J. exp. mar. Biol. Ecol., 111: 159–179.

Parson, T. R., Y. Maita & C. M. Lalli, 1984. A Manual of Chemical and Biological Methods for Seawater Analysis. Pergamon Press, Oxford: 171 pp.

Perez-Camacho, A., R. Gonzalez & J. Fuentes, 1991. Mussel culture in Galicia (N.W. Spain). Aquaculture 94 (2/3): 263–278.

Prieur, D., G. Mevel, J. L. Nicolas, A. Plusquellec & M. Vigneulle, 1990. Interactions between bivalve molluscs and bacteria in the marine environment. Oceangr. Mar. Biol. Annu. Rev. 28: 277–352.

Rodhause, P. G., C. M. Roden, M. P. Hensey & T. H. Ryan, 1984. Resource allocation in *Mytilus edulis* on the shore and in suspended culture. Mar. Biol. 84. 27–34.

Seed, R. & T. H. Suchanek, 1992. Population and community ecology of *Mytilus*. In Gosling, E. (ed.), The Mussel *Mytilus*: Ecology, Physiology, Genetics and Culture. Elsevier Science Publishers: 87–169.

Smaal, A. C. & M. R. van Stralen, 1990. Average annual growth and condition of mussels as a function of food source. Hydrobiologia 195: 179–188.

Thompson, R. J., 1984. Production, reproductive effort, reproductive value and reproductive cost in a population of blue mussel *Mytilus edulis* from a subarctic environment. Mar. Ecol. Prog. Ser. 16: 249–257.

Wilson, J. H., 1977. The growth of *Mytilus edulis* from Carlingford Lough. Ir. Fish. Invest. Ser. B (Mar.), 17: 1–15.

Hydrobiologia **484:** 111–120, 2002.
O. Vadstein & Y. Olsen (eds), Sustainable Increase of Marine Harvesting: Fundamental Mechanisms and New Concepts.

Response on scallop culture to enhanced nutrient supply by experimental fertilisation of a landlocked bay

Kjell Inge Reitan[1], Gunvor Øie[1], Olav Vadstein[3] & Helge Reinertsen[2]
[1]*SINTEF Fisheries and aquaculture, N-7465 Trondheim, Norway*
[2]*Norwegian University of Science and Technology, Department of Botany, N-7491 Trondheim, Norway*
[3]*Norwegian University of Science and Technology, Trondhjem Biological Station, N-7491 Trondheim, Norway*
Tel: +47-73-59-77-29. Fax: +47-73-59-63-63. E-mail: kjell.i.reitan@sintef.no

Key words: Scallops, *Pecten maximus*, nutrient supply, phytoplankton, primary production, growth rate, tissue content, shellfish

Abstract

The effect of an enhanced nutrient supply to coastal waters of a landlocked bay, Hopavågen in Central Norway, on the phytoplankton production and biomass, and on growth of scallops (*Pecten maximus*) was studied in 1997–1999. Nitrogen, silicon and phosphorous (N:Si:P = 16:8:1, atomic) were added daily between May and October in 1998 at a level of 0.4 mg P m^{-3} day^{-1}. The concentration of nutrient addition was doubled in 1999 during the same period. High addition of nutrients (1999) resulted in a significantly higher phytoplankton biomass in the summer period, expressed as chlorophyll *a* content, than without nutrient (1997) and low nutrient (1998). The respective mean chlorophyll *a* levels were 2.4 in 1999, 1.6 in 1998 and 1.2 $\mu g\ l^{-1}$ in 1997. The mean primary production during the summers generally increased with the addition of nutrients from an average level of 320 mg carbon m^{-2} day^{-1} in 1997 to 1200 mg carbon m^{-2} day^{-1} in 1999. Scallops placed at 10 m depth in Hopavågen showed an increase in growth rate of the outer scallop shell in the period July–September from 0.16% day^{-1} in 1997 to 0.53% day^{-1} in 1998. Scallops grown in an unfertilised control station in the fjord outside Hopavågen had a significantly lower growth rate than those grown in the fertilised water of Hopavågen. The results showed decreased growth rate with increasing shell sizes. However, for all size groups studied a higher growth rate of the scallops was observed when nutrients were added to the bay. The tissue dry weight content of scallops grown in Hopavågen was 2–4 times higher than in the control scallops.

Introduction

Recruitment and growth of shellfish in natural habitats are generally dependent on the concentration and the quality of the food particles in the sea. The main natural food for scallops is phytoplankton, especially single-celled algae, but also particulate organic material, bacteria and other micro-organisms are ingested (Fegley et al., 1992). More than 150 different algae and animals species were found in the digestive tract of 1-year-old *Patiepecten yessoensis*, mainly consisting of diatoms and protozoa (Miuklich & Tsikhon-Lukanina, 1981). Selective or preferential ingestion of certain cells by bivalves is believed to be a result of the various sizes and shapes of the microalgae (Shumway et al., 1985), and bivalves are reported to select algae due to nutritional quality and chemoreception stimuli (Ward & Targett, 1989; Ward et al., 1992).

The growth of scallops also depends on the water flow and the temperature (Broom & Mason, 1978; Bayne & Newell, 1983; Cahalan et al., 1989; Wildish et al., 1992). However, food availability is reported to be a major factor influencing the growth rate of scallops (Orensanz, 1984; Stockton, 1984; Mac Donald & Thompson, 1985). Furthermore, scallops (*Pecten maximus*) have shown different growth rate and survival in different locations such as fertilised shallow ponds and open sea sites (Andersen et al., 1993). Very few

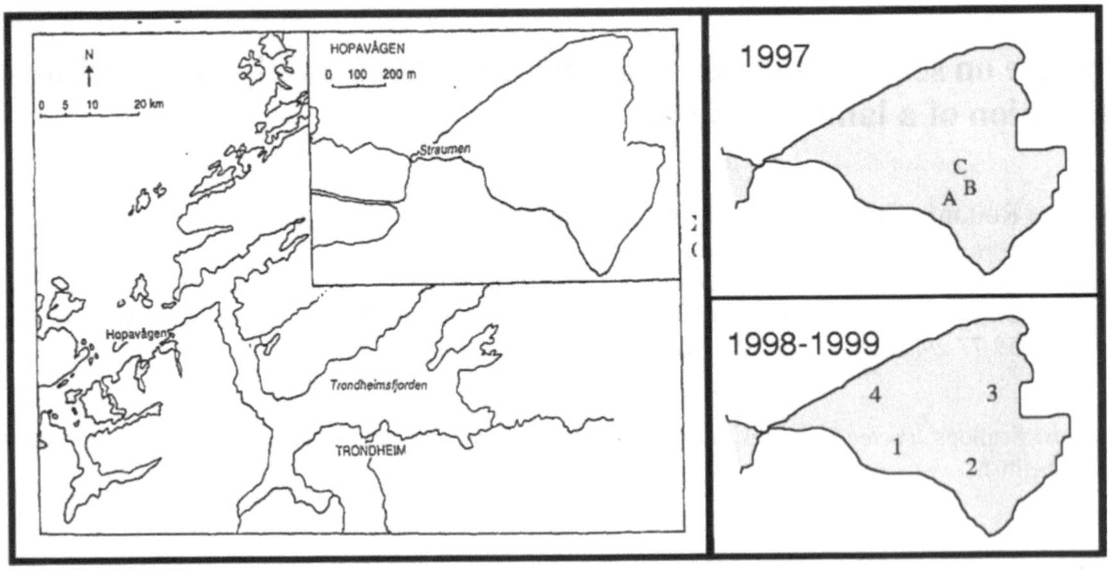

Figure 1. The landlocked bay, Hopavågen, located at the coast outside Trondheimsfjorden with the stations where the scallops were placed. In 1997 scallops were placed at 5 m (A), 10 m (B) and 15 m (C) depths. In 1998 and 1999 the scallops were placed at 10 m depth at four different stations in Hopavågen (1–4). In addition, one group of scallops was placed at 10 m depth outside the bay.

studies have quantified the food resources available to suspension feeders at various times of the year (Bayne & Newell, 1983) and its effect on growth.

Increasing production yield of fish and shellfish in marine systems is reported with increasing primary production (Nixon & Pilson, 1983). In this study the effect of enhanced nutrient supply on phytoplankton biomass and production, and further cascading effects on the growth of juvenile scallops were studied in an open landlocked bay. An experimental nutrient addition, providing a gradient in nutrient input, was carried out over three years. Effects on the phytoplankton biomass and primary production caused by the fertilisation are discussed in regard to its effects on the scallop cultivation.

Material and methods

Location of the experiment

The landlocked bay, Hopavågen, is located on the coast of central Norway. Hopavågen is a relatively large system (370 000 m^2 and 6.66 10^6 m^3), and a narrow, shallow channel connects the bay with the coast (Fig. 1). The mean depth is 18 m and the greatest depth is 31 m (Marion, 1996). The tidal water exchange is about 20–40% per day in the upper layer and about 2–4% below the threshold level. The retention time of the water increases with the distance from the inlet (Nygård, 1998). The input of antropogenic nutrients to Hopavågen is negligible.

Experimental design

No nutrients were added to Hopavågen in 1997. In 1998 and 1999 nitrogen, silicon and phosphorus (N:Si:P of 16:8:1, atomic ratio) were added during every tide cycle from May to early October. From 10–40 min. after the start of rising tide, nutrients were added to the inlet through a diffuser system. A tracer experiment revealed complete mixing of the nutrients within 2 h (unpublished results). The nitrogen component consisted of 1:1(atomic ratio) of NO_3:NH_4^+ . In 1998 P was added at a rate of 0.4 mg m^{-3} day^{-1}, while the dose was doubled in 1999.

Juvenile scallops from a scallop farm (Taroskjell AS, Roan) were placed in net baskets, with a diameter of 50 cm and height of 12 cm, in Hopavågen at different stations from July to November in each of the years 1997–1999 (Fig. 1). During the first year, 30 scallops (38–42 mm shell height) were placed at 5, 10 and 15 m depth at station 2. In 1998 and 1999, groups of 58–86 scallops (24–34 mm shell height) were placed at the different stations at 10 m depth. In addition, scallops were placed at a control station on the coast outside Hopavågen, in similar numbers as in Hopavågen.

Table 1. Current direction and current velocity in Hopavågen at 5 and 10 m depths at station 2 in 1997, and at 10 m depth at stations 1, 3, 4 and outside the bay (control) in 1998

	Station 2		Stations			
	5 m	10 m	1	3	4	Control
Main current, degree	198	236	72	91	70	134
Flow in the main direction (cm s^{-1})	0.5	1.2	0.4	0.1	0.3	0.5
Total current velocity (cm s^{-1})	1.5	1.8	1.2	1.1	1.2	1.2
Max. current velocity (cm s^{-1})	8.0	8.0	8.0	4.0	5.0	4.0
% of total current, 0-1 cm s^{-1}	52	31	72	68	73	64

Analytical methods

The biomass of algae, measured as a concentration of chlorophyll *a*, and the primary production in Hopavågen were registered every week from June to October during all years (1997–1999). Mixed water samples were taken from the euphotic zone (0–10 m) with a 2 m long Ramberg Sampler. Chlorophyll *a* was determined by a modified method of Parson et al. (1984) by filtering exact volumes of seawater on GF/F filters (Whatman) and extracting in 90% acetone (24 h at 4 °C). The content of chlorophyll a was quantified by fluorometry, using a Turner fluorometer.

Primary production was estimated with the ^{14}C-method. Water was filled in light and dark 50 ml tissue culture bottles (Nunc, total volume 62 ml), and 3.6 μCi of ^{14}C-bicarbonate was added (The International Agency for ^{14}C Determination, Denmark). The bottles were incubated *in situ* at 0.5, 1, 3, 5, 7, 9, 11 and 13 m depth for 3–4 h. In addition two dark bottles were incubated at 3 and 9 m depth. After termination of the incubation the bottles were kept in the dark, and filtration was begun within 30 min and completed within 90 min. Water was filtered with a low vacuum through filters of various pore sizes. To reduce errors caused by cell breakage, the filters were sucked dry but not rinsed (Goldman and Dennett 1985). Inorganic ^{14}C on filters was removed by HCl fumes, and incorporated ^{14}C was determined by liquid scintillation counting (Ultima Gold XR, Packard, and Packard Tricarb model 1900 Scintillation counter). Photosynthesis rates were calculated from incorporated ^{14}C and concentration of dissolved inorganic carbon. The samples were filtered on 0.45 μm mixed cellulose filters. Primary production per day was calculated by multiplying with a light factor defined as total daily insolation divided by insolation during the incubation period (mol m^{-2} s^{-1}, 400–700 nm). This assumption was supported by results from diel experiments.

In 1997 the temperature was registered every 30 min at station 2 at 5, 10 and 15 m depths, using a Tinytag Temperature logger (Gemini Data Loggers Lts, UK). Due to the loss of some temperature loggers in 1998 the temperature reported is based on weekly registrations at station 2, at10 m depth. In 1999 the temperature was registered at 10 m depth every 30 min at all stations including the control station.

The water current was registered continuously during 14 days (September – October) at 5 and 10 m depths at station 2 in 1997, and at 10 m depth in the other stations in 1998.

Every month the shell height was measured with a slide calliper, and the survival of the scallops was registered. The specific increase in shell height was calculated by the equation:

$$\mu = \ln(H_2 / H_1) / (t_2 - t_1). \quad (1)$$

H_1 and H_2 are the shell heights at two registrations and $t_2 - t_1$ is the number of days between the registrations. If more than two height determinations were involved, the increase was calculated by linear regression of ln *H versus* time. The daily height increase (DHI) expressed as per cent per day was calculated as:

$$\mathrm{DHI} = (e^{\mu} - 1)\,100. \quad (2)$$

All scallops were harvested in November 1998 and 1999, and the wet weight, the dry weight and the ash-free dry weight of the tissue were determined. For the determination of dry weight, samples were dried for 3 days at 60 °C. The ash-free dry weight was measured after combustion at 600 °C for 12 h.

Results are presented as mean values and standard errors of the mean (SE) or standard deviation, both in tables and figures. Significant differences between groups were tested using ANOVA followed by a Student–Newman–Keuls-Test. Statistical ana-

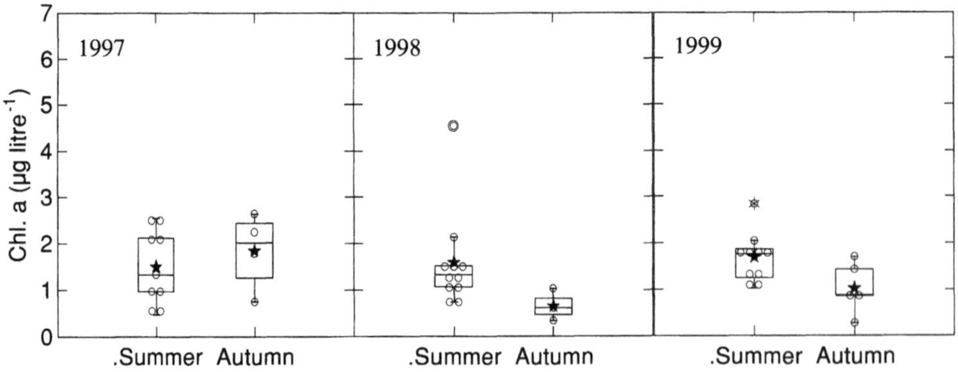

Figure 2. The biomass of phytoplankton measured as chlorophyll *a* (μg l^{-1}) of the in-streaming seawater (without added fertiliser) to Hopavågen in 1997, 1998 and 1999. The chlorophyll *a* content is shown for the summer (early July–September 19) and autumn (September 20–mid November). The star shows the average value. Boxes indicate 50% of all observations.

Table 2. The temperature (average and range) at different depths and stations in Hopavågen during summer (June 30–September 19) and autumn (September 20–late November) of 1997, 1998 and 1999

	Depth (m)	Station	Summer		Autumn	
			Average	Range	Average	Range
1997	5	2	15.3	10.7–20.0	18.5[a]	18.0–19.2[a]
	10	2	12.7	7.2–18.2	17.9[a]	17.5–18.1[a]
	15	2	6.4	5.5–7.5	7.7[a]	7.5–8.0[a]
1998	10	2	11.7	9.3–13.0	–	–
1999		1	11.5	9.6–14.0	10.8	8.5–12.5
		2	11.5	9.6–13.4	10.8	8.5–12.5
		3	11.5	9.7–13.8	10.8	8.6–12.3
		4	11.4	9.7–14.1	10.7	8.3–12.6
		Control	11.9	9.8–14.3	11.0	8.7–12.2

[a]Temperature registered in the period September 20–October 10.

lyses were performed with the programme SPSS for Windows (SPSS Inc., Chicago).

Results

Water exchange and water current

Both total current velocity and flow in the main direction was higher at 10 m than at 5 m depth in Hopavågen (Table 1). The total current velocity was low for all four stations at 10 m depth. The current was influenced by the tide and was strongest at stations 1 and 4. The direction of the current was relatively stable at station 2, whereas the current fluctuated at station 3.

The current velocity at the control station outside Hopavågen was at the same level as in Hopavågen. At this station the velocities were also affected by the tide, but the direction was relatively stable.

Temperature

The water temperature increased gradually during the growth season in 1997 with a rather high average temperature in the autumn at both 5 and 10 m depths (17.9 and 18.5 °C respectively, Table 2). In the summers of 1998 and 1999, 1.0–1.2 °C lower temperatures were observed at 10 m depth compared to 1997. There were no significant differences in the average summer temperature at 10 m depth in Hopavågen among the different stations and between 1998 and 1999. However, during the summer and autumn of 1999 the average temperature was 0.2–0.5 °C higher at the control station than in Hopavågen.

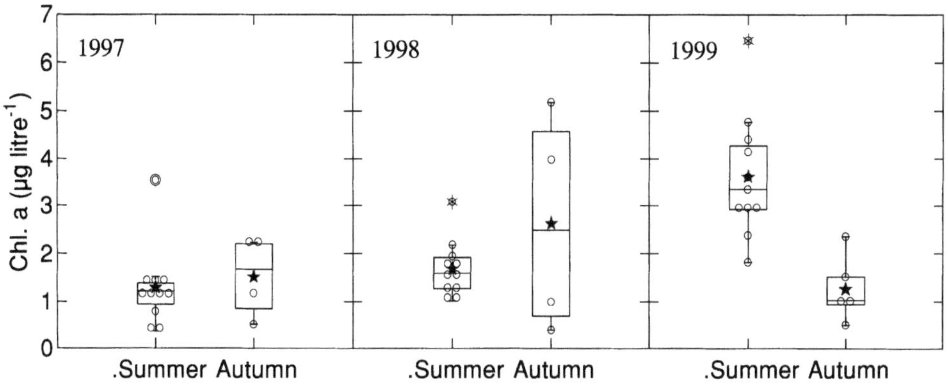

Figure 3. The biomass of phytoplankton (mixed sample, 0–10 m depth) measured as chlorophyll *a* ($\mu g\ l^{-1}$) in 1997 (before addition of fertiliser), 1998 (addition of low amount of fertiliser) and 1999 (addition of high amount of fertiliser). The chlorophyll *a* content is shown for the summer (early July–September 19.) and autumn (September 20 –mid November). The star shows the average value. The boxes indicate 50% of all observations.

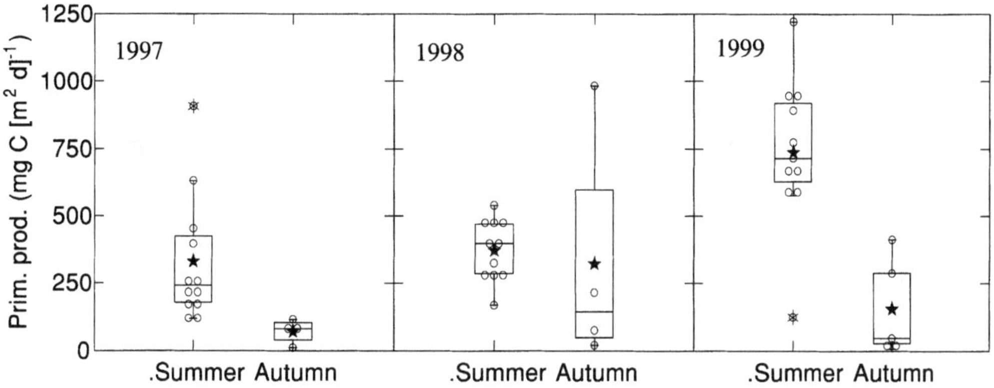

Figure 4. Primary production (mixed sample, 0–10 m depth) in 1997 (before addition of fertiliser), 1998 (addition of low amount of fertiliser) and 1999 (addition of high amount of fertiliser). The chlorophyll *a* content is shown for the summer (early July–September 19) and autumn (September 20–mid November). The star shows the average value. 50% of all observations are indicated in the square.

Biomass of phytoplankton and primary production

There were no differences in the mean biomass of phytoplankton, expressed as chlorophyll a, in the in-streaming tidal water to Hopavågen in the summers of 1997–1999 (Fig. 2). Comparable levels of chlorophyll *a* were also observed inside Hopavågen during the summer of 1997, when no nutrients were added (Fig. 3). The addition of a low amount of nutrients to Hopavågen gave a 30% increase in mean chlorophyll *a* content in the summer of 1998 compared to 1997 (not statistically significant). At the high dose of nutrients added in 1999, a 180% increase in average chlorophyll *a* level was observed during summer when compared to 1997.

No significant differences among years were observed for the chlorophyll *a* level during the autumn in Hopavågen (Figs 2 and 3). The same was valid for chlorophyll *a* levels during autumn in the in-streaming water.

The low addition of nutrients in 1998 resulted in 12% higher mean primary production in the summer compared to 1997, and the high addition of nutrients in 1999 gave a 125% increase in summer compared to 1997 (Fig. 4). The average primary production during the summer of 1999 at the control station was 468 ± 66 mg C m^{-2} d^{-1} (not shown).

No significant differences were observed in primary production during the autumns of the 3 years.

Growth and survival of scallops

Significantly higher growth in shell height during summer was observed in scallops at 5 and 10 m depths than at 15 m depth, being 0.17 and 0.10% day^{-1}, respectively (Fig. 5). The survival of the scallops at the

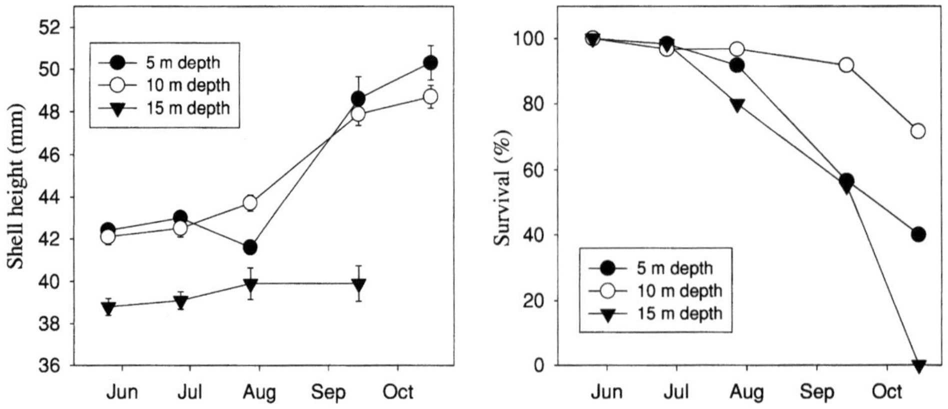

Figure 5. Growth (left panel) and survival (right panel) of scallops (30–40 mm) in Hopavågen in 1997 before the addition of extra nutrients.

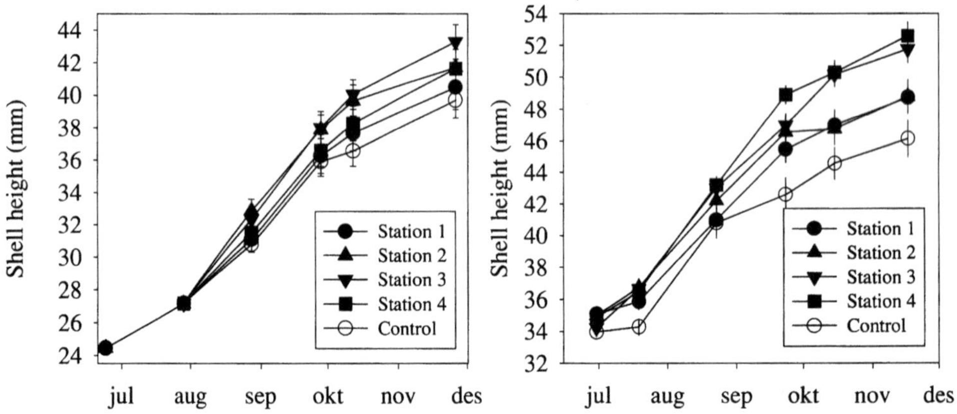

Figure 6. Growth of scallops in 1998 (left panel) and 1999 (right panel) when nutrients were added at low and high rates, respectively.

end of the growth period was highest at 10 m depth (>75%) and was significantly lower at 5 and 15 m depths (Fig. 5). Registration of the size of the dead scallops showed no difference to those that were living (not shown). Based on the results obtained in 1997 the scallops were placed at 10 m depth at different stations in Hopavågen and at the control station outside in 1998 and 1999 (Fig. 1).

A higher growth rate of scallops was generally observed during summer compared to autumn (Figs 5 and 6, Table 3). The scallops had a significantly lower growth in 1997, when no fertiliser was added, than during the following years. At the end of the summer in 1998 the scallops in the inner part of Hopavågen (stations 2 and 3) were significantly larger ($p < 0.05$) than at the control station. From September 1999 onwards all groups of scallops in Hopavågen were significantly larger than those at the control station (Fig. 6). The differences in shell height between station 3 in Hopavågen and the control station were higher especially in 1999 compared to 1998. The sur-

Table 3. The growth rate of scallop (%, day^{-1}) at 10 m depth in summer (early July–mid September) and autumn (mid September - late November) before (1997) and after the addition of nutrients (1998 and 1999)

		Growth periods	
		Summer	Autumn
1997	Station 2	0.16 ± 0.02	0.04 ± 0.04
1998	Control	0.44 ± 0.03	0.08 ± 0.02
	Station 1	0.46 ± 0.03	0.16 ± 0.17
	Station 2	0.53 ± 0.03	0.19 ± 0.14
	Station 3	0.53 ± 0.03	0.22 ± 0.14
	Station 4	0.48 ± 0.03	0.19 ± 0.16
1999	Control	0.29 ± 0.05	0.15 ± 0.03
	Station 1	0.33 ± 0.03	0.13 ± 0.01
	Station 2	0.35 ± 0.02	0.09 ± 0.02
	Station 3	0.38 ± 0.03	0.17 ± 0.06
	Station 4	0.42 ± 0.02	0.14 ± 0.03

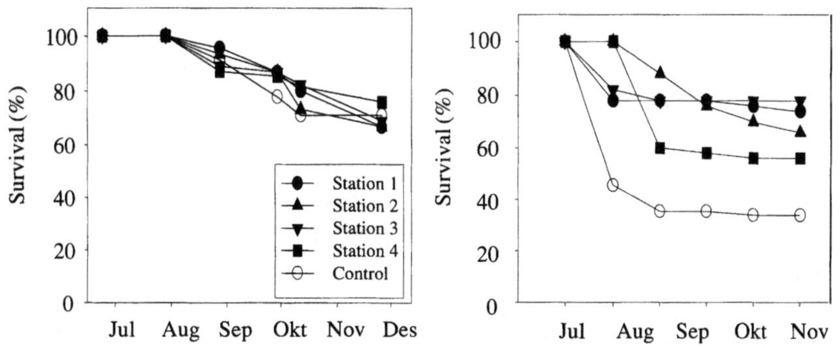

Figure 7. Survival rate of scallops in 1998 (left panel) and 1999 (right panel) when nutrients were added at low and high rates, respectively.

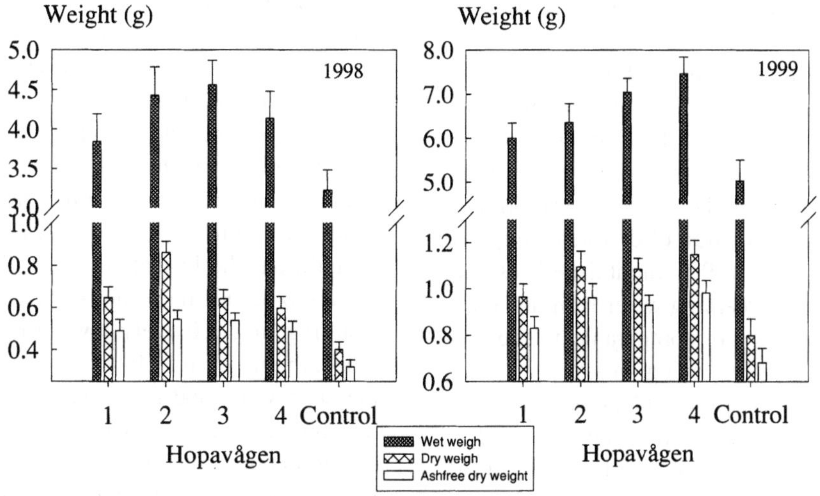

Figure 8. Wet weight, dry weight and ash-free dry weight in tissue of scallops at the end of the growth season in 1998 (left panel) and 1999 (right panel).

vival of scallops until the end of the growth period in 1998 was 70–78% in all groups (Fig. 7). In 1999 some mortality was observed until the first registration time in the control group, and almost no mortality for the rest of the period. For the rest of the groups the survival at the end of the experiment was 60–78%.

The growth rate (daily height increase) of scallops at 10 m depth during summer of 1997 was significantly lower than in the following years, being 0.16 and 0.33–0.53% day^{-1}, respectively (Table 3). In summer 1998 and 1999, the scallops in the inner part of Hopavågen had a significantly higher growth rate than those kept at the control station. During autumn, (mid-September to November) the growth rates were low, and no significant differences in growth rate were found among the different years and stations (Table 3).

The tissue wet weight and dry weight were significantly higher for scallops kept in Hopavågen than at the control station both in 1998 and 1999 ($p < 0.05$,

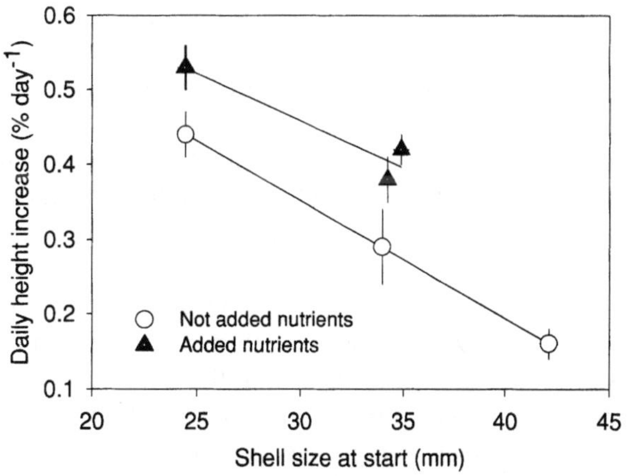

Figure 9. Daily height increase plotted against initial size of scallops grown at 10 m depth. The groups without added nutrients were the group in 1997 and the control station in 1998 and 1999 ($y = -0.0159\pm0.0001 + 0.828\pm0.002$, r^2=0.999), and the groups with added nutrients are the stations in Hopavågen in 1998 and 1999 ($y = -0.0127\pm0.0024 + 0.840\pm0.007$, r^2=0.933).

Fig. 8). Similar results were found also for ash-free dry weight.

Discussion

The experimental addition of nutrients to Hopavågen (1998 and 1999) gave a higher daily height increase during summer than no added nutrients for similar sizes of scallops. This indicates a positive effect of the nutrient addition on growth of scallops (Fig. 9). Increasing size of scallops gave reduced growth rate for both added and no added nutrients. The initial shell height of the scallops was higher in 1997 than in 1998 and 1999.

The addition of nutrients in 1998 gave a 27% increase in the average biomass of phytoplankton in summer compared to 1997. Furthermore the doubled addition in 1999 gave a 180% increase in the biomass during the summer compared to 1997. The addition of nutrients gave a higher fraction of diatoms compared to no added nutrients in 1997 (unpublished results). The biomass of the in-streaming water from the coast was similar in all years and comparable to the biomass in Hopavågen in 1997, being on average 1.5 μg chlorophyll *a* l^{-1}. Earlier studies of Norwegian coastal waters have shown a chlorophyll *a* content of 2 $\mu g\,l^{-1}$, which correspondes to 2 cells μl^{-1} of diatoms (Erga & Heimdal, 1984). Assuming that the algae in Hopavågen had a similar chlorophyll *a* content, the average numbers of cells were below 2 μl^{-1}, which is food limiting for scallop juveniles (Cahalan et al., 1989). Small increases in the biomass of phytoplankton will therefore have a positive effect on the growth of scallops. More registrations of low phytoplankton biomass were made in the in-streaming water and during the summer of 1997 than in 1998 and 1999 in Hopavågen. These differences indicate a higher and more stable food supply for the scallops when nutrients were added.

A higher average primary production during the summer 1999 was observed than in 1998, which was somewhat higher than in 1997. This shows an increased primary production with the addition of nutrients. The increased primary production suggests an increased specific growth rate of the algal population as the production increased more than the biomass. Increased growth rate is shown to affect the chemical composition of nutrient limited alga cells positively, i.e., their content of lipids, carbohydrates and proteins (Reitan et al., 1994). The nutrient quality of different algae has been shown to affect the growth of oyster spat (Laing & Millican, 1986) and the high nutritional value of diatoms for scallop larvae is well documented (Whyte et al., 1989, Thompson et al., 1994).

The effect of extra nutrient supply to Hopavågen on the primary production and the biomass of phytoplankton was probably strongest in the inner parts of Hopavågen. Hopavågen is connected to the open coast through a narrow, shallow channel (Fig. 1), and the water is exchanged through this channel by the tide. The tide is mixed with the surface layer down to approximately 10 m depth (Nygård, 1998). A simulation experiment suggests that the residence time of the water in the upper layer increases with distance from the inlet, meaning that the effect of the nutrient supply on the primary production was stronger in the inner parts of the bay (Stations 2 and 3).

The different stations in Hopavågen and the control station had low a current velocity with no differences in the flow velocity. All stations had a current velocity lower than 1 cm s^{-1} between 31 and 73% of the time. In addition, the baskets used for the scallops are reported to reduce the water current inside the baskets, resulting in a relatively low current for the scallops (Claereboudt et al., 1994). Low flow rates will limit the supply of food to the scallops and thereby limit the food availability (Wildish & Kristmanson, 1985). The total current velocity at the different stations was low compared to areas where scallops normally live. However, the similar current velocities at the control station and the stations in Hopavågen demonstrate that the velocity cannot explain the observed differences in growth between the control station and the nutrient-added bay.

The growth of the scallops in Hopavågen during 1997 was higher at 5 and 10 m than at 15 m depth. This was probably due to a combination of low water temperature at 15 m depth and less food particles. The average temperature at 10 m depth during the summers of 1997 to 1999 varied between 11.4 and 12.7 °C for all stations, with higher temperatures in 1997 than in 1998 and1999. The average temperature in 1999 was higher at the control station than at the stations in Hopavågen. It has been shown that scallops (*Argopecten irradians*) have a markedly decreased clearance rate at temperatures around 5 °C (Kirby-Smith, 1970). Higher temperatures are reported to result in higher growth rate in shell height for scallops fed at ad libitum (Kleinman et al., 1996). Laing (2000) has suggested that temperature is the environmental factor that most often limits the growth of scallops in the sea, as increasing temperature resulted in increasing algal consumption. Nevertheless, since the higher temperature at the control station did not result in higher growth rates, the temperature effects can not explain the differences in growth rate among the stations and years. Our data strongly suggest that the scallops were food limited in these areas.

The wet weight of the tissues at the end of the experiment was significantly higher in scallops from stations 2 and 3 compared to the control. Both the dry weight and the ash-free dry weight content was significantly higher in scallops grown in the Hopavågen under fertilised conditions compared to the unfertilised control station. The stronger effect on the somatic growth than on the growth in shell height is in accordance with earlier reports, showing somatic growth to be more sensitive than shell growth to environmental conditions such as temperature and food conditions (Bricel & Shumway, 1991).

In conclusion, the results suggest food limited growth under natural conditions and a positive effect of the addition of fertiliser on the growth of juvenile scallops. This was due to an increased food supply to the scallops followed by the fertilisation. Further studies should be done to demonstrate the generality of our conclusions. Moreover, a more detailed evaluation of nutrient addition, and their effects on phytoplankton species, composition and nutritional quality is required.

Acknowledgements

We thank Christian Beylich, Knut Arve Johansen, Jørund Larsen, Jan Ove Evjemo, Svenn Ove Linde and Kjersti Andresen for help with the accomplishment of the experiments. We also thank Dr Katja Hoehne Reitan for valuable comments on the manuscript. The study was supported financially by the MARICULT programme.

References

Andersen, S. & K. E. Naas, 1993. Shell growth and survival of scallop (*Pecten maximus* L) in a fertilised shallow seawater pond. Aquaculture 110: 71–86.

Bayne, B. L. & R. C. Newell, 1983. Physiological energetics of marine molluscs. In Saleuddin, A. S. M & K. M. Wilbur (eds), The Mollusca, Vol. 4. Academic Press Inc., New York: 407–515.

Bricelj, V. M. & S. Shumway, 1991. Physiology: energy aquisition and utilization. In Shumway, S. (ed.), Scallops: Biology, Ecology and Aquaculture. Elsevier Science Publishers: 333–337.

Broom, M. J.& J. Mason, 1978. Growth and spawning in the pectinid Chlamys opercularis in relation to temperature and phytoplankton concentration. Mar. Biol. 47: 277–285.

Cahalan, J. A., S. E. Siddal & M. W. Luckenbach, 1989. Effects of flow velocity, food concentration and particle flux on growth of juvenile bay scallops *Argopecten irradians*. J. exp. mar. Biol. Ecol. 129: 45–60.

Claereboudt, M. R., J. H. Himmelman & J. Cote, 1994. Field eval uation of the effect of current velocity and direction on the growth of the giant scallop, Placopecten magellanicus, in suspended culture. J. exp. mar. Biol. Ecol. 183: 27–39.

Erga, S. R. & B. R. Heimdal, 1984. Ecological studies on the phytoplankton of korsfjorden, western Norway. The dynamics of a spring bloom seen in relation to hydrographical conditions and light regime. J. Plankton Res. 6: 67–90.

Fegley, S. R., B. A. MacDonald & T. R. Jacobsen, 1992. Short-term variation in the quantity of seston available to benthic suspension feeders. Estaur., Coast. Shelf Sci. 34: 393–412.

Kirby-Smith, W. W., 1970. Growth of the scallops, Argopecten irradians concentricus (Say) and Argopecten gibbus (Linne), as influenced by food and temperature. Ph.D Thesis, Duke University, Durham, North Carolina, 126 pp.

Kleinman, S., B. G. Hatcher & R. E. Scheibling, 1996. Growth and content of energy reserves in juvenile sae scallops, *Placopecten magellanicus*, as a function of swimming frequency and water temperature in the laboratory. Mar. Biol. 124: 629–635.

Laing, I. & P. F. Millican, 1986. Relative growth and growth efficiency of Ostrea edulis L. spat fed various algal diets. Aquaculture 54: 245–262.

Laing, I., 2000. Effect of temperature and ration on growth and condition of king scallop (*Pecten maximus*) spat. Aquaculture 183: 325–334.

MacDonald, B. A. & R. J. Thompson, 1985. Influence of temperature and food availability on the ecological energetic of the gigant scallop *Placopecten magellanicus*. I. Growth rates of shell and somatic tissue. Mar. Ecol. Prog. Ser. 25: 279–294.

Marion, P. van., 1996. Ecological studies in Hopavågen, a landlocked bay at Agdenes, Sør Trøndelag, Norway. Gunneria 71: 39 pp.

Mikulich, L. V. & Y. A. Tsikhon-Lukanina, 1981. Food of scallop. Oceanology 21: 633–635.

Nixon, S. W & M. E. Q. Pilson, 1983. Nitrogen in estuarine and coastal marine ecosystems. In Carpenter, E. J. & D. G. Capone (eds), Nitrogen in the Marine Environment. Academic: 565–648.

Nygård, E., 1998. Estimation of the water exchange in Hopavågen (In Norwegian). NIVA report LNR 3861-98, 40 pp.

Orensanz, J. M., 1984. Size, environment, and density: the regulation of scallop stock and its management implications. In Jamieson G. S. & N. Bourne (eds), North Pacific Workshop on Stock Assessment and Management of Invertebrates. Can. Sp. Publ. Fish. aquat. Sci. 92, 195–227.

Pason, T. R., Y. Maita & C. M. Lalli, 1984. A Manual of Chemical and Biological Methods for Seawater Analysis. Pergamon Press, Oxford: 171 pp.

Reitan, K. I., J. R. Rainuzzo & Y. Olsen, 1994. Effect of nutrient limitation of fatty acid and lipid content of marine microalgae. J. Phycol. 30: 972–979.

Shumway, S. E., T. L. Cucci, R. C. Newell & C. M. Yentsch, 1985. Particle selection, ingestion and absorption in filter-feeding bivalves. J. exp. mar. Biol. Ecol. 91: 77–92.

Stockton, W. L., 1984. The biology and ecology of the epifaunal scallop *Adamussium colbecki* on the west side of McMurdo Sound, Antarctica. Mar. Biol. 78: 171–178.

Thompson, P. A., M.-x. Guo & P. J. Harrison, 1994. Influence of irradiance on the nutritional value of two phytoplankton species fed to larval Japanese scallops (*Patinopecten yessoensis*). Mar. Biol. 119: 89–97.

Ward, J. E. & N. M. Targett, 1989. Influence of marine microalgal metabolites on the feeding behaviour of the blue mussel *Mytilus edulis*. Mar. Biol. 101: 313–321.

Ward, J. E., H. K. Cassell & B. A. MacDonald, 1992. Chemoreseption in the sea scallop Placopecten magellanicus (Gmelin). I. Stimulatory effects of phytoplankton metabolites on clearance and ingestion rates. J. exp. mar. Biol. Ecol. 163: 235–250.

Whyte, J. N. C., N. Bourne & C. A. Hodgson, 1989. Influence of algal diets on biochemical composition and energy reserves in *Patinopecten yessoensis* (Jay) larvae. Aquaculture 78: 333–348.

Wildish; D. J. & D. D. Kristmanson, 1985. Control of suspension feeding bivalve production by current speed. Helgolander Meeresunters. 39: 237–243.

Wildish, D. L., D. D. Kristmanson & A. M. Saulinier, 1992. Interactive effects of velocity and seston concentration of giant scallop feeding inhibition. J. exp. mar. Biol. Ecol. 155: 161–168.

Hydrobiologia **484:** 121–131, 2002.
O. Vadstein & Y. Olsen (eds), Sustainable Increase of Marine Harvesting: Fundamental Mechanisms and New Concepts.

Coastal upwelling, primary production and mussel growth in the Rías Baixas of Galicia

F.G. Figueiras, U. Labarta & M.J. Fernández Reiriz
CSIC, Instituto de Investigacións Mariñas, Eduardo Cabello, 6, 36208-Vigo, Spain
Tel.: + 34-986-231930, ext 372. Fax: +34-986-292762. E-mail: paco@iim.csic.es

Key words: coastal upwelling, phytoplankton, primary production, mussel growth and physiology, Rías Baixas, Galicia, NW Spain

Abstract

The Rías Baixas are four flooded tectonic valleys located on the northwest Iberian Peninsula that support a mussel production of about 250 × 10^6 kg y^{-1} from 3337 mussel rafts. Mussel production in this region is the highest in Europe and one of the most intensive in the world, giving employment to 9000 people directly and 20 000 indirectly. The causes of this high mussel production are discussed through the analysis of published and some unpublished information. The interaction between coastal upwelling and the circulation patterns in the Rías, which channel the 3-dimensional variability of the open ocean into a 2-dimensional system, promotes a massive response in the productivity of phytoplankton populations inside the Rías, even during weak upwelling events along the coast. Coastal upwelling in the area typically occurs between March–April and September–October. The mean value of gross primary production during the whole upwelling season is ~1.4 g C m^{-2} d^{-1}, although high sporadic values of ~4 g C m^{-2} d^{-1} may occur during upwelling relaxation events, when phytoplankton export to the coastal shelf is restricted. Mussel growth occurs mainly during the upwelling season. It is estimated that mussel harvest extracts ~10% of the primary production. The phytoplankton response to upwelling provides food of high quality (f ~0.5) that determines high absorption efficiency (~0.6), whereas the characteristics of the Rías maintains the seston concentration at levels (0.5 – 1.3 mg TPM l^{-1} and <5 mg Chl a m^{-3}) below the threshold of pseudo-faeces production. The physiological behaviour of mussels indicates that the high yield of mussel culture in the Rías of Galicia is a consequence of the particular characteristics of the seston.

Mussel culture in Galicia

The Rías Baixas are four flooded tectonic valleys on the northwest of the Iberian peninsula oriented in a northeast-southwest direction (Fig. 1). They are about 30 km long and 6 km wide in their middle part, the Ría de Muros-Noia being slightly shorter and the Ría de Arousa wider than the other three. These four Rías support a high density of mussel floating rafts.

The culture of *Mytilus galloprovincialis* Lamarck began in the Ría de Arousa in 1946 with 125 m^2 rafts with 500 hanging ropes 5 m long each. Nowadays, rafts are of 500 m^2 with the same number of ropes but 12 m long. During the past 50 years, the region has evolved to be the highest mussel production zone in Europe and one of the most intensive in the world. The present number of rafts in Galicia is 3337, with 97% concentrated in the four Rias Baixas (Table 1). The highest density is found in the Ría de Arousa, where 2.292 rafts (69% of the total rafts in Galicia) covers a surface of 27 km^2, which represents 11% of the total surface of the Ría. The estimated annual mussel production per raft is similar in the four Rías (~75 × 10^3 kg $raft^{-1}$ y^{-1}, Table 1), rendering a total estimated mussel production in Galicia of 250 × 10^6 kg y^{-1}. This estimated production agrees with the average production values provided by the farmers (290 × 10^6 kg y^{-1}) and the commercialised production (223 × 10^6 kg y^{-1}). Mussel production gives employment to 9000 people directly and to 20 000 indirectly and, is therefore an activity of great socio-economic importance for the region (Labarta, 1999; 2000). In this contribution we will discuss the causes that explain the high mussel production in this relatively small area.

Table 1. Mussel production in Galicia in 1997 after Labarta (1999)

Area of mussel Cultivation	Production per raft kg × 10^6 raft^{-1} y^{-1}	Number of rafts
Ria de Muros-Noia	69.7	118
Ria de Arousa	75.91	2292
Ria de Pontevedra	75.51	346
Ria de Vigo	72.18	478
Galicia	74.92	3337
Total estimated production		250 × 10^6 kg
Total production from mussel farmers		290 × 10^6 kg
Total commercialised production		223 × 10^6 kg
Mean		256 × 10^6 kg

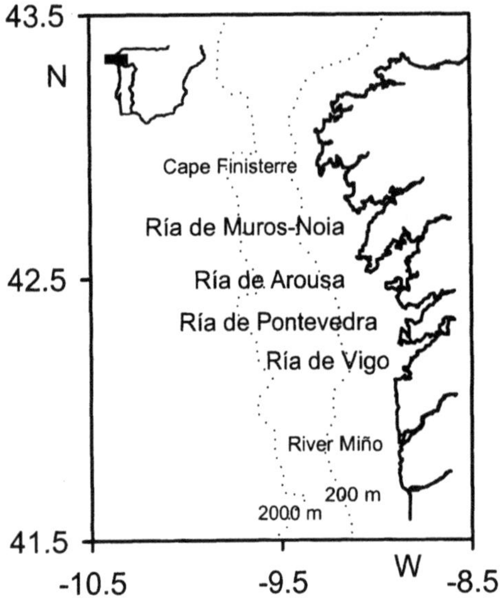

Figure 1. Map showing the location of the four Rías Baixas on the NW of the Iberian Peninsula.

Coastal upwelling and circulation patterns in the Rías

Galicia is located in the northern limit of the NW Africa upwelling system, and therefore in an area where the along-shore winds interact with the coastal topography to generate upwelling–downwelling dynamics (cross-shore Ekman transport) on the continental shelf. The evolution of the upwelling–downwelling index (Fig. 2b), a surrogate variable for the cross-shore Ekman transport that is estimated from geostrophic winds (Bakun, 1973), shows in the western Galician coast (Blanton et al., 1984) a characteristic seasonal variation (Fig. 2a). This seasonality mirrors the seasonally varying changes in the strength and position of the atmospheric pressure cells that govern the North Atlantic climatology, the Azores High and the Greenland Low, defining two wind-featured oceanographic seasons. The upwelling season occurs on average from March to September, when the weakening of the Greenland Low and strengthening and northward displacement of the Azores High promotes northerly winds. The rest of the year, due to the intensification of the Greenland Low, the winds are predominantly from the south and south-west, favouring the predominance of downwelling conditions.

However, this characteristic seasonal pattern of the upwelling–downwelling dynamics varies considerably. For instance, the transition from the downwelling to the upwelling season (spring transition) can occur in February or in April, while the autumn transition (from the upwelling to the downwelling season) shows lesser variability and occurs generally around October (Fig. 2a). The intensity of the seasonal signal also shows significant interannual variability. Seasonal maximum upwelling can occur in late spring or late summer, while seasonal maximum downwelling occurs in December or January (Fig. 2a). The interannual variability in the timing and intensity of the seasonal signal is due to the characteristic high-frequency variation of the wind patterns, the forcing of the Ekman transport dynamics. This high-frequency variability, which is associated with the short-term fluctuations of the large-scale climatology of the North Atlantic, is the most important source of variation of the upwelling–downwelling dynamics and generates, for instance, the occurrence of relaxation or oppos-

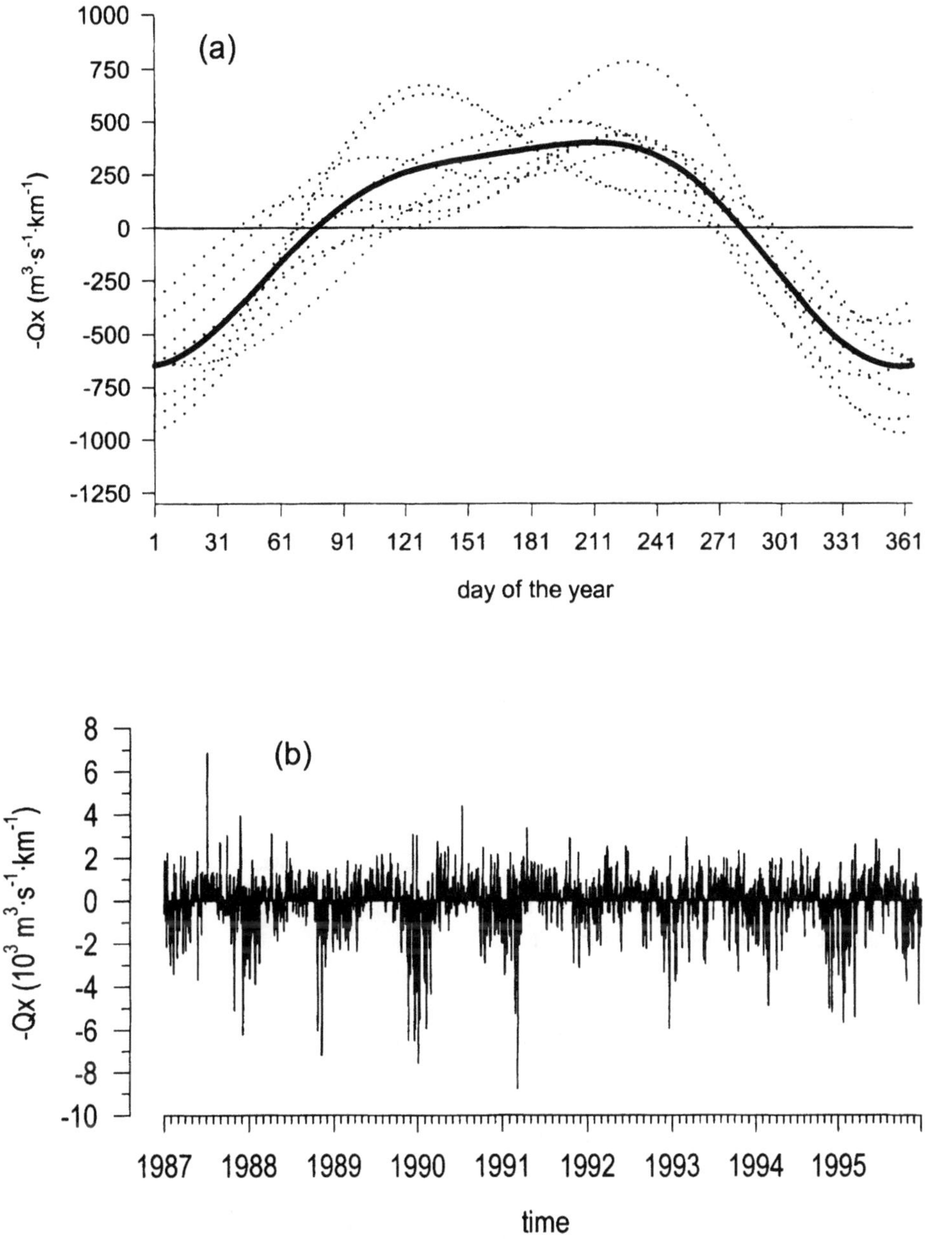

Figure 2. Upwelling index for a 9 year period (1987–1995). (a) Seasonal cycle of the upwelling index. This seasonal cycle is defined by the 1st and 2nd harmonics of the annual period extracted from the daily time series of the upwelling index for the period 1987–1995; average seasonal cycle – thick line; average for the single years – dotted lines. (b) Daily-time series of the upwelling index. Upwelling index, an estimate of cross-shore Ekman transport, was calculated according to Wooster et al. (1976):

$$-Q_x = \frac{\rho_{air} \cdot C_d \cdot |V| \cdot V_N}{\rho_{sw} \cdot f},$$

where ρ_{air} is the density of air, 1.22 kg m^{-3} at 15 °C; C_d is an empirical drag coefficient, 1.3×10^{-3}; $|V|$ is wind speed and V_N is the north component of wind speed deduced from surface pressure charts (three times d^{-1}) in front of Cape Finisterre (43 ° N, 11 ° W); ρ_{sw} is the density of seawater, ~1025 kg m^{-3}; and f is the Coriolis parameter at 43 °N latitude, 9.946×10^{-5} s^{-1}.

ite events within a given upwelling or downwelling season (Fig. 2b).

The hydrographic coastal features resulting from this seasonal wind pattern are well exemplified by the satellite picture in Figure 3. Cold (<15 °C) and nutrient rich upwelled water is found on the shelf during the summer upwelling, when off-shore filaments often develop (Joint et al., 2001). In contrast, warm water (>16 °C) is found at the surface on the shelf in winter. This warm water on the shelf is a consequence of the well-defined reversed surface circulation that takes place during winter in the area (e.g. Wooster et al., 1976; Fiúza, 1984; Ambar et al., 1986; Frouin et al., 1990; Pingree et al., 1999; Castro et al., 1997), when the Portugal Coastal Counter Current (PCCC) transport eastern North Atlantic central Water of subtropical origin from ~39 °N to 47 °N. The PCCC is maintained as a coastal feature during the downwelling season.

The dynamics on the coastal shelf interact with circulation in the Rías in different ways. Upwelled water on the coast, which is colder and saltier than the water in the Rías, forces a density-induced positive circulation pattern, characterised by the outflow of surface waters and a compensating inflow of upwelled waters at the bottom. Conversely, coastal downwelling causes a reversal of this circulation pattern, characterised by the intrusion of warm coastal waters at the surface, which accumulates in the Rías interior and finally flows towards the ocean through the bottom (e.g. Figueiras et al., 1994). The Rías act driving the 3-dimensional variability of the open ocean into a 2-dimensional system, where all oceanographic signals will be amplified. In fact, we can better understand some aspects of these ecosystems when we consider them as very efficient extensions of the shelf. Thus, a weak upwelling with no or minor effect on the coast may place nutrient-rich upwelled water at a depth of 50 to 60 m, which is enough to enter the Rías by positive estuarine circulation and cause a conspicuous response of phytoplankton. Upwelling relaxation and downwelling block the export of phytoplankton and favour the accumulation of organic matter inside the Rías and even re-introduce the material that has been exported to the shelf during previous upwelling episodes (e.g. Figueiras et al., 1994; Tilstone et al., 1994; Fermín et al., 1996; Tilstone et al., 2000).

A good example of the link between coastal upwelling and circulation in the Rías was obtained during fieldwork in the Ría de Vigo on September 1990, when a thermistor chain and near-surface cur-

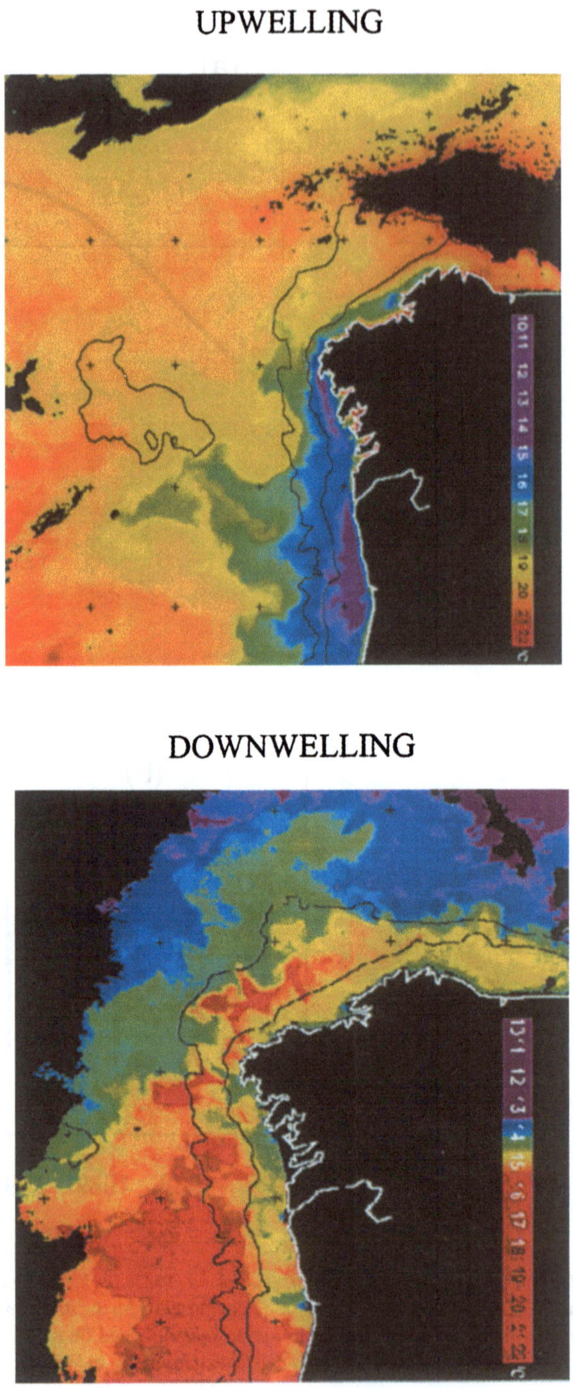

Figure 3. Typical AVHRR sea surface temperature images of the NW Iberian Peninsula during upwelling (summer, 3 August 1998) and downwelling (winter, 21 January 1998) situations. Courtesy of the RMS group of Plymouth Marine Laboratory.

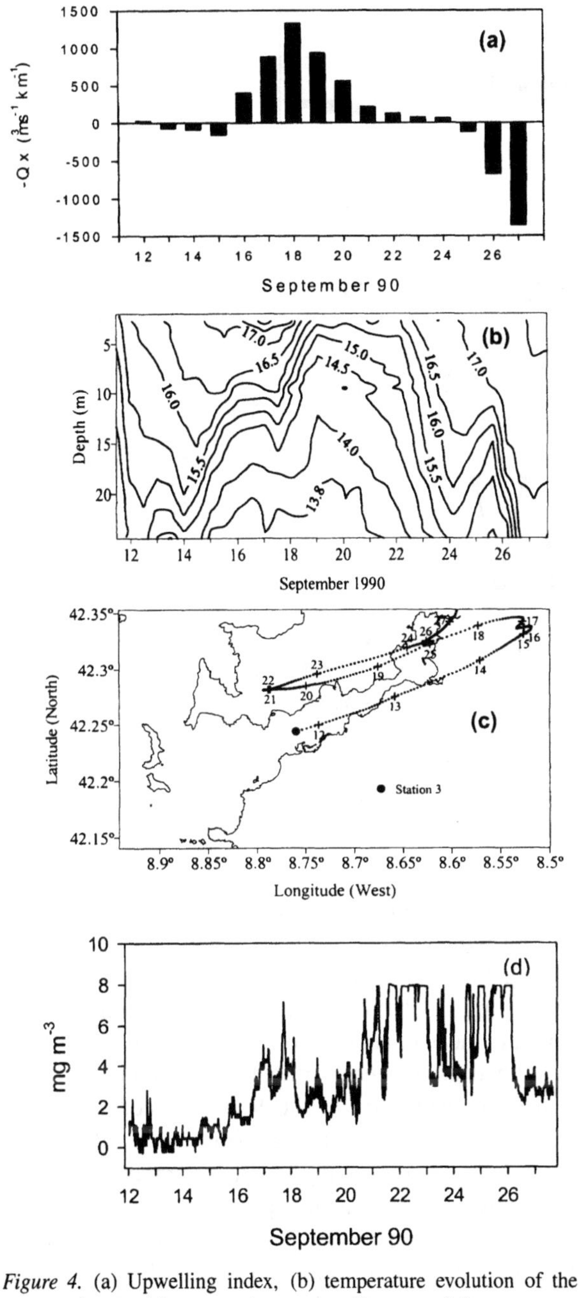

Figure 4. (a) Upwelling index, (b) temperature evolution of the water column, (c) progressive vector diagram of the near surface current and (d) time-evolution of the near surface chlorophyll concentration between 12 and 27 September 1990 in the Ría de Vigo. Thermistor chain, near surface current meter and near surface fluorometer were deployed at one station in the middle of the Ría de Vigo (station 3, c). Fluorometer readings were converted to chlorophyll concentrations after calibration against extracted chlrophyll. Redrawn from Figueiras et al. (1994).

rent meter and fluorometer was deployed in a station in the middle of the Ría (Fig. 4, see Figueiras et al., 1994 for more details). Variations in wind-driven coastal upwelling (Fig. 4a) produced a downwelling–upwelling–downwelling cycle in the Ría that was recorded by the thermistor chain (Fig. 4b). The slight downwelling between 12 and 15 September (Fig. 4a) at the beginning of the observations, forced the surface intrusion of coastal warm water (>16 °C) into the Ría (Fig. 4b). Later, the upwelling between 16 and 18 September provoked the bottom intrusion of cold water (<14 °C). The upwelling relaxation between 19 and 24 September was followed by a strong downwelling (26–27 September) that left a homogeneous water column of ~17 °C. This downwelling–upwelling–downwelling cycle had a clear signal in the residual surface current, as is showed by the progressive vector diagram of the near-surface current meter records (Fig. 4c). Downwelling episodes (12–15 and 22–27 September) caused the water flow into Ría, whereas during upwelling (17–22 September) water flowed to the southwest out of the Ría. During the two downwelling periods, inflowing water was transported to a distance beyond the inner limit of the Ría basin, which obviously is not possible, and consequently indicates that water was piled up towards the interior of the Ría. There were three periods with no net flow into or out of the Ría, which coincided with the transient situations between successive downwelling and upwelling: 15–17, 21–22 and 25–26 September. The covariance between upwelling index and convective residual fluxes in the Rías has also been pointed out in other studies (Rosón et al., 1997; Tilstone et al., 2000; Míguez et al., 2001). Vertical and horizontal convective fluxes are coupled with the highest vertical and horizontal transports occurring during strong upwelling and downwelling and the lowest during relaxation periods (Tilstone et al., 2000).

The surface chlorophyll (Chl *a*) concentration recorded at the same time from the moored fluorometer (Fig. 4d) was low, usually below 1 mg m^{-3}, during the first downwelling (12–15 September). During the upwelling period (15–18 September) the concentration rose at an average rate of 0.69 d^{-1} from 1 to ~4 mg m^{-3}, which is compatible with phytoplankton growing at near maximum rates. However, between 20 and 22 September, during upwelling relaxation, Chl *a* concentration increased at a mean rate >1 d^{-1} and remained high until 25 September. Chl *a* levels dropped abruptly on 26 September and remained low (~2.5 mg m^{-3}) until the end of the observations. The evolu-

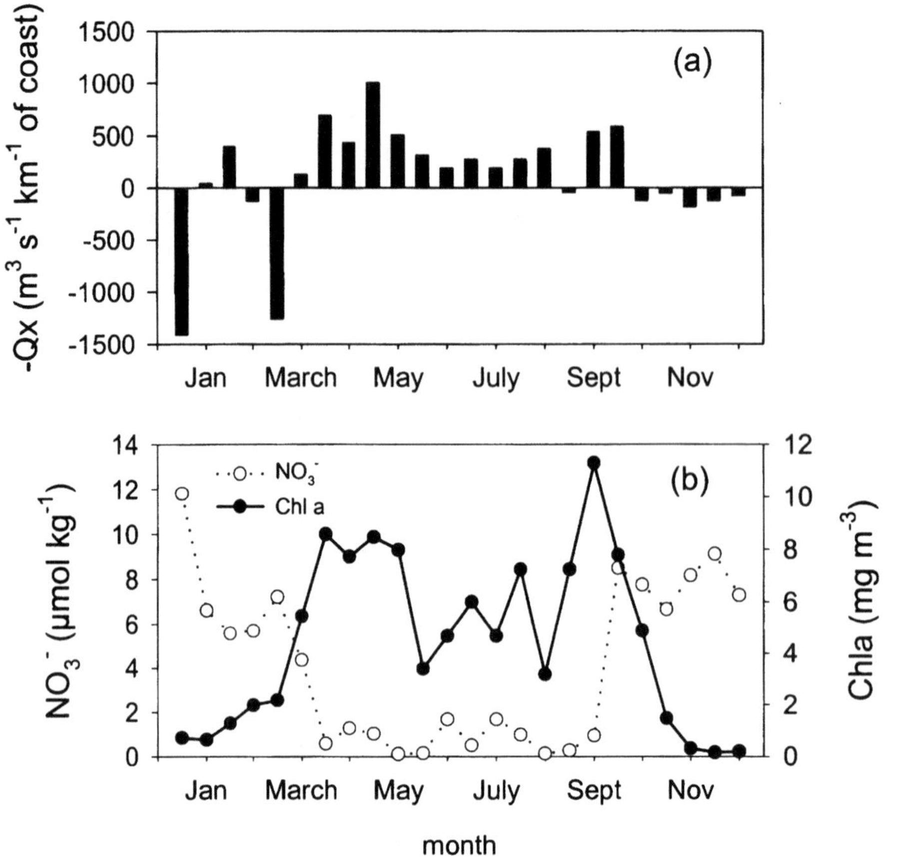

Figure 5. (a) Time-evolution of upwelling index and (b) nitrate and chlorophyll concentration (b) at the surface waters in the Ría de Vigo. Values are fortnightly means ($n = 4$). Sampling was done at station 3 in Figure 4c in 1991. Mirón & Figueiras (unpublished).

tion of Chl *a* concentration indicates that the reversal circulation of the Ría caused by upwelling relaxation followed by downwelling, favoured the phytoplankton accumulation, which was finally removed from the water column due to the persistence of downwelling until the end of the fieldwork. Similar situations of phytoplankton accumulation in the interior of the Rías during upwelling relaxations and low Chl *a* concentrations during strong downwelling have been corroborated by other field studies in the region (e.g. Tilstone et al., 1994; Fermín et al., 1996).

Phytoplankton

On the basis of the annual cycle, phytoplankton biomass expressed as Chl *a* concentration determined fluorometrically (Yentsch & Menzel, 1963) shows a bimodal pattern (Fig. 5b), with the maxima located at the transition periods of the upwelling index: spring and autumn respectively (Mirón & Figueiras, unpublished). Concentration is rather low (<1 mg m^{-3}) during winter, rises to mean values of $\sim$8 mg m^{-3} during the spring and autumn maxima and remains at $\sim$5 mg m^{-3} during summer. This temporal pattern and mean concentrations for each period has been described as a general rule for the annual cycle of Chl *a* concentration in the Rías deduced from a 6-year period of observations (Nogueira et al., 1997). There are differences in the oceanographic processes that cause the two seasonal peaks of chlorophyll. The spring maximum results from the phytoplankton response to the first upwelling episodes of the year (Fig. 5a), which coincide with the increase of the incident irradiance, and represents the characteristic spring bloom of shallow temperate ecosystems, composed mainly of large diatoms. (Figueiras & Niell, 1987). The autumn maximum occurs at the time of the seasonal transition from upwelling to downwelling (Fig. 5a), when intense reversal circulation takes place in the Rías (see Fig. 4). The high Chl *a* concentration during the autumn bloom is rather due to the accu-

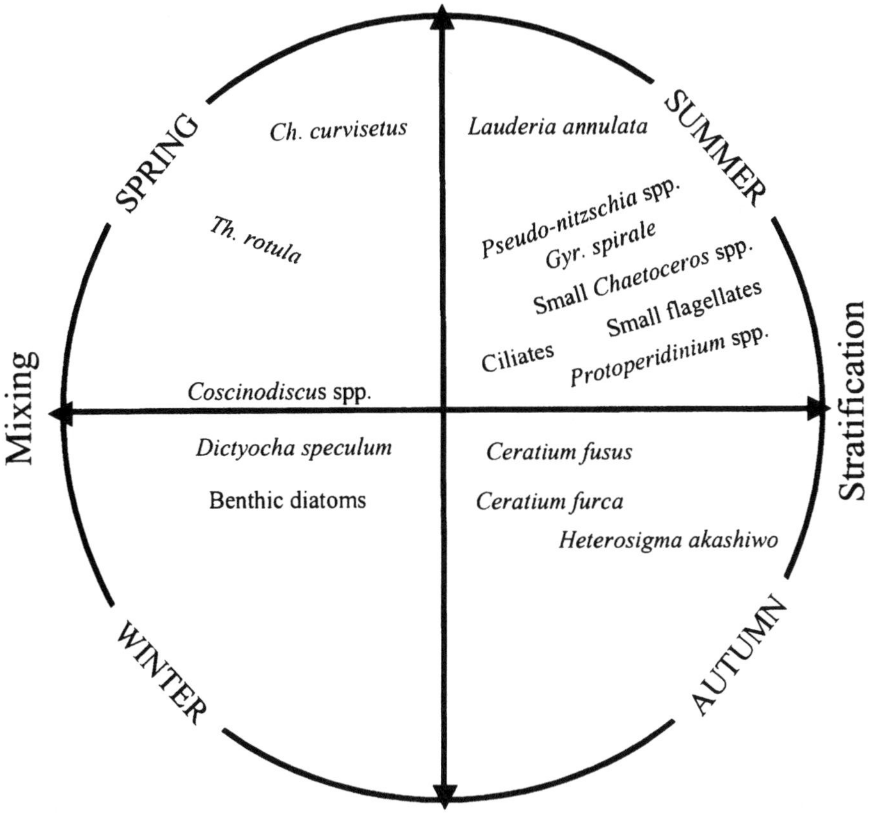

Figure 6. Annual phytoplankton cycle in the Ría de Pontevedra. The axes are the two first principal components of an analysis of phytoplankton abundance. Redrawn and simplified from Figueiras &Niell (1987).

mulation promoted by circulation reversal than to '*in situ*' phytoplankton growth. During this period, phytoplankton assemblages are composed mainly of mixed populations of diatoms and dinoflagellates.

The typical phytoplankton composition of the Rías (Figueiras & Niell, 1987) is shown in Figure 6. Large diatoms (e.g. *Thalassiosira rotula*, *Chaetoceros curvisetus*, *Lauderia annulata*) are abundant during the first spring upwelling events, when stratification of the water column is weak. In summer, when stratification is stronger, small diatoms (e.g. *Pseudo-nitzschia* spp., small *Chaetoceros* spp.,) and heterotrophic species (*Protoperidinium* spp., *Gyrodinium spirale*, ciliates) coexist. Potential red tide species usually appear at the end of summer, during the transition from upwelling to downwelling season, when circulation reversal is established (Figueiras & Niell, 1987; Figueiras et al., 1994; Fermín et al., 1996). It has been demonstrated that large dinoflagellates, many of them red tide species, can remain in the water column due to their swimming capacity to counteract the downward water velocity generated by the circulation reversal, whereas diatoms are removed from the water column (e.g. Fraga et al., 1988; Tilstone et al., 1994; Figueiras et al., 1995, 1996). Phytoplankton abundance during winter is comparatively lower and benthic species (*Paralia sulcata*, *Diploneis* spp., *Navicula* spp.) are often found in the water column. These studies allowed us to establish a correlation between diatoms biomass and upwelling index (Figueiras & Ríos, 1993). The correlation was high ($r = 0.92, n = 10$) when only the first part of the upwelling period (February–June) was considered. However, the correlation was lower ($r = 0.66, n = 25$) when all the upwelling season was included in the analysis. The reduction of the correlation found for the second half of the upwelling period (July–October) might be explained by the presence of heterotrophic plankton in the water column (Fig. 6), which have a rapid response time to any food availability.

One aspect that remained hidden until now was the importance of the nanoplankton in the Rías. Classically, it was considered that the microplankton fraction dominates the phytoplankton biomass, because the Rías are coastal and shallow environments. Nevertheless, nanoplankton fraction may also be important

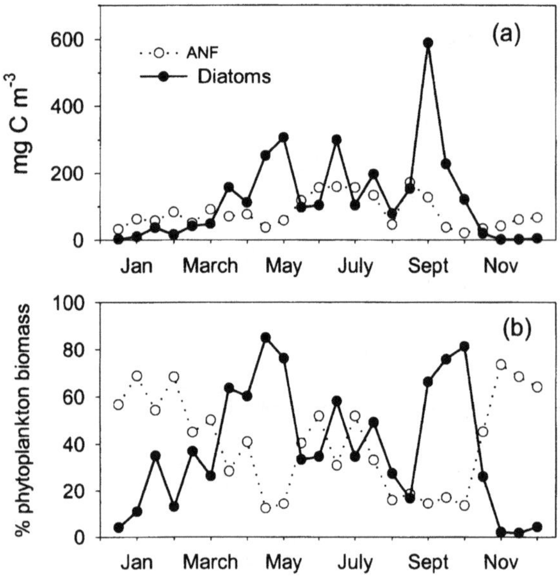

Figure 7. Carbon biomass of diatoms and autotrophic nanoflagellates (ANF) along an annual cycle (a) and respective contribution to the total autotrophic biomass (b) along an annual cycle in the Ría de Vigo. Values are fortnightly means ($n = 4$). Sampling was done at station 3 in Figure 4c in 1991. Diatoms were counted with an inverted microscope, their biovolumes calculated (Edler, 1979) and converted to cell carbon (Strathman, 1967). ANF were counted with an epifluorescence microscope and spherical biovolumes converted to cell carbon (Verity et al., 1992).

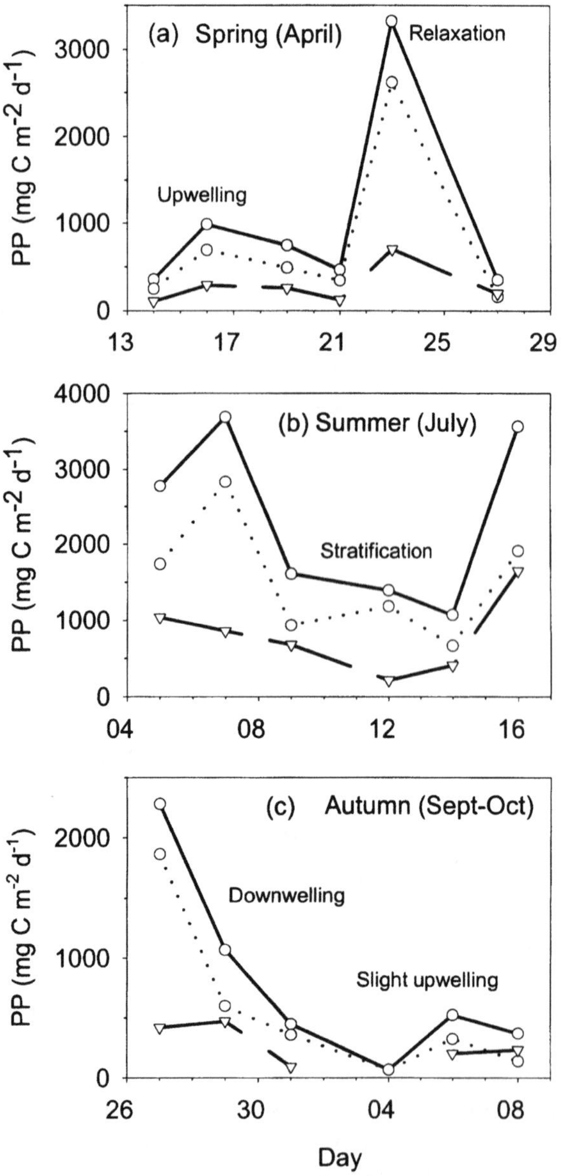

Figure 8. Daily depth-integrated primary production in the Ría de Vigo in spring (a), summer (b) and autumn (c) after Tilstone et al., (1999). Solid line: total phytoplankton; dotted line: phytoplankton fraction >20 μm; dashed line: phytoplankton fraction <20 μm.

in upwelling systems (Legendre & Rassoulzadegan, 1995). In this sense, more recent studies in the Ría de Vigo (Mirón & Figueiras, unpublished) indicate that nanoplankton is an important component of the phytoplankton biomass in the area (Fig. 7a). Diatoms significantly contribute to the two annual Chl *a* maxima of spring and autumn (Fig. 7a,b ~80% of total autotrophic biomass), whereas autotrophic nanoplankton is more important during winter (up to 60% of the total autotrophic biomass). Both components, nano- and microphytoplankton, are equally important during summer (Fig. 7b). This is an aspect that can be relevant for the mussel life cycle because nanoplankton may be the main food for mussel larvae. The body size ratio between mussel larvae (150 μm < Ø< 300 μm) and nanoplankton (2 μm < Ø< 20 μm) allows a predator-prey relationship (Hansen et al., 1994).

Primary production

Daily depth-integrated gross primary production (GPP) estimated from short time ^{14}C incorporation (Tilstone et al., 1999), during the upwelling season, as is usual in other upwelling systems, shows high short-term variability (Fig. 8) that is related with upwelling–downwelling and relaxation events. Relaxations during spring raise primary production (Fig. 8a) due to the accumulation of phytoplankton inside the Rías (Tilstone et al., 2000). Primary production during the summer stratified period is high during upwelling events (~3.5 mg C m^{-2}) and lower when upwelling relaxes (Fig. 8b). Autumn downwelling causes the abrupt drop of primary production (Fig. 8c). Net phytoplankton (>20 μm) is responsible for 66±13% of the mean primary production, but pico- and nanophytoplankton (<20 μm) can contribute up to 50% of total primary production during upwelling episodes that follow strong downwelling (Fig. 8c), when a large portion of net phytoplankton has been removed from the water column by circulation reversal during downwelling (see above). Comparing the three situations that may be defined during the upwelling season: downwelling–upwelling transition (spring), upwelling (summer) and upwelling–downwelling transition (autumn) mean primary production is higher during upwelling (~2.4 g C m^{-2} d^{-1}) than during the transition periods (~1 g C m^{-2} d^{-1}) despite the mean chlorophyll concentration being lower in summer (Fig. 5b). The higher chlorophyll-specific maximum photosynthetic rate in summer (Tilstone et al., 1999) is responsible for the differences in primary production.

Mean GPP during upwelling season in the Rías is ~1.4 g C m^{-2} d^{-1} (Fraga, 1976; Alvarez-Salgado et al., 1996; Tilstone et al., 1999) although higher values (2.5 g C m^{-2} d^{-1}) have also been reported (Moncoiffé et al., 2000).

Mussel production, food availability and physiological processes

The intensive mussel culture in the Rías Baixas of Galicia is expected to have a great influence on the food web. The high mussel biomass supported by the 3300 rafts suggest that this organism is the most important grazer in the Rías. Mussel growth is affected by a number of factors, among which food availability is determinant (Fernández-Reiriz et al., 1996). According to several authors (e.g. Frechette & Bouget, 1985; Smal & Stralen, 1990) the ingestion capacity is a function of phytoplankton concentration and current speed. Phytoplankton concentration establishes the maximum available food, and water velocity passing through the culture area determines the rate at which the food is supplied. As was shown previously, coastal upwelling–downwelling is the oceanographic process that is responsible for both factors. The first quantitative relationship between upwelling and mussel growth was published by Blanton et al. (1987). They found that mussel condition index in the Ría de Arousa, which is the percentage of the meat content of mussels harvested, was related ($r^2 = 0.66$) with the mean upwelling index from April to September for an 11 year period (1973–1983). More recently, Pérez-Camacho et al. (1995) have demonstrated that 98% of the variance of the growth rates in length and dry weight of cultured mussels in the Rías Baixas is explained by current velocity and Chl *a* concentration, with current velocity explaining 66–79% and Chl *a* concentration 33–19% of the variance in both growth rates, respectively. Covariation of Chl *a* concentration and current velocity is possible during upwelling, as shown previously. High current speeds can also be found during downwelling, but are associated with low Chl *a* concentrations. Upwelling relaxations, when current velocity is close to zero will lead to exhaustion of Chl *a* after a given period of time and, consequently, mussel growth will be limited by the low food availability.

Considering that mussel growth rate follows a seasonal pattern with the highest values during spring-summer (e.g., Pieters et al., 1980; Mallet et al., 1987; Babarro et al., 2000), which is the upwelling season, a simple calculation will allow us to estimate the fraction of primary production that is channelled to mussels. From the total mussel production of 172 × 10^6 kg y^{-1} in the Ría de Arousa (Table 1) and carbon content of mussel wet weight (~2.5%, Tenore et al., 1982; Pérez-Camacho & González, 1984) it follows that total carbon yield of mussels harvested in the Ría de Arousa is ~4.3 × 10^6 kg C y^{-1}. In addition, from the surface area of the Ría (230 km^2), the extension of the upwelling period (~180 days) and the mean GPP (1. 4 g C m^{-2} d^{-1}), it is deduced that the mean GPP for the upwelling season is ~5.796 × 10^7 kg C. Therefore, mussels incorporate ~7% of carbon produced in the Rías during the upwelling season, a value that rises to ~12% when net community production (~0.84 g C m^{-2} d^{-1}, Alvarez-Salgado et al., 1996), that is, GPP minus carbon respired in the water column (i.e. including mussels respiration), is considered.

Mussel growth in the Rías Baixas is not only dependent on food availability; mussel physiology also plays an important role. Fernández-Reiriz & Labarta (unpublished) have established that the determining factor of potential growth, or energy available for growth (SFG), is the absorbed food ration, which is of greater importance than the metabolic losses, which are rather low and constant for a standardised mussel. Moreover, Fernández-Reiriz et al. (2001) have shown that the feeding behaviour of *Mytilus galloprovincialis* responds to a series of functional relationships between physiologic rates related to energy incorporation and environmental variables that define food quantity (total particulate matter, TPM) and quality (the ratio *f* between particulate organic matter, POM, and TPM). Thus, the study demonstrates that the organic ingestion rate (OIR, mg org h^{-1}) for a standardised mussel of 40 mm and 300 mg is exponentially related with the concentration of TPM (mg l^{-1}):

$$OIR = 1.29\ (1 - \exp - 0.78\ TPM)$$
$$(n = 27;\ r^2 = 0.70;\ p < 0.01),$$

from which it is deduced that OIR saturates at concentrations of TPM higher than ~7 mg l^{-1}. In addition, the same authors (*loc. cit.*) have also found that the absorption efficiency AE is related with *f* in the following way:

$$AE = 0.95 - 0.18 \times 1/f$$
$$(n = 17;\ r^2 = 0.61;\ p < 0.01),$$

which indicates that the maximum AE (~0.6) is obtained for a food quality *f* of ~0.5. From these relationships and the mean TPM content (~0.88 ± 0.27 mg l^{-1}) and food quality value ($f = 0.52 \pm 0.05$) in the seawater under rafts during the upwelling season (Babarro et al., 2000) it is deduce that the absorption rate (AR = OIR × AE) of mussels in the Rías is ~0.4 mg org h^{-1}.

The physiological processes described here indicate that the high yield of mussel culture in the Rías of Galicia is a consequence of the seston characteristics, which in turn depend on the interaction between coastal upwelling and circulation of the Rías. The phytoplankton response to upwelling provides food of high quality (*f* ~0.5) that determines high absorption efficiency (~0.6), whereas the oceanographic characteristics of the Rías maintains the seston concentration at levels (0.5 – 1.3 mg TPM l^{-1} and < 5 mg Chl a m^{-3}) below the threshold of pseudo-faeces production. These two environmental factors influence the mussel physiology response and consequently the mussel growth rates (Fernández-Reiriz et al., 2001) in the Rías of Galicia. These growth rates are higher than those obtained in other ecosystems with higher seston concentrations and/or higher fluctuations of chlorophyll concentrations and of food quality (Pérez-Camacho & Román, 1979; Pérez-Camacho et al., 1995; Fernández-Reiriz et al., 1996; Babarro et al., 2000).

Acknowledgements

We thank the members of the IIM groups of Oceanography and Physiology, Nutrition and Culture of Mollusc Bivalves for their contributions. We are also very grateful to the Remote Sensing Group of the Plymouth Marine Laboratory for the AVHRR images and to Dr E. Nogueira for the computations of upwelling indices and for useful criticism of the manuscript.

References

Alvarez-Salgado, X. A., G. Rosón, F. F. Pérez, F. G. Figueiras & Y. Pazos, 1996. Nitrogen cycling in an estuarine upwelling system, the Ría de Arousa (NW, Spain).I. Short-time-scale patterns of hydrodynamic and biogeochemical circulation. Mar. Ecol. Prog. Ser. 135: 259–273.

Ambar, I., A. F. G. Fiúza, T. Boyd & R. Frouin, 1986. Observations of a warm oceanic current flowing northward along the coasts of Portugal and Spain during November–December 1983. EOS transactions, American Geophysical Union 67(44): 1054.

Babarro, J. M. F., M. J. Fernández-Reiriz & U. Labarta, 2000. Growth of seed mussel (*Mytilus galloprovincialis*): effects of the environmental parameters and seed origin. J. Shellf. Res. 19: 187–193.

Bakun, A., 1973. Coastal upwelling indices, west coast of North America, 1946–1971. NOAA Tech. Rep., NMFS SSRF-671, U.S. Dept. of Commerce, 103 pp.

Blanton, J. O., L. P. Atkinson, F. Fernández de Castillejo & A. Lavin Montero, 1984. Coastal upwelling off the Rias Bajas, Galicia, Northwest Spain, I: Hydrography studies. Rapports et Proc-verbeaux Reun. Cons. int. Explor. Mer. 183: 79–90.

Blanton, J. O., K. R. Tenore, F. Castillejo, L. P. Atkinson, F. B. Schwing & A. Lavin, 1987. The relationship of upwelling to mussel production in the rias on the western coast of Spain. J. mar. Res. 45: 497–511.

Castro, C. G., X. A. Álvarez–Salgado, F. G. Figueiras, F. F. Pérez & F. Fraga, 1997. Transient hydrographic and chemical conditions affecting microplankton populations in the coastal transition zone of the Iberian upwelling system (NW Spain) in September 1986. J. mar. Res. 55: 321–352.

Edler, L., 1979. Recommendations for marine biological studies in the Baltic Sea, Phytoplankton and Chlorophyll. Baltic Marine Biology. 5: 38 pp.

Fermín E. G., F. G. Figueiras, B. Arbones & M. L. Villarino, 1996. Short-time evolution of a *Gymnodinium catenatum* population in the Ría de Vigo. J. Phycol. 32: 212–221.

Fernández-Reiriz, M. J., U. Labarta & J. M. F. Babarro, 1996. Comparative allometries in growth and chemical composition of mussel (*Mytilus galloprovincialis* Lmk) cultured in two zones in the ria Sada (Galicia, NW Spain). J. Shellf. Res. 15: 349–354.

Fernández-Reiriz, M. J., U. Labarta & J. M. F. Babarro, 2001. Relaciones funcionales del comportamiento alimentario del mejillón (*Mytillus galloprovincialis*) en relación al seston natural en las rías gallegas. In Fernández Pato, C. A. et al. (eds), Acuicultura y Desarrollo Sostenible. Consejería de Ganaderia, Agricultura y Pesca, Santander: 296–300.

Figueiras F. G. & A. F. Rios, 1993. Phytoplankton succession, red tides, and the hydrographyc regime in the Rías Baixas of Galicia. In Smayda, T. J. & Y. Shimizu (eds), Toxic Phytoplankton Blooms in the Sea. Elsevier Science Publishers. EEUU: 239–244.

Figueiras F. G., E. Gómez, E. Nogueira & M. L. Villarino, 1996. Selection of *Gymodinium catenatum* under downwelling conditions in the Ría de Vigo. In Yasumoto, T., Y. Oshima & Y. Fukuyo (eds), Harmful and Toxic Algal Blooms. Intergovernmental Oceanographic Commission of UNESCO, Paris: 215–218.

Figueiras, F. G. & F. X. Niell, 1987. Composición del fitoplancton en la ría de Pontevedra (NO de España). Invest. Pesquera 51: 371–409.

Figueiras, F. G., K. Jones, A. M. Mosquera, X. A. Alvarez-Salgado, A. Edwards & N. MacDougall, 1994. Red tide assemblage formation in an estuarine upwelling ecosystem: Ría de Vigo. J. Plankt. Res. 16: 857–878.

Figueiras, F. G., T. Wyatt, X. A. Alvarez-Salgado & I. Jenkinson, 1995. Advection, diffusion, and patch development of red tide organisms in the Rías Baixas. In Lassus, P., G. Arzul, E. Erard, P. Gentin & C. Marcaillou (eds), Harmful Marine Algal Blooms. Technique et Documentacion-Lavoisier, Intercept Ltd., Francia: 579–584.

Fiúza, A. F. G., 1984. Hidrologia e dinâmica das aguas costeiras de Portugal. Ph D. Thesis, University of Lisbon, 294 pp.

Fraga, F., 1976. Fotosíntesis en la ría de Vigo. Invest. Pesquera 40: 151–167.

Fraga, S., D. M. Anderson, I. Bravo, B. Reguera, K. A. Steidinger & C. M. Yentsch, 1988. Influence of upwelling relaxation on dinoflagellates and shellfish toxicity in Ría de Vigo, Spain. Estuar. coast. Shelf Sci. 27: 349–361.

Fréchette, M. & E. Bouget, 1985. Energy flow between the pelagic and benthic zones: Factors controlling particulate organic matter available to an intertidal mussel bed. Can. J. Fish. aquat. Sci. 42: 1158–1165.

Frouin, R., A. F. G. Fiúza, I. Ambar & T. J. Boyd, 1990. Observations of a poleward surface current off the coasts of Portugal and Spain during winter. J. Geophys. Res. 95: 679–691.

Hansen, B., P. K. Bjørnsen & P. J. Hansen, 1994. The size ratio between plankton predators and their prey. Limnol. Oceanogr. 39: 395–403.

Joint, I., M. Inall, R. Torres-Almarza, F.G. Figueiras, X.A. Alvarez-Salgado & M. Woodward, 2001. Two lagrangian experiments in the Iberian upwelling system: tracking an upwelling event and an off-shore filament. Progr. Oceanogr. 51: 221–248.

Labarta, U., 1999. La Acuicultura en Galicia. A Economía Galega. Serie Informes anuais n° 13, Informe 1997/1998. Instituto de Desarrollo de Galicia. Universidad de Santiago-Fundación Caixa Galicia.

Labarta, U., 2000. La Acuicultura en Galicia. A Economía Galega. Serie Informes anuais n° 14, Informe 1998/1999. Fundación Caixa Galicia.CIEF Centro de Investigación Económica y Financiera.

Legendre, L. & F. Rassoulzadegan, 1995. Plankton and nutrient dynamics in marine waters. Ophelia 41: 153–172.

Mallet, A. L., C. E. A. Carver, S. S. Coffen & K. R. Freeman, 1987. Winter growth of the blue mussel *Mytilus edulis* L.: importance of stock and site. J. exp. mar. Biol. Ecol. 108: 217–228.

Míguez, B. M., L. Fariña-Busto, F. G. Figueiras & F. F. Pérez, 2001. Succession of phytoplankton assemblages in relation to estuarine hydrodynamics in the Ría de Vigo: A box model approach. Scient. Mar. 65: 65–76.

Moncoiffé, G., X. A. Alvarez-Salgado, F. G. Figueiras & G. Savidge, 2000. Seasonal and short time-scale dynamics of microplankton community production and respiration in an inshore upwelling system. Mar. Ecol. Prog. Ser. 196: 111–126.

Nogueira, E., F. F. Pérez & A. F. Ríos, 1997. Seasonal patterns and long-term trends in an estuarine upwelling ecosystem (Ría de Vigo, NW Spain). Estuar. coast. Shelf Sci. 44: 285–300.

Pérez-Camacho, A. & R. Gonzalez, 1984. La filtración del mejillón (*Mytilus edulis* L.) en laboratorio. In Actas do Primeiro Seminario de Ciencias do Mar, As Rías Galegas, Cuadernos da Area de Ciencias Marinas, Seminario de Estudos Galegos, 1: 427–437.

Pérez-Camacho, A. & G. Román, 1979. Estudio del mejillón y de su epifauna en los cultivos flotantes de la Ría de Arosa. II. Bol. Inst. Esp. Oceanogr. 5: 21–42.

Pérez-Camacho, A., U. Labarta & R. Beiras, 1995. Growth of mussel (*Mytilus edulis galloprovincialis*) in cultivation raft. Influence of seed source, cultivation site and food availability. Aquaculture 138: 349–362.

Pieters, H., J. H. Kluytmans, D. I. Zandee & G. C. Cadee, 1980. Tissue composition and reproduction of *Mytilus edulis* in relation to food availability. Neth. J. Sea Res. 14: 349–361.

Pingree, R. D., B. Sinha & C. R. Griffiths, 1999. Seasonality of the European slope current (Goban Spur) and ocean margin exchange. Cont. Shelf Res. 19: 929–975.

Rosón, G., X. A. Alvarez-Salgado & F. F. Pérez, (1999) Carbon cycling in a large coastal embayment, affected by wind-driven upwelling: short-time-scale variability and spatial differences. Mar. Ecol. Prog. Ser. 176: 215–230.

Smaal, A. C. & M. R. van Stralen, 1990. Average annual growth and condition of mussels as a function of food source. Hydrobiologia 195: 179–188.

Strathmann, R., 1967. Estimating the organic carbon content of phytoplankton from cell volume or plasma volume. Limnol. Oceanogr. 12: 411–418.

Tenore, K. R. & 14 co-authors, 1982. Coastal upwelling in the Rías Baixas, NW Spain. Contrasting the benthic regimes of the Ría de Arousa and Muros. J. mar. Res. 40: 701–772.

Tilstone, G. H., B. M. Míguez, F. G. Figueiras & E. G. Fermín, 2000. Diatom dynamics in a coastal ecosystem affected by upwelling: coupling between species succession, estuarine circulation and biogeochemical processes. Mar. Ecol. Prog. Ser. 205: 23–41.

Tilstone, G. H., F. G. Figueiras & F. Fraga, 1994. Upwelling-Downwelling Sequences in the generation of red tides in a coastal upwelling system. Mar. Ecol. Prog. Ser. 112: 241–253.

Tilstone, G. H., F. G. Figueiras, E. G. Fermín & B. Arbones, 1999. Significance of nanophytoplankton photosynthesis and primary production in a coastal upwelling system (Ría de Vigo, NW Spain). Mar. Ecol. Prog. Ser. 183: 13–27.

Verity, P. G., C. Y. Roberton, C. R. Tronzo, M. G. Andrews, J. R. Nelson & M. E. Sieracki, 1992. Relationships between cell volume and the carbon and nitrogen content of marine photosynthetic nanoplankton. Limnol. Oceanogr. 48: 357–366.

Wooster, W. S., A. Bakun, & D. R. Mclain, 1976. The seasonal upwelling cycle along the eastern boundary of the North Atlantic. J. mar. Res. 34: 131–141.

Yentsch, C. S. & D. W. Menzel, 1963. A method for the determination of phytoplankton chlorophyll and phaeophytin by fluorescence. Deep Sea Res. 10: 221–231.

Hydrobiologia **484:** 133–148, 2002.
O. Vadstein & Y. Olsen (eds), Sustainable Increase of Marine Harvesting: Fundamental Mechanisms and New Concepts.

Effects of the large-scale uncontrolled fertilisation process along the continental coastal North Sea

Franciscus Colijn[1], Karl-Jürgen Hesse, Norbert Ladwig & Urban Tillmann[2]
Westcoast Research and Technology Centre, Hafentörn, 25761 Büsum, Germany
[1]*Present address: GKSS, Institute for Coastal Research, Max Planck Strasse 1, 21502 Geesthacht, Germany*
E-mail: colijn@gkss.de
[2]*Present address: Alfred Wegener Institute for Polar Research, Bremerhaven, Germany*

Key words: eutrophication, North Sea, phytoplankton, macrobenthos, fishes

Abstract

In this paper, effects of eutrophication in selected compartments of the North Sea ecosystem are discussed, encompassing the possibly positive effects of nutrient enrichment. Based on a variety of studies, impacts on biomass of phytoplankton, macrozoobenthos, microzooplankton, shrimps and fishes and productivity are presented. Enhanced nutrient concentrations and loadings can be observed in several coastal areas of the North Sea. As a result, increases in the concentration, production and changes in the species composition was observed in the phytoplankton. In addition, there are some indications for an increased biomass of macrozoobenthos, whereas an increase in microzooplankton can only be assumed from mesocosm experiments. A concomitant increase of higher trophic levels such as shrimps and fishes, as observed in some coastal regions of the North Sea, is difficult to link directly to eutrophication due to a lack of conclusive field observations showing the causality of the changes. That the large fertilisation process in the North Sea has led to a series of changes is, however, without doubt. The answer, to what extent these can be claimed as being harmless, positive or negative from the anthropogenic point of view, is hampered by the lack of good assessment criteria for marine ecosystems and requires a thorough analysis of all compartments involved by means of long-term-series long enough to discriminate between man-made and natural variability.

Introduction

During the last decades a large-scale fertilisation process has been initiated in the North Sea through the input of large amounts of plant nutrients such as phosphorus and nitrogen compounds. Riverine nutrient inputs from European rivers into the continental coastal water of the North Sea increased from 202 ktN/y and 15 ktP/y in 1950 to 918 ktN/y and 100 ktP/y in 1980 (Radach et al., 1990). Efforts in nutrient reduction, intensified during the last decade, have resulted in decreasing phosphorus inputs in most of the countries bordering the North Sea. The effect is less obvious for nitrogen compounds, because of the major importance of diffuse sources inputs (Fig. 1 A–B).

On the basis of a selected number of papers, we want to discuss the extent to which this fertilisation has been effective on different levels of the trophic foodweb, both qualitatively and quantitatively. Because the topic encompasses so many aspects, we will concentrate only on some of the burning points, thereby using the diagram originally drawn by Gray, as presented in the introduction of the book 'Eutrophication in Coastal Marine Ecosystems' by Jørgensen and Richardson (1996) (Fig. 2).

So far the negative effects of eutrophication have attracted the most interest. Within the framework of the Norwegian Maricult programme (Olsen, this volume) an attempt has been made to see what positive effects have been observed and where they occurred.

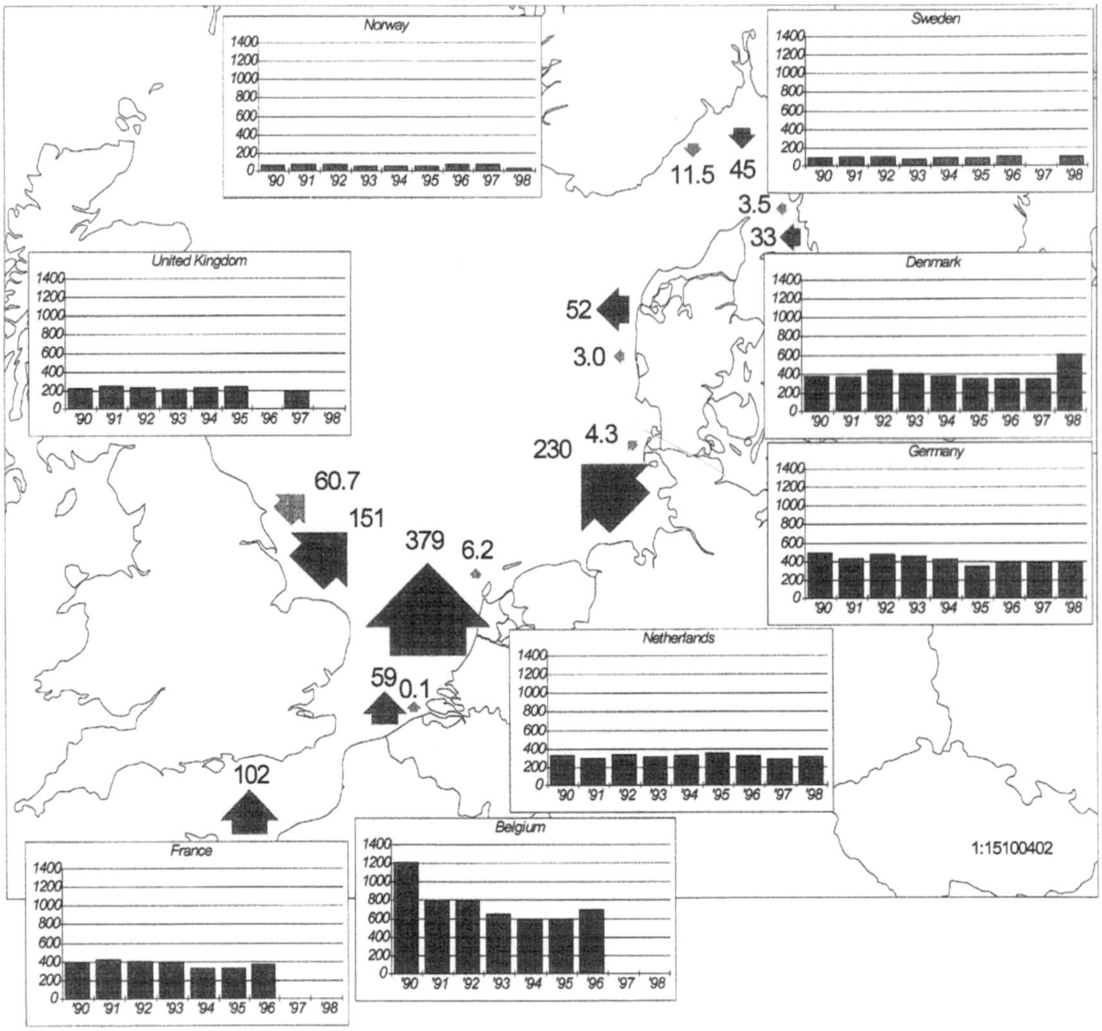

Annotations
Belgium: channel Gent-Terneuzen lacking; France: Seine only; Netherlands: data including polder effluents; Norway: without tributaries; Sweden: without minor rivers and tributaries; United Kingdom: data not complete.

Figure 1. (A) Mean of total annual riverine (dark grey arrows) and direct (light grey arrows) nitrogen inputs [kt N/a] to OSPAR subregion II (Greater North Sea) over the period 1990–98 and annual means of total nitrogen concentrations [μmol N/l] of the riverine discharge calculated from input data and flow rate (RID database).

Positive effects encompass a stimulation of biomass of certain trophic levels, of enhanced productivity or a change in the food web structure, which might positively influence food conversion in the trophic web. The parts discussed here as indicated in Figure 2 are: increase of phytoplankton biomass, benthos and fish standing stocks and changes in species composition as a primary effect. For this purpose we have selected those papers that claim to describe eutrophication effects. The link to eutrophication, however, is not always very clear and decisive, since a causal relationship is hardly ever proven. This also holds true for long-term changes, which nowadays are often coupled to climatic variability.

The number of published papers related to eutrophication of the marine waters of Europe and the Mediterranean Sea over the period 1988–1997 has recently been elaborated by Vidal et al. (1999). They listed a total number of papers dedicated to eutrophication of the North Sea of about 170. Their numbers clearly

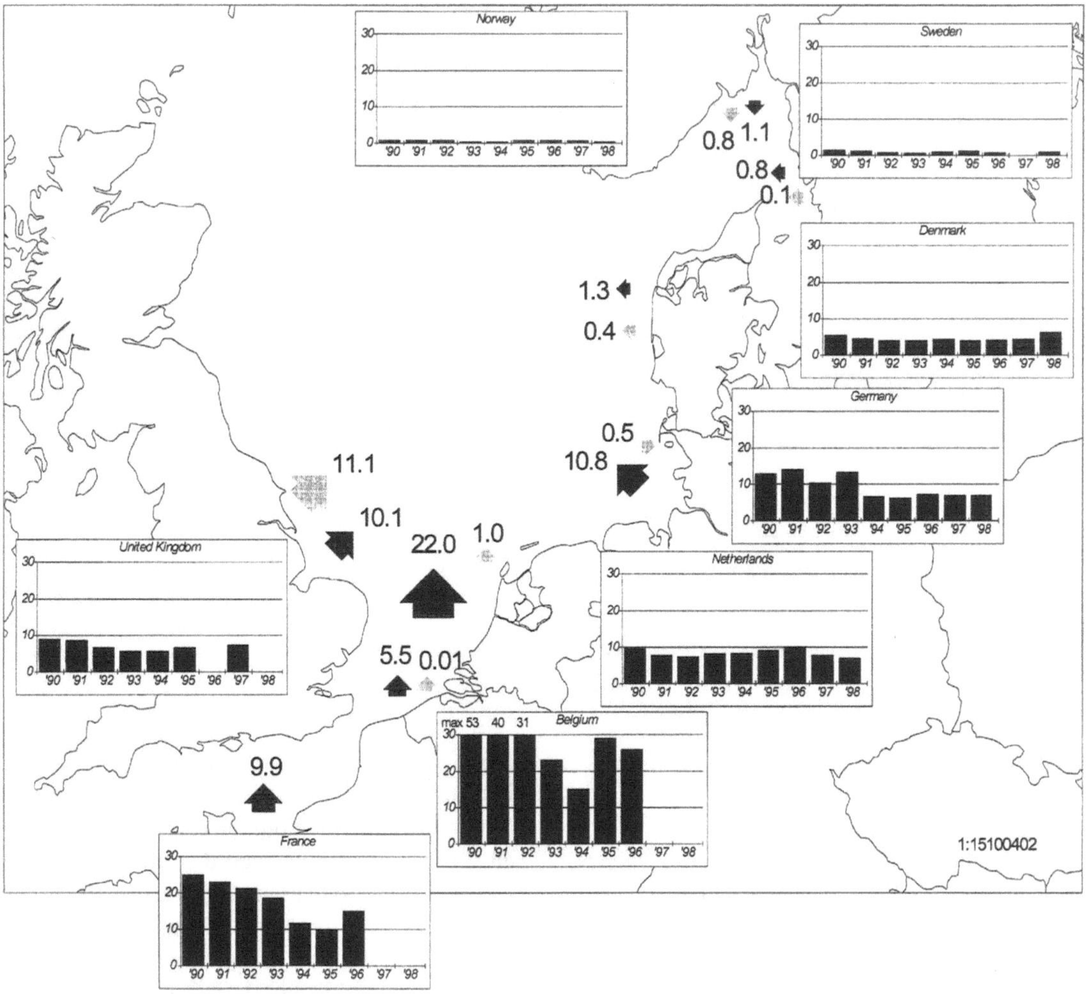

Annotations
Belgium: channel Gent-Terneuzen lacking; France: Seine only; Netherlands: data including polder effluents; Norway: without tributaries; Sweden: without minor rivers and tributaries; United Kingdom: data not complete.

Figure 1. (B) Mean of total annual riverine (dark grey arrows) and direct (light grey arrows) phosphorus inputs [kt P/a] to OSPAR subregion II (Greater North Sea) over the period 1990–98 and annual means of total phosphorus concentrations [μmol P/l] of the riverine discharge calculated from input data and flow rate (RID database).

show the increasing interest over the last decade in eutrophication matters and probably a saturation level can presently be observed. The authors have also presented data to show that eutrophication research is biased towards the North Sea region. However, a clear interest for the Mediterranean Sea has evolved (Vidal et al., 1999). The small amount of papers about the Norwegian Sea and the N.E. Atlantic is not due to a lack of interest, but simply to the fact that those areas are not considered as eutrophied (or in political terms are 'non-problem areas') and are therefore not studied for eutrophication effects.

Our bibliographies reveal over 100 recent titles alone for eutrophication-related studies in the North Sea area (Ladwig, unpubl.). Some of the material used for this review has been compiled in a study on the development of eutrophication criteria in sensitive areas for the German Environmental Agency.

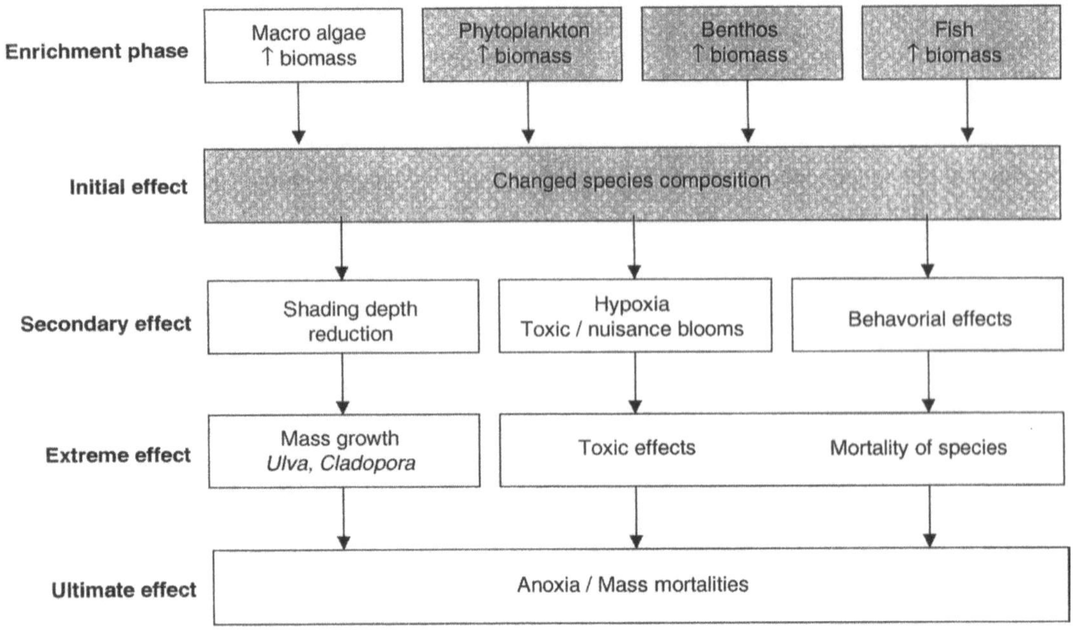

Figure 2. Schematic presentation of eutrophication effects: grey shadowed parts are discussed in detail in the text (changed after Jørgensen & Richardson, 1996).

Phytoplankton biomass and changes in species composition

One of the first studies that showed long-term changes in pelagic systems is the Continuous Plankton Recorder Survey. As an indication of the phytoplanktonic biomass 'greenness', the colour of the silk nets from the recorder, is used. This index has been shown to increase since 1985, but with a high degree of variability for which no simple explanations are available. The CPR does not enable us to correlate these variances directly with other biological or abiotic parameters (Reid et al., 1998), because they are not measured in the CPR itself. Only indirect correlations can be made, using data from other studies.

Analyses made on the basis of the CPR data for smaller parts of the North Sea show differences within the compartments investigated: the most southerly compartments generally show higher phytoplankton biomass values than the more northerly compartments. Differences in timing of blooms also occur as well as long-term changes in the duration of high phytoplankton biomass levels. The method has several limitations, because the areas potentially affected by eutrophication are not well covered during the preponderantly offshore sampling strategy of the CPR tows. Furthermore there is some evidence of a strong climatic influence on the decadal phytoplankton variability in the North Sea as assessed by the CPR surveys. Major changes in biomass and the relative importance of dinoflagellates and diatoms on a wider scale are attributed to variations in atmospheric forcing, as indicated by the North Atlantic Oscillation index (NAO). It is difficult to discriminate among these effects and eutrophication-based changes. Therefore we inspected preferably data from those sites which are directly under the influence of freshwater discharge. These are the well-known eutrophied coastal areas of the European continent: the Dutch, German and Belgian coastal zones, where a number of large European rivers flow into the North Sea.

In a recent paper, Philippart et al. (2000) have analysed the long-term data set for the Marsdiep, a tidal inlet to the Wadden Sea, close to the island of Texel. This long-term station sampled at high tide is considered as being representative for the Dutch coastal zone, but is affected by the Wadden Sea system as well, which in turn is flushed by a tributary of the River Rhine. The data set, which was collected by Cadée & Hegeman, comprises phytoplankton species abundance and composition for a 20-year period (1974–1994) (see refs. in Philippart et al., 2000). The nutrient fluxes in this area have also been measured or calculated based on hydrographic models for the

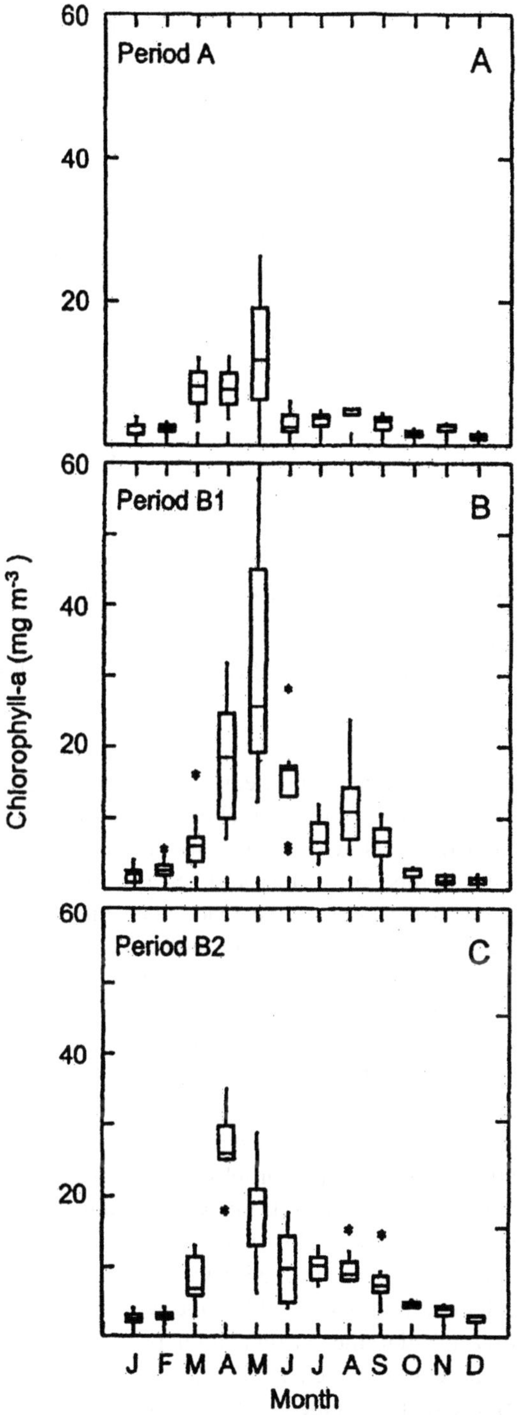

Figure 3. Box-and-whisker plots of seasonal (monthly averaged) time series of chlorophyll-*a* concentrations (mg m^{-3}) of the phytoplankton community in Marsdiep from 1974 to 1994. (A) period 1974–1976; (B) period 1978–1987, (C) period 1988–1994 (from Philippart et al., 2000).

Wadden Sea. The time series was split into three periods, according to the nutrient fluxes. The starting years 1974–1976 (Fig. 3 A) show monthly averaged chlorophyll-*a* concentrations which were highest during the spring bloom in May with a mean slightly above 10 mg m^{-3}; during the second period (Fig. 3 B, 1978–1987) a doubling of the chlorophyll amount occurs in spring with an extended growing season towards the summer, which consequently also exhibits elevated concentrations. During the last period (Fig. 3 C, 1988–1994) median values slightly decreased but phytoplankton biomass remained distinctly higher when compared to the start of the study period.

The nutrient data revealed that there was a shift from a phosphorus-controlled system in the first period to nitrogen control during the second period. A reshift towards phosphorus control occurred after 1987. There was a strong correlation between phytoplankton community structure and TN:TP ratios. Under eutrophic N controlled conditions large diatom forms such as *Rhizosolenia* and *Thalassiosira* spp. proliferated, but flagellates also increased. A Principal Component Analysis (PCA) revealed groups with a clearly different species composition, reflecting the three trophic periods.

The dominance of larger species is supported by theoretical studies i.a. Stolte & Riegman (1995).These large phytoplankton species would then fuel benthic communities after sedimentation.

Several other studies principally support the view given in the paper of Philippart et al. (2000). The interesting point in this study is that the authors link several of the observations to experimental studies on cultures, thereby strengthening the theoretical framework of understanding eutrophication effects in an ecological concept (cf. Sommer et al., 2002).

Studies by Lancelot's group in Belgium have shown that the length of the growing season and the biomass level are strongly correlated to the input of riverine nutrients (Lancelot et al., 1997; Rousseau, 2000). Because of the dominance of *Phaeocystis* and its intense blooms this species has attracted strong interest over the last decade. In their well-known paper, Cadée & Hegeman (1986) showed that the frequency and duration of *Phaeocystis* blooms off the Dutch coast has distinctly increased from 1973 until 1985. Analysis of long-term data (1982–1991) along the German east Frisian coast did not support this trend, but revealed an increase of *Phaeocystis* bloom intensity (Hanslik et al., 1998).

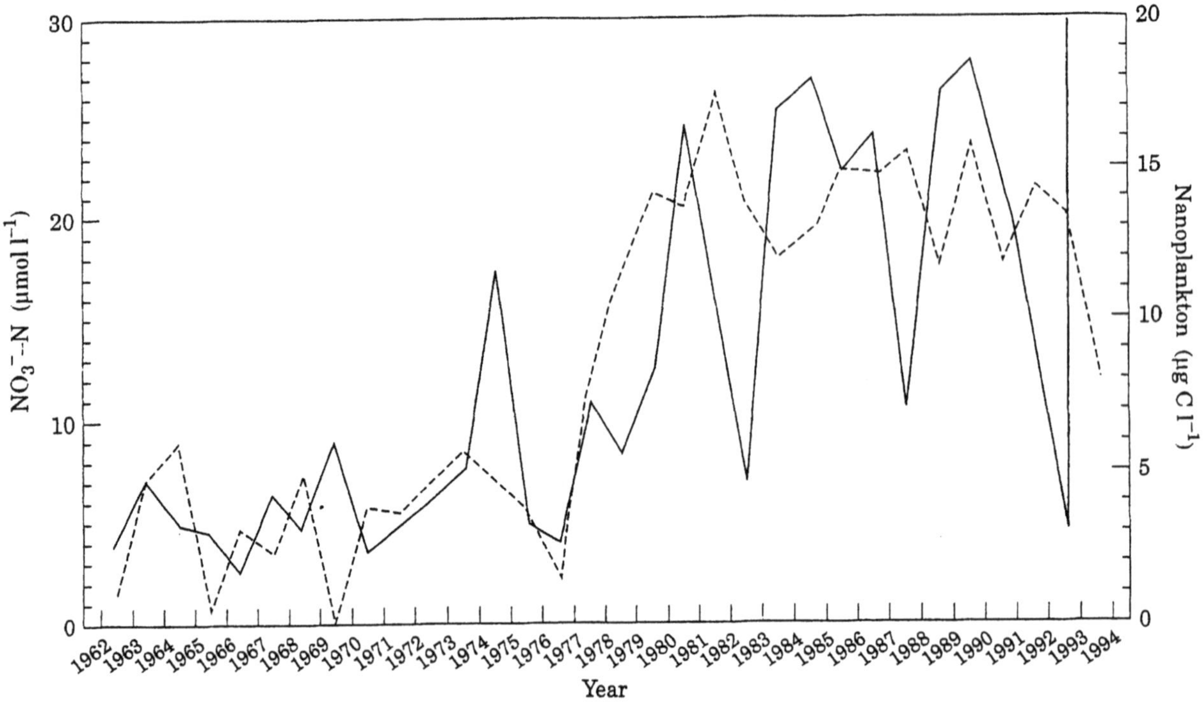

Figure 4. Median values of nanoplankton biomass and nitrate concentrations during winter at Helgoland Roads, 1962–1994 (from Hickel, 1998) (—— = nitrate; - - - - = nanoplankton).

On the basis of long-term data from the permanent station Helgoland Roads, a similar effect to those observed along the Dutch coast has been found for the more offshore part of the German Bight: apart from an increase in total phytoplankton, measured as carbon through estimates of the biovolume, a sudden shift towards flagellates has occurred at the end of the 1970s. This has been attributed to eutrophication, though a causal link to nutrient inputs or concentrations in the sense of an effect-response relationship could not be statistically demonstrated. Figure 4 (Hickel, 1998) combines the median values of the nanoplankton biomass and nitrate concentrations during winter from 1962 to 1994 at Helgoland. An overall increase has been observed in the period 1978 to 1980. Apart from explanations linked to eutrophication, the author suggests a combined effect of several factors, including climatic ones. Obviously the large-scale changes in the North Sea which could be related to the NAO may also interfere with the more subtle eutrophication effects at Helgoland Roads. Hypothesis testing is difficult because the station where the phytoplankton analyses are performed is under the variable influence of the Elbe outflow. The consideration of the German Bight as a highly eutrophied sea area is mostly based on the rather well-known loadings and concentrations of inorganic nutrients. However, long-term and large-scale information about the relevant biological parameters with respect to the eutrophication process in this part of the North Sea is largely missing, partly because of the inadequate intensity and design of specific monitoring activities. Accidental observations of eutrophication symptoms in the area encompass the occurrence of exceptional phytoplankton blooms, oxygen deficiencies and related events (Table 1), which demonstrate the sensitivity of the area to elevated nutrient inputs. Analysis of long-term data from Danish coastal waters reveals a close relationship between chlorophyll-*a* concentration and the amount of nitrogen in the water, being most pronounced in summer (Christensen, 1998). Considering the mean flushing time of 33 days in the German Bight (Lenhart & Pohlmann, 1997), the winter situation is poorly linked to the eutrophication process during the growth season, which is strongly dependent on the complex hydrographic structures and nutrient retention capacities in the coastal area. A main feature in this context is the formation of thermohaline fronts in the inner German Bight, which affect the transport of matter and stimulate phytoplankton mass development and accumulation (Hesse et al., 1995). The usual assessment of the eutrophication status merely based

Table 1. Compilation of exceptional plankton blooms in the German Bight

Species	Year	Concentration [cells/l]	Reference
Ceratium furca	Aug (1981)	$0.5 \cdot 10^6$	Hickel (1983)
	Aug 1987	$0.1 \cdot 10^6$	Westernhagen et al. (1989)
	Jul 1993	$1 \cdot 10^6$	Hagmeier et al. (1995)
Ceratium fusus	Jul 1982	dense stocks	Hagmeier (pers. comm.)
	Aug 1984	$1 \cdot 10^6$	Hagmeier (1985)
Coscinodiscus concinnus	Apr 1985	$0.35 \cdot 10^6$	Gerlach (1990)
	Mar 1996	twice as usual	Hickel (1997)
Coscinodiscus wailesii	1991	480	Hagmeier (1992)
	1993	650	Hagmeier et al. (1995)
	1994	2,000	Hagmeier et al. (1995)
Pseudonitzschia pungens	1992	$1.7 \cdot 10^6$	Hagmeier (1992)
Gymnodinium mikimotoi	Aug 1968	$3 \cdot 10^6$	Hickel et al. (1971)
	Aug 1979	$1 \cdot 10^6$	Doerffer & Amann (1984)
	Jul 1993	124,000	Hagmeier et al. (1995)
Gymnodinium chlorophorum	Aug 1990	$100 \cdot 10^6$	Hagmeier (1991)
	Aug 1990	$1 \cdot 10^6$	Nehring et al. (1995)
Dinophysis norvegica	Aug 1987	10,000	Westernhagen et al. (1989)
Dinophysis acuminata	Aug 1984	$0.16 \cdot 10^6$	Nehring et al. (1995)
Fibrocapsa japonica	Jul 1997	$0.3 \cdot 10^6$	Rademaker et al. (1998)
Chatonella verruculosa	May 1998	high	Göbel (pers. com.)
Prorocentrum triestinum	1993	51,000	Hagmeier et al. (1995)
	1994	30,000	Hagmeier et al. (1995)
Phaeocystis globosa	regularly	$200 \cdot 10^6$	divers
Myrionecta rubra	Jun 1984	$0.06 \cdot 10^6$	Hagmeier (1985)
	Jul 1994	$3 \cdot 10^6$	Hagmeier et al. (1995)
Noctiluca scintillans	regularly	70,000	divers

on winter concentrations of inorganic nutrients seems thus questionable.

Primary production

In their review of primary production numbers for the North Sea, van Beusekom & Diel (1994) have shown that there are large regional differences, which seem to be dependent on the trophic gradients. A combined picture (Fig. 5) from different sources has been compiled by de Vries et al. (1998). It shows a positive productivity gradient along with rising concentrations of dissolved inorganic nitrogen in different parts of the North Sea.

Apart from an increase in phytoplankton biomass, one might expect an increase of the growth rate along with a better nutrient supply only if the phytoplankton community suffered from nutrient limitation. That the yield has increased in many coastal regions can

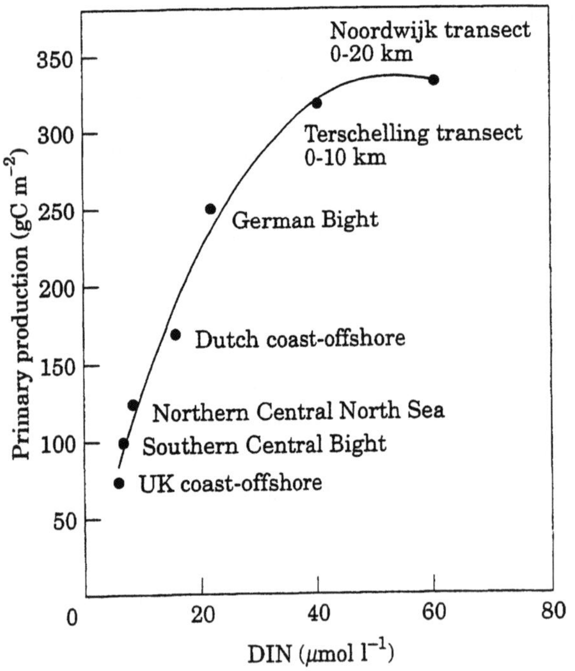

Figure 5. Relation between annual primary production and winter DIN concentrations within the North Sea (after de Vries et al., 1998).

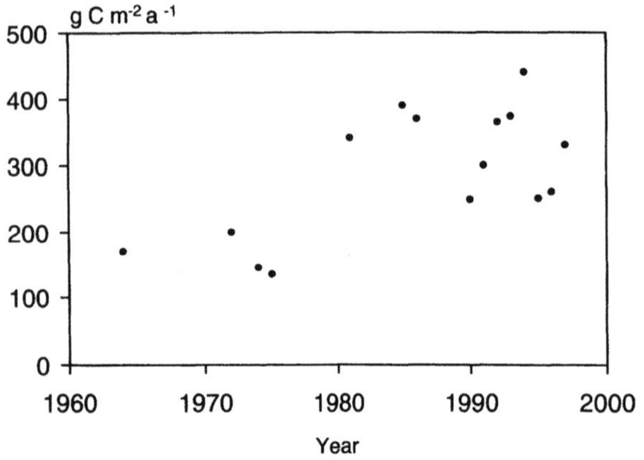

Figure 6. Annual primary production of phytoplankton for the Marsdiep area (Cadée & Hegeman, 1993; Cadée, pers. comm.).

be suggested from most long-term datasets. However, concerning primary production there are only few long-term estimates available. One is that of Richardson & Heilmann (1995) for the Kattegat area, which strongly suggests an increase in primary production over the last decade. In the Marsdiep Cadée & Hegeman (1993; recent data, pers. comm) showed a clear increase of the primary production from 1950 until the late nineties (Fig. 6).

Even though a reduction in nutrient loadings, especially in phosphate inputs has occurred since the nineties, the data reveal that the primary production in the Marsdiep area has hardly been reduced; this certainly has something to do with the role of light limitation, as recently presented in a model by Cloern (1999), as well as with the adaptive capability of the phytoplankton community to different nutrient ratios.

It seems that in the Wadden Sea – despite the progress in phosphorus reduction – nutrients are still plentiful enough (in relation to light availability) to sustain primary production at such a high level. This is partially due to the temporal storage capacity of Wadden Sea sediments with respect to phosphate, and the subsequent release of large amounts of phosphorus from the sediments during summer (Hesse et al., 1992; Dick et al., 2000; Ladwig et al., in press). In the northern German Wadden Sea, this export may be on average a factor of 6 higher than the phosphate input via the river Elbe at that time of the year.

Recently, Tillmann et al. (2000) studied potential light and nutrient limitation in the northern German Wadden Sea, at a site close enough to the Elbe river to be under the direct estuarine impact. Effects of DIN, phosphate and silicate were compared with light limitation, defined as a balanced value between average light availability in the mixed water column ($E_{average}$) and the light requirement at saturating intensity (E_k) of the actual phytoplankton community. The measurements show that light limitation is the most important factor affecting primary production in the northern Wadden Sea, whereas nutrient limitation, defined by k_s values from the literature occurs only occasionally (Fig. 7).

Because of the pronounced difficulties in the interpretation of field data especially with respect to natural variability and man-made impacts, scientists have started to combine field investigations with laboratory studies and/or mesocosms to mimic the field situation under controlled conditions. Such studies have been performed within the EU/COMWEB project and also in the Netherlands in the Eastern Scheldt (Prins et al., 1999) in order to investigate the effects of nutrient reductions. It is interesting to see that some scientists have performed mesocosm experiments with the aim to analyse the effects of nutrient additions, whereas others try to deduce the effects of nutrient reductions. The types of results are, however, mostly quite comparable. In mesocosm studies within the EU/COMWEB project Duarte et al. (2000) investigated the development of different phytoplankton size fractions in relation to different nutrient inputs. They found a clear shift from picoplankton-dominated to

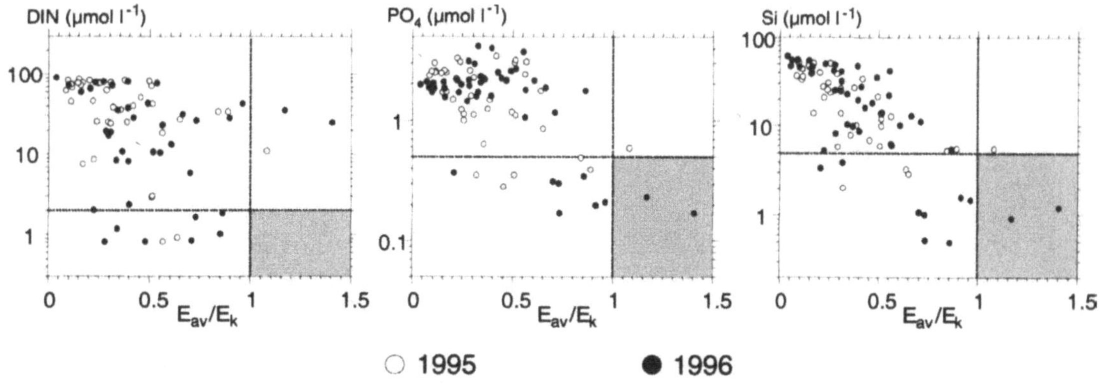

Figure 7. Plot of E_{av}/E_k with ambient nutrient concentrations. The dotted area defines the region for which $E_{av}/E_k > 1$ ('light saturation') and nutrient concentrations are below published half-saturation constants for nutrient uptake of natural populations (from Tillmann et al., 2000).

Table 2. Primary production in mesocosms (g C m^{-2}) under different loadings (after Prins et al. 1999)

Loading	Spring (85 days)	Summer (119 days)
High	82	80
Medium	72	18
Low	44	11

microplankton-dominated phytoplankton with rising nutrient loadings, supporting the observations made by Philippart et al. (2000). Due to differential grazing, the evolution of the total biomass was enhanced only at relatively high nutrient loadings (2.54 and 0.12 μmol N and P d^{-1} or higher), whereas the production increased very strongly.

In their studies on effects of nutrient reductions in Dutch coastal waters, Prins et al. (1999) simulated reduction in nutrient loadings (N, P) by 50% (medium) and 75% (low), respectively, in comparison to the ambient concentrations in the Dutch coastal waters in 1995 (high). Si-loadings were at equal rates. The treatments resulted in large differences in chlorophyll-*a* content and these biomass differences (Fig. 8) were reflected in the measured primary production in the mesocosms: the cumulative primary production over two different periods clearly shows that especially the primary production in the mesocosms in summer was very low in the'medium' and 'low' loaded systems, but much more elevated in the 'high' loaded assay, which corresponded to the actual ambient conditions (Table 2). In spring the differences were less, but the gradient in nutrient loadings reflected the primary production values quite well. In all the experiments diatoms formed the main biomass component as compared to dinoflagellates and nanoflagellates. In general, their biomass exceeded the others by a factor of five. However, dinoflagellate development was enhanced along with rising N- and P-loading.

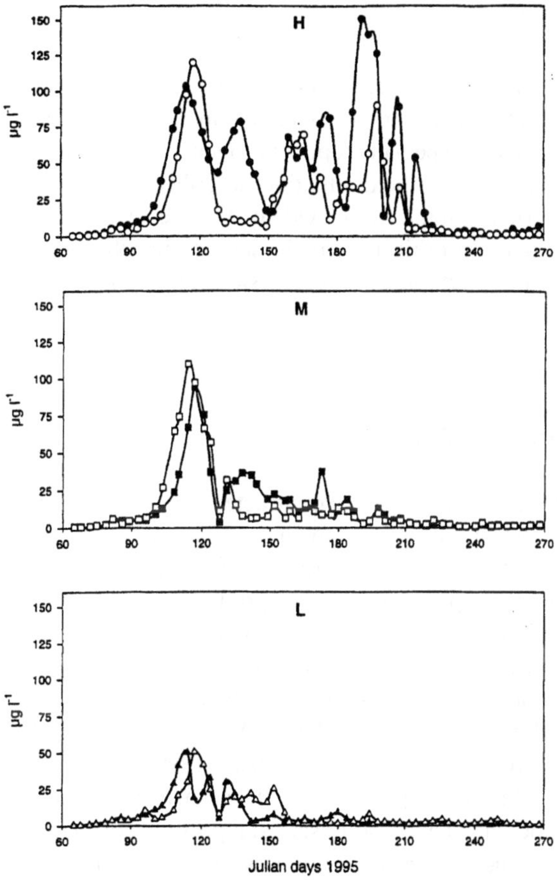

Figure 8. Mesocosm chlorophyll-*a* concentrations in duplicate treatments High, Medium and Low representing different loading regimes (after Prins et al., 1999).

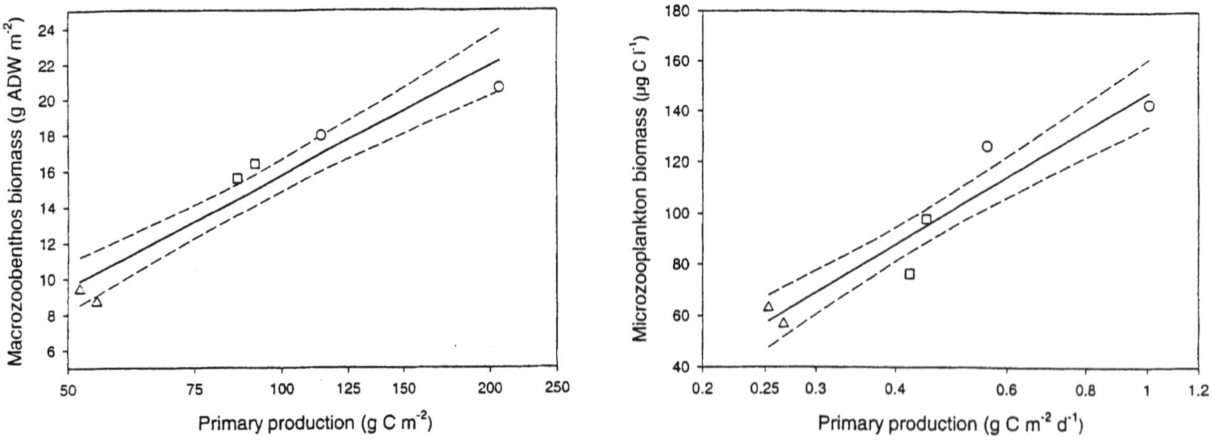

Figure 9. Relation between microzooplankton biomass and primary production (right panel), and between macrozoobenthos biomass and primary production (left panel) under different nutrient loading regimes (circles: High; squares: Medium; triangles: Low) (after Prins et al., 1999).

There are other mesocosm studies with natural sea water which end up with massive flagellate development, especially upon nitrogen enrichment, thus reflecting the field situation described above for the German Bight. It is evident that the availability of silica is a critical factor in determining the outcome of the phytoplankton succession (cf. Sommer, 1994; Rousseau, 2000).

Macrozoobenthos

Macrozoobenthos have attracted relatively much interest in eutrophication studies, but more because of negative extreme or ultimate effects than as an indicator group for an enhanced production. Such negative effects encompass massive kills of macrozoobenthos during periods of oxygen deficiency, especially in stratified bottom waters (German Bight: Rachor & Albrecht, 1983; Dethlefsen & Westernhagen, 1983; Fjords in Denmark: Møhlenberg, 1999).

In their experiments with mesocosms Prins et al. (loc. cit.) have also observed a positive correlation of the macrozoobenthos biomass and the microzooplankton biomass as well with primary production. At least under the conditions of these experiments the relations seem to be linear (Fig. 9).

Beukema & Cadée (1997) analysed field data from the Dutch Wadden Sea based on long-term studies and observed positive correlations between food supply and the biomass of macrozoobenthos (Fig. 10). The authors stated that food availability when tested on individual stations only affects benthic populations, which are not stressed by other environmental conditions like exposure to waves or high sediment mobility.

Another example is also available from long-term data and recurrent spatial surveys in the German Bight (Salzwedel et al., 1985; Rachor, 1990; Schröder & Knust, 1998). A clear tendency towards increasing biomass levels can be observed when data from the years 1966, 1975 and 1983 are compared to the early macrozoobenthos survey of Hagmeier (1925) in 1923/24. This primarily holds true for mud and muddy-sand habitats. A more recent survey from 1995 supports this situation (Thatje & Gerdes, 1997).

There are indications that the response of macrozoobenthos to increasing organic loads in the coastal North Sea fits quite well with the Pearson–Rosenberg model (1978), which describes an increase in biomass along with a better food supply and, with ongoing enrichment, a shift to opportunistic species and a subsequent decrease due to harmful effects such as hypoxia, anoxia and immune stress, caused by acidification. A shift to small, opportunistic forms has been observed in sedimentation areas, which are especially susceptible to eutrophication, such as the mud hole in the German Bight east of Helgoland (Rachor pers. comm.).

Zooplankton, shrimps and fish

It is difficult to find clear evidence for eutrophication effects in the pelagic food chain of the North Sea. Fransz et al. (1992) describe the density of adult copepods (*Temora longicornis*) in the Marsdiep for a 20-year period (Fig. 11). They conclude that an increase of the adult population has occurred, which is

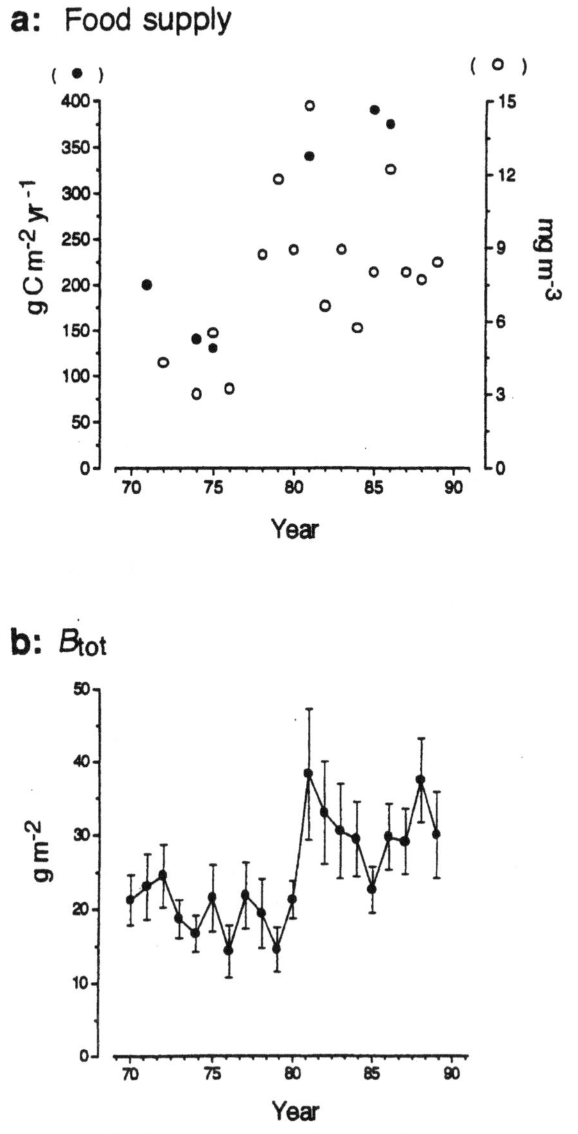

Figure 10. Changes in food supply (a) and biomass (b) of macrozoobenthos in the Dutch Wadden Sea (from Beukema & Cadée, 1997).

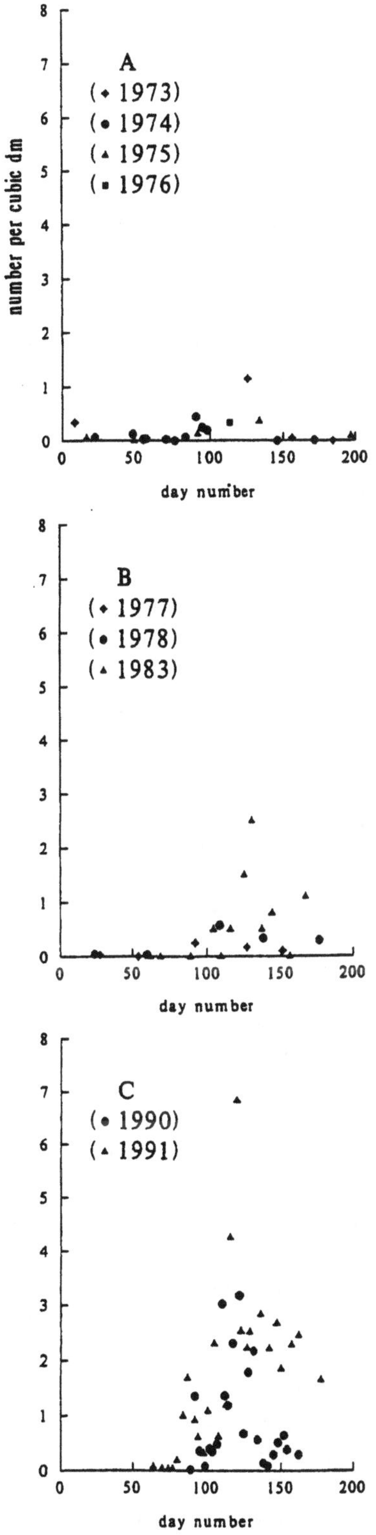

Figure 11. Abundance of *Temora longicornis* adults in Dutch coastal waters in three different periods A, B and C (from Fransz et al., 1992).

linked to the eutrophication of these coastal waters (see above).

Boddeke (1996) studied the distribution and growth rate of brown shrimps in Dutch coastal waters and concluded that during the period of nutrient reductions shrimp stocks were negatively affected. In Figure 12 the distribution of commercial shrimps (size above 54 mm) off the Dutch coast in the period 1980–1982 is compared to that of 1991–1993, at a time when phosphate reduction was effective. The author concluded that in 1991–1993 food availability limited shrimp recruitment and densities further offshore. However,

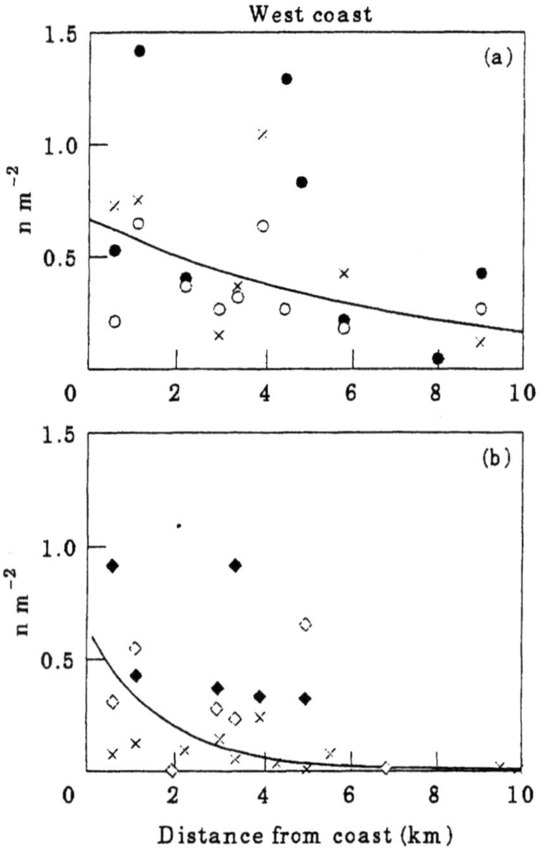

Figure 12. Relation between densities of commercial shrimps (>54 mm) and distance from the shore in two different periods (a: 1980–1982; b: 1991–1993), symbols represent different years (from Boddeke, 1996).

without actually knowing the food availability and requirements this assumption remains speculative.

Because of the complexity in the relationship between eutrophication and fisheries, it is difficult to untangle the interactions. Nielsen & Richardson (1996) compared the fisheries yield in the Kattegat region with changes in primary production, and concluded that the observed increase in production fits with the increased fish yield (Fig. 13).

Rijnsdorp & van Leeuwen (1996) showed that there is a positive significant correlation between the annual length increment of North Sea plaice and the nutrient level (based on ambient phosphate concentrations). However, as other factors strongly affect the growth rates as well, such as temperature and seabed disturbance, no definite conclusion with respect to eutrophication can be drawn from this investigation.

These and many other observations, like the compilation of Micheli (1999), support our simplified picture that, according to the former food-chain concept of Steele, primary production is often positively correlated with fish production, resulting in yields in the order of less than 1% of the primary production, depending on the physical and biological characteristics of the system. However, the enhanced formation of organic material deriving from anthropogenic nutrient inputs may result in oxygen deficiency in the bottom layer, to acidification of the water and increased concentrations of carbon dioxide (hypercapnia). Altogether these are factors which may promote the loss of spawning grounds, emigration of fish or morbidity and mortality of less motile species. Some of these effects already occur at moderate hypoxic conditions. It has been shown that the feeding rate and the growth rate of some flatfish (flounder) decreases significantly at 50% oxygen saturation (Breitburg pers., comm.).

Other effects of oxygen deficiency encompass modifications of the predator–prey interactions, such as the loss of suitable prey organisms for demersal fish in the aftermath of benthic mass mortalities (Hickel et al., 1989) and – as a top-down effect – modifications in the grazing pressure on fish larvae under low oxygen conditions.

In contrast to the situation in the world's main hypoxic areas as, e.g. the Gulf of Mexico, where decreases of summer spawning demersal fish species were observed (Chesney et al., 2000), no such observations exist for the North Sea. Due to the physical properties of this shallow area, periods of oxygen deficiency are limited in space and time. Major events occurred in coastal regions of the German Bight in the years 1981–83, 1989, and 1994 (Fig. 14). It is of note that these observations were mostly a byproduct of research projects focusing on other topics and were therefore rather accidental. Because of the lack of adequate monitoring activities, no information exists on the temporal and spatial variability of hypoxia, especially in the offshore regions of the North Sea.

Conclusions

The balance between the positive and negative effects of nutrient enrichment on yields of wild fish and aquaculture is as yet poorly understood. A main difficulty remains the discrimination between effects of anthropogenic origin and natural variability, especially with respect to management concepts of fisheries resources.

There is sound evidence that positive effects of increased nutrient inputs on fish recruitment and harvest are strongly dependent on the degree of nutrient limitation in the system and on the physical properties

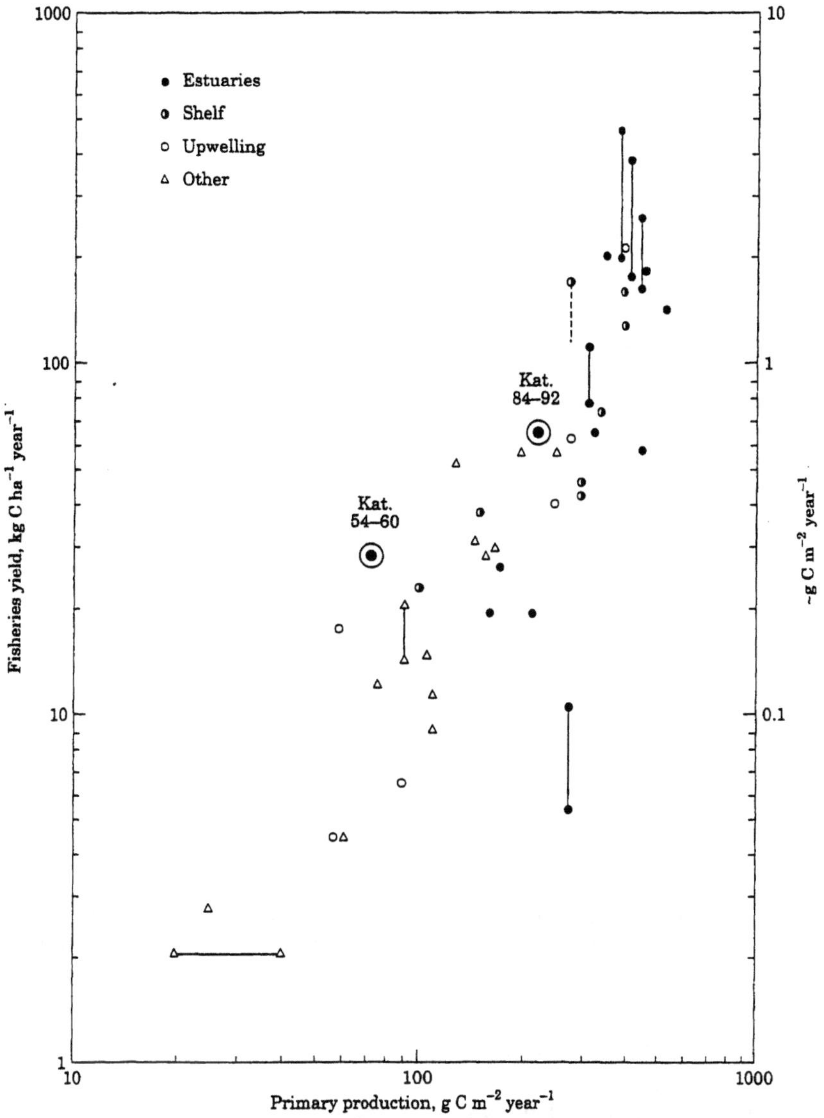

Figure 13. Fisheries yield as a function of primary production in general and the Kattegat in particular (from Nielsen & Richardson, 1996).

of the sea area. It is very likely that, in oligotrophic seas, moderate nutrient enrichment will be of profit to top predators and that in well-mixed areas characterised by high flushing rates effects of low oxygen on fish and shellfish are of minor importance. However, special attention should be given to changes in N/P/Si ratios, which are decisive for the species composition. A rising N/P-ratio may result in the proliferation of undesirable blooms such as *Phaeocystis* (Riegman, 1995), or in the stimulation of algal toxin formation, at least in a few toxic species belonging to the generus *Alexandrium* (Anderson et al., 1990; John & Flynn, 2000).

In the early *in situ* fertilisation experiments in the Scottish sea lochs Craiglin and Kyle Scotnish (Gross et al., 1950), it has been shown that eutrophication, up to a certain level, may be of benefit to benthos and fish yield. The critical thresholds of nutrient overload leading to the manifold observed deterioration of the system has to be determined site-specifically, according to the eutrophication susceptibility of the receiving water body, with primary focus on the different hydrodynamics. First concepts encompass sensitivity factors relating nutrient inputs to residence time (Quinn et al., 1989). Other supporting factors related to the site-specific distribution of light and nutrient resources, such as stratification parameters, critical depth and remineralisation intensity, have to be considered in the future.

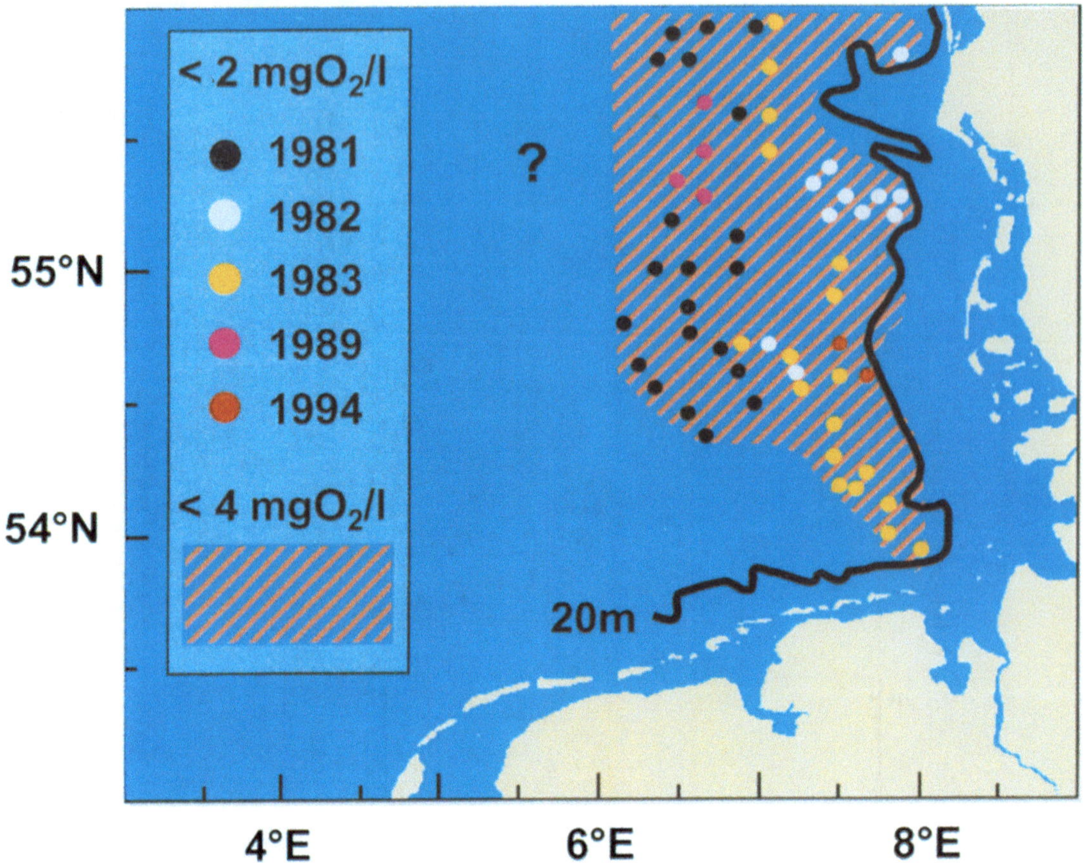

Figure 14. Oxygen deficiency areas in the German Bight 1981–1994 (after Rachor & Albrecht, 1983; Dethlefsen & Westernhagen, 1983; Niermann 1990; KUSTOS project 1994). The hatched area was temporarily affected by low oxygen concentrations in at least one of the years considered.

Turbulent diffusion and stratification intensity are especially critical not only to the oxygen regime and organic carbon flux in the bottom layer, but also to the proliferation of dinoflagellate blooms, which own the main potential as toxin producers. It has been shown that turbulence may inhibit growth in a variety of dinoflagellate species independent of nutrient and light availability (Estrada & Berdalet, 1997).

In keeping with this, in turbulent regions such as the Wadden Sea the phytoplankton community is largely dominated by diatoms, in spite of high concentrations in both inorganic and organic dissolved nitrogen and phosphate compounds in relation to silicate. Absolute amounts of inorganic nutrients are usually enough so as not to limit primary production, with exception of short periods of phosphate and silicate limitation in spring, and nitrogen and silicate limitation in summer.

For the German Bight, efforts in nutrient point source reduction undertaken over the last decade have resulted in salinity standardised (34 PSU) winter concentrations of phosphorus close to those observed by Kalle 1935/36 (ibid., 1937; Gaul, 2000). Despite of successful P-reduction, the situation of an altered phytoplankton community characterised by a shift towards flagellates and an overall increase in biomass continues to persist in the coastal area of the German Bight. This may result in a changed edibility of phytoplankton and impact the coupling with other trophic levels. There is increasing evidence, however, that the observed changes have a multicausal origin and cannot be traced back to eutrophication alone, but also to changing atmospheric forcing circulation patterns related to the North Atlantic Oscillation (NAO). Considering the non-linear behaviour of nature, it seems unlikely that the system will return to its former state through further nutrient reduction.

Acknowledgements

We thank the MARICULT team for the invitation to present this paper during the first MARICULT conference in Trondheim in June 2000. Furthermore, we thank the Umweltbundesamt for support for N. Ladwig during this study and the BMBF for support to U. Tillmann in the SYKON project. M. Reckermann is acknowledged for support in editing and printing several figures.

References

Anderson, D. M., D. M. Kulis, J. J. Sullivan, S. Hall & C. Lee, 1990. Dynamics and physiology of saxitoxin production by the dinoflagellates *Alexandrium* spp. Mar. Biol. 104: 511–524.

Beukema, J. J. & G. C. Cadée, 1997. Local differences in macrozoobenthic response to enhanced food supply caused by mild eutrophication in a Wadden Sea area: Food is only locally a limiting factor. Limnol. Oceanogr. 42(6): 1424–1435.

Beusekom, J. van & S. Diel-Christiansen, 1994. A synthesis of phyto- and zooplankton dynamics in the North Sea environment. World Wildlife Fund, UK.

Boddeke, R., 1996. Changes in the brown shrimp (*Crangon crangon* L.) population off the Dutch coast in relation to fisheries and phosphate discharge. ICES J. Mar. Sci. 53: 995–1002.

Cadée, G. C. & J. Hegeman, 1986. Seasonal and annual variation in *Phaeocystis pouchetii* (Haptophyceae) in the westernmost inlet of the Wadden Sea during the 1973 to 1985 period. Neth. J. Sea Res. 20: 29–36.

Cadée, G. C. & J. Hegeman, 1993. Persisting high levels of primary production at declining phosphate concentrations in the Dutch coastal area (Marsdiep). Neth. J. Sea Res. 31: 147–152.

Chesney, E. J., D. M. Baltz & R. G. Thomas, 2000. Louisiana estuarine and coastal fisheries and habitats: Perspectives from a fish's eye view. Ecol. Applic. 10 (2): 350–366.

Christensen, P. B. (ed.), 1998. The Danish Marine Environment: has action improved its state? Havforsking fra Miljøstyrelsen 62: 1–120.

Cloern, J. E., 1999. The relative importance of light and nutrient limitation of phytoplankton growth: a simple index of coastal ecosystem sensitivity to nutrient enrichment. Aquat. Ecol. 33: 3–16.

Dethlefsen, V. & H. v. Westernhagen, 1983. Oxygen deficiency and its effects on bottom fauna in the eastern German Bight 1982. ICES C. M. 1983/L:2.

Dick, S., U. H. Brockmann, J. E. E. van Beusekom, B. Fabiszisky, M. George, U. Hentschke, K.-J. Hesse, B. Mayer, T. Nitz, T. Pohlmann, K. Poremba, K. Schaumann, W. Schönfeld, A. Starke, U. Tillmann & G. Weide, 1999. Exchange of matter and energy between the Wadden Sea and the coastal waters of the German Bight – Estimations based on numerical simulations and field measurements. Dtsch. Hydrogr. Z. 51 (2/3): 181–219.

Doerffer, R. & V. Amann, 1984. The development of the horizontal distribution of a North Sea phytoplankton bloom. Special meeting on causes, dynamics and effects of exceptional marine blooms and related events, ICES, Copenhagen, 4–5 October 1984.

Duarte, C. M., S. Agusti & N. S. R. Agawin, 2000. Response of a Mediterranean phytoplankton community to increased nutrient inputs: a mesocosm experiment. Mar. Ecol. Progr. Ser. 195: 61–70.

Estrada, M. & E. Berdalet, 1997. Phytoplankton in a turbulent world. In Marrasé, C. et al. (eds), Lectures on Plankton and Turbulence. Scientia Marina 61 (Suppl. 1): 125–140.

Fransz, H. G., S. R. Gonzalez, G. C. Cadée & F. C. Hansen, 1992. Long-term change of *Temora longicornis* (Copepoda, Calanoida) abundance in a Dutch tidal inlet (Marsdiep) in relation to eutrophication. Neth. J. Sea Res. 30: 23–32.

Gaul, H., 2000. MURSYS – Nährstoffkonzentrationen in der küstennahen Zone und in der Deutschen Bucht. http://www.bsh.de/Meeresumweltschutz/Mursys/

Gerlach, S. A., 1990. Nitrogen, phosphorus, plankton and oxygen deficiency in the German Bight and in Kiel Bay – final report on the project 'Eutrophication of the North Sea and the Baltic Sea'. Kieler Meeresforsch. Sonderh. Vol 7.

Gross, F., S. R. Nutman, D. T. Gauld & J. E. G. Raymont, 1950. A fish cultivation experiment in an arm of a sea loch I–V. Proc. Roy. Soc. Edinb. B 64: 1–135.

Hagmeier, A., 1925. Vorläufiger Bericht über die vorbereitenden Untersuchungen der Bodenfauna der Deutschen Bucht mit dem Petersen-Bodengreifer. Ber. dt. wiss. Komm. Meeresforschung 1: 247–272.

Hagmeier, E., 1985. Mikroplankton in der deutschen Bucht. Jahresb. 1984 der Biologischen Anstalt Helgoland.

Hagmeier, E., 1991. Mikroplankton in der deutschen Bucht. Jahresb. 1990 der Biologischen Anstalt Helgoland.

Hagmeier, E., 1992. Phytoplankton bei Helgoland im Sommer 1991. Jahresb. 1991 der Biologischen Anstalt Helgoland: 45–46.

Hagmeier, E., 1992. Zum Auftreten der Kieselalge *Coscinodiscus wailesii* im Sommer 1991. Jahresb. 1991 der Biologischen Anstalt Helgoland: 46 pp.

Hagmeier, E., W. Hickel, P. Mangelsdorf & K. Treutner, 1995. Räumliche Ausbreitung von Mikroplankton-Populationen und Nährstoffen. Zweijahresbericht 1993/94 der Biologischen Anstalt Helgoland: 20–23.

Hanslik, M., J. Rahmel, M. Bätje, S. Knieriemen, G. Schneider & S. Dick, 1998. Der Jahresgang blütenbildender und toxischer Algen an der niedersächsischen Küste seit 1982. Umweltbundesamt Texte, 1998-3.

Hesse , K.-J., U. Hentschke & U. H. Brockmann, 1992. A synoptic study of nutrients and phytoplankton characteristics in the German Wadden Sea with respect to eutrophication. In Colombo, G. et al. (eds), Marine Eutrophication and Population Dynamics. Proc. 25th EMBS, Olsen & Olsen, Fredensborg: 45–53.

Hesse, K.-J., U. Tillmann, S. Nehring & U. Brockmann, 1995. Factors controlling phytoplankton distribution in coastal waters of the German Bight (North Sea). In Eleftheriou, A., A. D. Ansell & C. J. Smith (eds), Biology and Ecology of Shallow Coastal Waters. Olsen & Olsen, Fredensborg, Denmark: 11–22.

Hickel, W., 1983. *Ceratium furca* – 'red tide' in der Deutschen Bucht (August 1981). Jahresb. 1982 der Biologischen Anstalt Helgoland.

Hickel, W., 1997. Räumliche Ausbreitung von Mikroplankton-Populationen und Nährstoffen. Zweijahresber. der Biologischen Anstalt Helgoland 1995/1996: 20–22.

Hickel, W., 1998. Temporal variability of micro- and nanoplankton in the German Bight in relation to hydrographic structure and nutrient changes. ICES J. Mar. Sci. 55: 600–609.

Hickel, W., E. Hagmeier & G. Drebes, 1971. *Gymnodinium* blooms in the Helgoland Bight (North Sea) during August, 1968. Helgol. Wiss. Meeresunters. 22: 401–416.

Hickel, W., E. Bauerfeind, U. Niermann & H. von Westernhagen, 1989. Oxygen deficiency in the south-eastern North Sea: Sources and biological effects. Ber. Biol. Anst. Helgol. 4: 143.

Jørgensen, B. B. & K. Richardson (eds), 1996. Eutrophication in Coastal Marine Ecosystems. Coastal and Estuarine Studies 52, American Geophysical Union, Washington.

John, E. H. & K. J. Flynn, 2000. Growth dynamics and toxicity of *Alexandrium fundyense* (Dinophyceae): the effect of changing N:P supply ratios on internal toxin and nutrient levels. Eur. J. Phycol. 35: 11–23.

Kalle, K., 1937. Nährstoff-Untersuchungen als hydrographisches Hilfsmittel zur Unterscheidung von Wasserkörpern. Annalen der Hydrographie und Maritimen Meteorologie. 65: 1–18.

Ladwig, N., K.-J. Hesse, F. Colijn & U. Tillmann, in press. Have German Wadden Sea waters changed their eutrophic state over the last 10 years? ICES Mar. Sci. Symp.

Lancelot, C., V. Rousseau, G. Billen & D. van Eeckhout, 1997. Coastal eutrophication of the Southern Bight of the North Sea: assessment and modelling. NATO-ASI Series 439–454.

Lenhart, H. J. & T. Pohlmann, 1997. The ICES box approach in relation to results of a North Sea circulation model. Tellus 49A: 139–160.

Micheli, F., 1999. Eutrophication, fisheries, and consumer-resource dynamics in marine pelagic ecosystems. Science 285 (5432): 1396–1398.

Møhlenberg, F., 1999. Effect of meteorology and nutrient load on oxygen depletion in a Danish micro-tidal estuary. Aquat. Ecol. 33 (1): 55–64.

Nehring, S., K.-J. Hesse & U. Tillmann, 1995. The German Wadden Sea: a problem area for nuisance blooms? In Lassus, P. et al. (eds), Harmful Marine Algal Blooms: 199–204.

Nielsen, E. & K. Richardson, 1996. Can changes in the fisheries yield in the Kattegat (1950–1992) be linked to changes in primary production? ICES J. Mar. Sci. 53: 988–994.

Niermann, U. 1990. Oxygen deficiency in the south eastern North Sea in summer 1989. ICES C.M. 1990: 18 pp.

Olsen, Y., 2002. MARICULT Research Programme: background, status and main conclusions. Hydrobiologia 484/Dev. Hydrobiol. 167: 1–10.

Pearson, T. H. & R. Rosenberg, 1978. Macrobenthis succession in relation to organic enrichment and pollution of the marine environment. Oceanogr. Mar. Biol. Ann. Rev. 16: 229–311.

Philippart, C. J. M., G. C. Cadée, W. van Raaphorst & R. Riegman, 2000. Long-term phytoplankton-nutrient interactions in a shallow coastal sea: algal community structure, nutrient budgets, and denitrification potential. Limnol. Oceanogr. 45(1): 131–144.

Prins, T. C., V. Escaravage, L. P. M. J. Wetsteyn, J. C. H. Peeters & A. C. Smaal, 1999. Effects of different N- and P-loadings on primary and secondary production in an experimental marine ecosystem. Aquat. Ecol. 33: 65–81.

Quinn, H. A., J. P. Tolson, C. J. Klein, S. P.Orlando & C. Alexander, 1989. Strategic assessment of near coastal waters. Susceptibility of east coast estuaries to nutrient discharges: Passamaqouddy Bay to Chesapeake bay. Summary Rep., Strategic Assessment Branch. NOS/NOAA, Rockville, MD: 37 pp.

Rachor, E., 1990. Changes in sublittoral zoobenthos in the German Bight with regard to eutrophication. Neth. J. Sea Res. 25(1/2): 209–214.

Rachor, E. & H. Albrecht, 1983. Sauerstoff-Mangel im Bodenwasser der Deutschen Bucht. Veröff. Inst. Meeresforsch. Bremerh. 19: 209–227.

Radach, G., W. Schönfeld & H. Lenhart, 1990. Nährstoffe in der Nordsee Eutrophierung, Hypertrophierung und deren Auswirkungen. In Lozan, J. L. et al. (eds), Warnsignale aus der Nordsee. Parey, Berlin.

Rademaker, M., M. Reckermann, U. Tillmann, A. Tillmann, F. Colijn, W. Zevenboom & P. Houpt, 1998. *Fibrocapsa japonica* and *Heterosigma akashiwo*: new observations. Harmful Algae News 17: 8–10.

Reid, P. C., M. Edwards, H. G. Hunt & A. J. Warner, 1998. Phytoplankton change in the North Atlantic. Nature 391: 546.

Richardson, K. & J. P. Heilmann, 1995. Primary production in the Kattegat: past and present. Ophelia 41: 317–328.

Riegman, R., 1995. Nutrient-related selection mechanisms in marine phytoplankton communities and the impact of eutrophication on the planktonic food web. Wat. Sci. Technol. 32: 63–75.

Rijnsdorp A. D. & P. I. van Leeuwen, 1996. Changes in growth of North Sea plaice since 1950 in relation to density, eutrophication, beam-trawl effort, and temperature. ICES J. mar. Sci. 53: 1199–1213.

Rousseau, V., 2000. Dynamics of *Phaeocystis* and diatom blooms in the eutrophicated coastal waters of the Southern Bight of the North Sea. Ph.D. Thesis , Free University Brussels.

Salzwedel, H., E. Rachor & D. Gerdes, 1985. Benthic macrofauna communities in the German Bight. Veröff. Inst. Meeresforsch. Bremerh. 20: 199–267.

Schröder, A. & R. Knust, 1998. Long term trends in demersal fish and benthic invertebrates. In Lindeboom, H. & S. J. de Groot (eds), The Effects of Different Types of Fisheries on the North Sea and the Irish Sea Benthic Ecosystems. NIOZ 1998-1.

Sommer, U., 1994. Are marine diatoms favoured by high Si:N ratios? Mar. Ecol. Progr. Ser. 115: 309–315.

Sommer U., H. Stibor, A. Katechakis, F. Sommer & T. Hansen, 2002. Pelagic food web configurations at different levels of nutrient richness and their implications for the ratio fish production:primary production. Hydrobiologia 484/Dev. Hydrobiol. 167: 11–20.

Stolte, W. & R. Riegman, 1995. The effect of phytoplankton cell size on transient state nitrate and ammonium uptake kinetics. Microbiology 141: 1221–1229.

Thatje, S. & D. Gerdes, 1997. The benthic macrofauna of the inner German Bight: present and past. Arch. Fish. Mar. Res. 45(2): 93–112.

Tillmann, U. , K.-J. Hesse & F. Colijn, 2000. Planktonic primary production in the German Wadden Sea. J. Plankt. Res. 22(7): 1253–1276.

Vidal, M., C. M. Duarte & M. Carmen Sanchez, 1999. Coastal eutrophication research in Europe: Progress and Imbalances. Mar. Poll. Bull. 38(10): 851–854.

Vries, I. de, R. N. M. Duin, J. C. H. Peeters, F. J. Los, M. Bokhorst & R. W. P. M. Laane, 1998. Patterns and trends in nutrients and phytoplankton in dutch coastal waters: comparison of timeseries analysis, ecological model simulation, and mesocosm experiments. ICES J. Mar. Sci. 55: 620–634.

Westernhagen, H. v., W. Hickel, E. Bauerfeind, U. Niermann, 1987. Sauerstoff-Defizit in tiefen Rinnen der Deutschen Bucht: Ursachen und biologische Auswirkungen. BMFT Abschlussbericht MFU 05384.

Westernhagen, H. v., W. Hickel, E. Bauerfeind, U. Niermann, 1989. Sauerstoff-Zehrung im Bodenwasser der Deutschen Bucht. BMFT Abschlussbericht MFU 05514.

Hydrobiologia **484:** 149–166, 2002.
O. Vadstein & Y. Olsen (eds), Sustainable Increase of Marine Harvesting: Fundamental Mechanisms and New Concepts.

Littoral macrofauna (secondary) responses to experimental nutrient addition to rocky shore mesocosms and a coastal lagoon

Patrik Kraufvelin[1,*], Hartvig Christie[2] & Marianne Olsen[3]
[1]*Environmental and Marine Biology, Åbo Akademi University, Akademigatan 1, FIN-20500 Turku/Åbo, Finland;*
[2]*Norwegian Institute for Nature Research (NINA), P.O. Box 736 Sentrum, N-0105 Oslo, Norway*
[3]*Norsk Hydro, Research Centre Porsgrunn, P.O. Box 2560, N-3901 Porsgrunn, Norway (*Author for correspondence) Fax. +358-2-2153428. E-mail: patrik.kraufvelin@abo.fi*

Key words: coastal eutrophication, rocky shores, macrofauna, BACIP, mesocosm, whole-ecosystem experiment

Abstract

Macrofauna secondary responses to controlled eutrophication within two MARICULT/MAST-III projects, EULIT and COMWEB, are compared. EULIT utilises a nutrient gradient established in eight hard bottom mesocosms, whereas the data from COMWEB originate from a whole-ecosystem study – the case of experimental nutrient addition to Hopavågen lagoon. In both systems, nutrient addition started in May 1998, after initial studies of the background macrofauna communities, thus allowing application of Before-After-Control-Impact-Pairs techniques (BACIP). The main objectives have been to investigate the macrofauna responses to eutrophication in littoral rocky shore ecosystems and to evaluate if similar responses could occur in mesocosm and 'field' systems, despite their inherent differences. Apart from a distinct increase in numbers of *Littorina littorea* L. and some indications of increased abundance of the genus *Jaera*, no significant changes, caused by 2.5 years of nutrient addition, could be detected in the mesocosm fauna. It is interesting, however, that these two possibly stimulated animal groups have two things in common: (1) both belong to the rather few mobile taxa, which are not flushed out of the system through the mesocosm outlets, (2) both feed on microalgae and green algae in the upper littoral zone, i.e., algal groups that have shown the clearest response to the nutrient addition. In Hopavågen, 1.5 years of nutrient addition only caused modest plant and animal responses in the rocky shore ecosystem, although there was a markedly increased settlement of *Mytilus edulis* L. at one lagoon site. This increase was probably due to elevated levels of digestible particles (increased phytoplankton production) caused by the nutrient addition. There were also some indications of increased production of filamentous algae at this same lagoon site. All in all, both studies demonstrate only minor responses to increased nutrient levels within the littoral community (both plants and animals). One explanation to this may be that longer time-scales are needed in order for clear-cut changes to occur. Several observations also indicate that wave exposure (significant in both systems) may largely modulate the impact of increased nutrient load on the structure of littoral communities. Another explanation is the high degree of stability among littoral macroalgae-dominated communities with internal biological regulation factors (like grazing) possibly being able to counteract effects.

Introduction

Increased nutrient inputs have led to a general eutrophication of the European coastal zone, with most notorious effects in the North Sea and the Baltic Sea (Baden et al., 1990; Bonsdorff et al., 1997ab). While the understanding of the responses to increased nutrient levels generally is adequate for the pelagic zone, seagrass dominated ecosystems and soft sediments (Duarte, 1995), present knowledge of the response of rocky shore ecosystems is largely qualitative. In rocky littoral communities, high nutrient levels are generally considered to stimulate the growth rate of photosynthetic organisms and to increase the primary production of the ecosystem. An increased production will, however, not only be visible as an increased growth rate of the existing algae, but also as a replacement of slow-growing species by faster-growing species (Bokn et al., 1992; Pedersen & Borum, 1996). This changed species composition of benthic algae (primary effect) may result in changes in abundance and quality of food (type of species and nutrient content) for herbivores as well as habitat changes for the littoral fauna (Seed & O'Connor, 1981; Edgar, 1990) and may in turn cause changes in the animal community (secondary effects). Nevertheless, the critical nutrient load needed to exceed the buffering capacity of rocky shore ecosystems and cause either increased production or undesirable effects is presently unknown. Direct experimental evidence of causality in connection with eutrophication of rocky shore communities, as well as information on the significance of the spatial and temporal scales involved, have to our knowledge not been published previously. Quantitative knowledge of the eutrophication process derived from controlled large-scale and long-term experimentation may serve to remedy this situation.

Rocky shores, dominated by macroalgal beds, are important components of northern seas and have a high primary production, which in turn supports a complex and highly productive animal community (Moore & Seed, 1985). The fauna of these ecosystems consists of both sessile and motile invertebrates attached or associated to both the rocky bottom and the algae. The macrofauna community is mainly composed of filter-feeders (mussels, barnacles, serpulid polychaets, bryozoans) herbivores and detrivores (gastropods, amphipods, isopods, sea-urchins) and carnivores (polychaets, crabs, starfish and fish). As for the herbivores, which depend on the algal growth, the filter-feeders and detrivores in benthic algal communities are also dependent on the organic particles produced by the benthic algae (Robertson & Lucas, 1983; Duggins & Eckman, 1994), in addition to the particles of planktonic origin. Thus the major components of the fauna are expected to be dependent on the development of the algal community and to respond to nutrient addition subsequently to responses within plant communities.

Mesocosms may be useful tools for establishment of causal relationships, since they allow controlled experimentation and it is possible to replicate experimental units and to randomly disperse treatments and controls (Lundgren, 1985; Gearing, 1989; Clark & Cripe, 1993; Lawton 1995). At the same time every mesocosm investigation will suffer from its intermediate position – caught in the middle between realism and control (Kraufvelin, 1998, 1999). Full-scale experiments (whole-ecosystem studies), on the other hand, operate at a higher, more realistic, spatial scale and should thus be better suited for ecosystem management tasks and for understanding and prediction of the ecology of ecosystems (Carpenter, 1999). However, each approach has inherent weaknesses and no approach alone is able to provide a complete understanding of how natural ecosystems function. Therefore the problems would ideally be met by a combination of several approaches (Drenner & Mazunder, 1999; Huston, 1999).

The present study aims to produce novel scientific knowledge by combination of two long-term and large-scale MARICULT/MAST-III experiments, EULIT (Effects of Eutrophicated Seawater on Rocky Shore Ecosystems Studied in Large Littoral Mesocosms) and COMWEB (Comparative Analysis of Food Webs Based on Flow Networks: Effects of Nutrient Supply on Structure and Function of Coastal Plankton Communities). Littoral macrofauna responses along a nutrient gradient established in eight hard bottom mesocosms at Marine Research Station Solbergstrand (EULIT) 1996–2000 will be compared to littoral macrofauna responses in a whole-ecosystem context – the case of experimental nutrient addition to Hopavågen lagoon in central Norway (COMWEB) 1997–1999. In both systems, addition of extra nutrients started in May 1998, after initial studies of the background macrofauna communities thus allowing application of a Before-After (BACI) analytic (e.g., Stewart-Oaten et al., 1986, 1992). The main objectives of this study have been to investigate secondary responses to eutrophication in rocky shore ecosystems and evaluate if similar responses could occur in mesocosm tanks and in the 'field', despite the inherent differences of these two systems and differences in the magnitude of the added nutrient levels. Within EULIT, many other studies have been accomplished on the rocky shore ecosystem (e.g., Bokn et al., 2002), whereas within COMWEB/Hopavågen whole-ecosystem experiment, which mainly has been focused on planktonic communities (see other Hopavågen papers of this issue), the present study is the only one concerned with littoral communities (both animals and plants).

Material and methods

Solbergstrand rocky shore mesocosms

The mesocosm facility is situated at the Marine Research Station Solbergstrand by the Oslofjord (59° 37′ N, 10° 40′ E), southeast Norway, and belongs to the

Norwegian Institute for Water Research (NIVA). The Solbergstrand mesocosm and the main methods have previously been described by Bakke (1990), Bokn et al. (1998, 2001, 2002) and Kraufvelin et al. (submitted), but some information relevant to this paper will be recapitulated below.

Eight concrete land-based mesocosms were set up in early summer 1996. The communities were established by transplantation of small boulders with attached macroalgae and associated animals from the Oslofjord to steps built at the eastern side of each mesocosm. During the initiation phase, natural community development was allowed to contribute to the flora and fauna assemblages. During this pilot period, all eight mesocosms acted as controls and the methodology and experimental design for EULIT were employed to test the efficiency and to provide a background data set. Nutrients were added as a mixture, which consisted of 14.3 mol N as NH_4NO_3 and 0.9 mol P as H_3PO_4 and had a N/P mol ratio of 16/1, along a geometrical gradient corresponding to 1, 2, 4, 8, 16 and 32 μmol l^{-1} N (hereafter referred to as concentration 01, 02, 04, 08, 16 and 32) as well as 0.06, 0.12, 0.25, 0.5, 1.0 and 2.0 μmol l^{-1} P above background levels. Two basins continued to serve as controls without extra nutrient addition (00). The mean summer background nutrient levels in the water inlet (from May to September each year 1998–2000) were about 8 μmol l^{-1} nitrogen and 0.2 μmol l^{-1} phosphorus (Bokn et al., 2003). Since nutrient addition started in May 1998, the total experimental period consisted of two background years and 2.5 years with elevated nutrient levels in August 2000. The mesocosms have a water volume of 7–13 m^3 depending on tide level and they are fed with water from 1 m depth in the Oslofjord, which also acts as a source for spores, zygotes and larvae. The mean water residence time has been about three hours. A tidal regime with 36 cm amplitude has been maintained and waves have been generated constantly (18 strokes per minute). The maximum depth of the mesocosms is 1 m. To ensure that the physical and chemical properties in the mesocosms satisfy the experimental design, oxygen concentration, temperature and salinity have been recorded continuously and nutrients on an average weekly basis both in the water intake and in the outflow water (Bokn et al., 2002).

The abundance and biomass of littoral macrofauna in each basin were normally investigated during three campaigns each year (spring, summer and autumn). With 'littoral' in this paper we refer to the intertidal zone and the uppermost 30 cm of the subtidal zone. The amount of large motile and all sessile animals was determined by counting inside a frame (grid) system on the steps in all eight basins. Smaller mobile animals (macrofauna and meiofauna) were estimated by collection of artificial substrates, consisting of three ropes, one stone and one petri dish, tied closely together. This substrate was able to catch an intermediate fraction of the animals normally present in the most dominating mesocosm algal types (green, red and brown algae). Four + four replicate substrates at two different depths/steps (one intertidal and one subtidal step just below the tidal zone) were exposed for 2 days in each basin. Three artificial substrates on each step were analysed for fauna content non-destructively, whereas one artificial substrate from each step was sampled destructively for more detailed analysis of animal diversity as well as for analysis of meiofauna. The number of individuals per basin could be estimated by extrapolation from artificial substrates to algal communities with knowledge of the total amount of algae and previous information about the distribution of animals among the major algal groups as well as in the artificial substrates. Animal biomass was estimated from standardised weights of the different species/taxa.

Hopavågen coastal lagoon

The effects of enhanced nutrient supply to coastal waters were also examined in a landlocked bay, Hopavågen, located in central Norway (63° 35′ N, 09° 30′ E). Hopavågen has a surface area of 400 000 m^2, a total volume of 5.4 million m^3, and a volume of euphotic waters of 3.7 million m^3. The depth of the main basin is 22–32 m, and the bottom is relatively flat. The average depth is 20 m. The water exchange (between the landlocked bay and the coastal waters outside) is primarily controlled by tidal forces. The tidal range outside Hopavågen varies between 75 and 225 cm. Water exchanges through a narrow and shallow inlet and can easily be monitored. Daily water exchange is variable and dependant on at least tidal cycle and air pressure, but averages 0.61 $\pm$ 0.22 million m^3 or 11 $\pm$ 4% of the total volume day^{-1}. This corresponds to 19% of the productive waters per day. Due to the narrow inlet, the tidal range in Hopavågen is very limited compared to that outside – 0.3–1.0 m. The volume of freshwater runoff from the areas surrounding Hopavågen is restricted. Hopavågen is not affected by pollution.

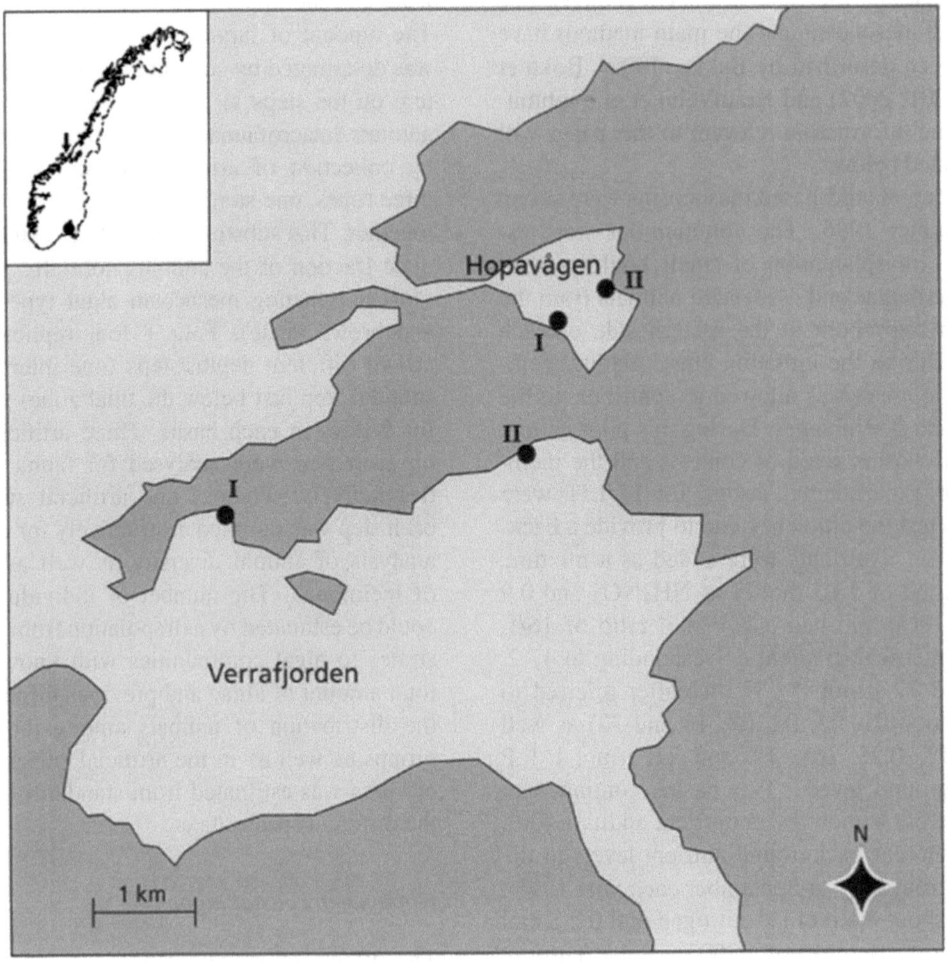

Figure 1. Map of Hopavågen and Verrafjorden with the investigated areas. On the inset map of Norway, the Hopavågen area 63° 35′ N, 09° 30′ E) is marked with an arrow and Marine Research Station Solbergstrand (59° 37′ N, 10° 40′ E) by a black circle.

The experiment took place over a 3-year period, where 1997 and the spring of 1998 provided data for an undisturbed situation. Artificial eutrophication during the first year was initiated 5 May 1998 and was terminated 12 October 1998. During the second year (May–October 1999) the dose was doubled. Nutrients were added at every rising tide (approximately twice a day) into the narrow inlet where water is exchanged with the coastal water outside. To assure thorough mixing of the added nutrients, the nutrients were added during the early period of water inflow. The dose of nutrients corresponded to 0.014 μmol P l^{-1} day^{-1} 1998 (0.028 μmol P l^{-1}day^{-1} in 1999), and with a 14:5.4:1 molar ratio N:Si:P. The dosing was calculated on the basis of sedimentation rates measured in 1996 and 1997, with the aim of increasing the nutrients entering Hopavågen by 100% (200% in 1999). N was supplied as NH_4NO_3, P as H_3PO_4 and Si as waterglass; $Na_3Si_2O_7$, all nutrients in aqueous solutions. When compared to the Solbergstrand mesocosms, Hopavågen thus represents a much lower nitrogen addition, with its 0.2 μmol N l^{-1} (1998) and 0.4 μmol N l^{-1} (1999) added, than any of the Solbergstrand mesocosms (all $\geq$1 μmol N l^{-1}). The halted nutrient addition to Hopavågen during winter is an additional difference, since the dosing was continuous at Solbergstrand.

To facilitate the ability to discover changes in the littoral community, several different investigations were initiated. *In situ* qualitative and quantitative analysis of littoral macroalgae and macrofauna were done in the tidal zone (and in the uppermost 30 cm of the subtidal zone) at two localities inside Hopavågen (Hopavågen I and II) and two localities in the nearby Verrafjorden (Verrafjorden I and II), which acted as reference sites (Fig. 1). The macroalgae associations and the sessile or slow-moving large fauna of the four localities were investigated non-destructively by use of frames (10 replicate squares, 0.5 × 0.5 m) by direct counts or estimated coverage at two vertical levels in the littoral zone (upper/mid littoral and low littoral/upper sublittoral zone). The position of the first square was marked at all localities to enable investigation of the same areas during subsequent sampling. This was done because random sampling of communities with highly patchy distribution gives too high variability between sampling dates. Square analyses were performed in October 1997, in April, May, August and October 1998, as well as in July and October 1999, thus generating a time series with three measurements from the Before period and four measurements from the After period. Quantitative and qualitative data of mobile herbivore and detrivore invertebrates, associated to the macroalgae, were also achieved from three replicate algal samples collected lowest in the tidal zone at each locality. This was done by 'seaweed washing', after which each plant was standardised to a displacement volume of 500 ml. *Fucus vesiculosus* L. was chosen at three localities, whereas at Hopavågen II only *Halidrys siliquosa* (L.) Lyngbye could be found in 1997 and 1998. In 1999 *Fucus vesiculosus* could be found at all stations and this year *Fucus vesiculosus* was also sampled at Hopavågen II, in addition to *Halidrys*. This sampling was performed in October 1997, in August and October 1998, as well as in July and October 1999.

Statistical analysis

BACIP one-way ANOVA was the major univariate technique used for hypothesis testing, but also regression analysis was utilised. At Solbergstrand, differences in abundance of some key-species, e.g., *Jaera* spp. and *Littorina littorea*, along the nutrient gradient were of special interest. The BACIP design proposed here means that repeated sampling takes place 'simultaneously' in the two controls and each of the six impacts (treatments) before and after the start of nutrient dosage. The analytical approach, suggested by Stewart-Oaten et al. (1986, 1992), is to calculate the differences between the mean values for controls and each of the treatments on each sampling date (termed 'delta') and to test by one-way ANOVA

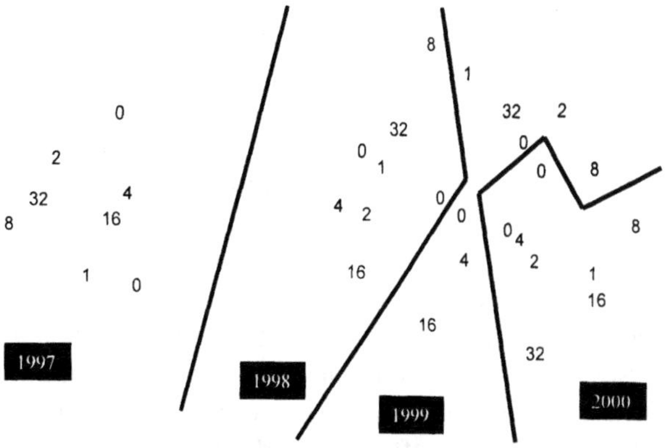

Figure 2. NMDS-ordination of Solbergstrand mesocosms (labelled as the nitrogen concentration added). The ordination is based on species/biomass listings (fourth root transformed values) as investigated by frame analysis and artificial substrates, stress, 0.12. Results from the sampling in late summer 1997, 1998, 1999 and 2000 are presented. Solid lines were drawn to separate the samples from different years from each other.

whether the means of the deltas change from before to after periods. The ANOVAs were performed after checking for normality with Kolmogorov Smirnov's test and homogeneity of variances with Levene's test (Sokal & Rohlf, 1995). When necessary, values were $\log_{10}(x+100)$ transformed or $\log_{10}(x+1000)$ transformed. Addition of 100 or 1000 was necessary to deal with negative differences (deltas) between treatments and controls. The BACIP for *L. littorea* was built on a series of seven Before measurements and six After measurements, in which July 1998 was the last Before measurement and September 1998 the first After measurement (Kraufvelin et al., submitted). For the genus *Jaera*, five Before measurements and eight After measurements were utilised (due to later start of the sampling in the Before period, but more frequent sampling in the After period). BACIP techniques were also applied on Hopavågen frame data, with special interest to percentage cover of *Mytilus edulis* and filamentous macroalgae (arcsin transformed values) and *L. littorea* abundance, coming up with three Before measurements and four After measurements for all three variables.

Also multivariate statistical techniques were applied. These were non-parametric methods as recommended for biological data by Clarke (1993). To obtain a reasonable balance between dominant and rare species, all data were transformed prior to analysis. A fourth root transformation was used at Solbergstrand, whereas Hopavågen data were square root transformed. Spatial and temporal differences among Solbergstrand mesocosms (animal biomass data) were tested by use of a two-way crossed analysis of similarity without replication (ANOSIM2), after which non-metric multidimensional scaling (NMDS) was used for graphical presentation of between or within group similarities. The differences between localities and development over time for animal community structure (abundance data) at Hopavågen and Verrafjorden were tested by a two-way crossed ANOSIM and also presented graphically by NMDS. All univariate analyses were carried out on SPSS 7.5 for Windows, whereas multivariate analyses were carried out using the PRIMER software (Clarke & Warwick, 1994; Carr, 1996).

Table 1. BACIP one-way ANOVA on *Jaera* spp. abundance using the five measurements until July 1998 as Before values and the eight measurements from September 1998 until August 2000 as After values. Original values were transformed by $\mathrm{Log}_{10}(x + 1000)$ in order to deal with unequal variances. Both uncorrected significance values and values corrected by a sequential Bonferroni are given: $^{*}0.01 < P \leq 0.05$;$^{**}\ 0.001 < P \leq 0.01$

Concentration	df	*F*	Uncorrected sign. level	Corrected sign. level
01	1, 11	0.035	0.8558 ns	0.8558 ns
02	1, 11	10.132	0.0083**	0.0498*
04	1, 11	6.813	0.0243*	0.1215 ns
08	1, 11	1.001	0.3386 ns	0.6772 ns
16	1, 11	4.850	0.0500*	0.2000 ns
32	1, 11	2.600	0.1351 ns	0.4053 ns

Results

Solbergstrand mesocosms

The mesocosms hosted at least 80 species/taxonomic groups of macrofauna during the study period (Appendix 1). Of these taxa, >40% were crustaceans (33 taxa) and almost 30% were molluscs (22 taxa). Other taxa include: bristle worms (six taxa); sea anemones and fish (three taxa each); bryozoans, echinoderms and ascidians (two taxa each); sponges, flatworms, ribbon worms, oligochaets, arachnids, insects and sea-spiders (one taxon each). The meiofauna of the mesocosms consisted mainly of Copepoda, Ostracoda, Nematoda and Halacaridae. Regarding the plants, 47 species of macroalgae were registered in the mesocosms during the study period. Of these, 21 species were red algae, 16 species were brown algae and 10 species were green algae, although most of the plant biomass was made up by *Fucus vesiculosus* and *Fucus serratus* L. The plant community is more thoroughly outlined by Bokn et al. (2002).

There were no overall effects of the nutrient addition on animal biodiversity and biomass in the Solbergstrand mesocosms (as analysed by frames and artificial substrates). A two-way crossed ANOSIM without replication on animal biodiversity data (using species/biomass listings) after the start of nutrient addition (July and September 1998, May, August, October 1999 as well as August 2000), gives a *P* value of 0.397. Temporal (seasonal/annual) differences are very clear, however, with a *P* value less than 0.001. The changes in animal community structure with time (or an ongoing ecological succession) in all eight basins are illustrated by an NMDS-ordination using animal species/biomass data 1997–2000 with frame counts and estimations from artificial substrates combined (Fig. 2). Only results from the late summer sampling each year are presented to facilitate comparisons between years and basins. The only indication of a response to the nutrient addition in this figure (note that 1997 samples were background samples in the absence of extra nutrient addition), is the fact that the two controls cluster out at one side of the year 2000 subgroup while the three highest concentrations (08, 16 and 32) cluster out at the opposite end, but this is probably purely coincidental. As a whole, no effects of nutrient addition occurred in macrofauna community structure, no matter which variable that was examined (species composition, species/abundance data or species/biomass data). Regarding meiofauna, which had a very insignificant biomass compared to that of the macrofauna, there were neither any differences along the nutrient gradient.

Nevertheless, when looking into single animal populations, significant changes due to the nutrient addition can be found. Both the small isopod of the genus *Jaera* and the common periwinkle, *Littorina littorea* (L.) show stimulated abundance values along the gradient. In addition, there were indications of stimulated (although not statistically significant) abundance values along the nutrient gradient for the dominating amphipod *Gammarus locusta* L.

For *Jaera*, the stimulated abundance is significant when tested by a BACIP one-way ANOVA (with five Before and eight After measurements) for the low nutrient additions 02 (uncorrected $P = 0.008^{**}$) and 04 ($P = 0.024^{*}$), as well as the second highest nutrient addition 16 ($P = 0.050^{*}$). After correction for the

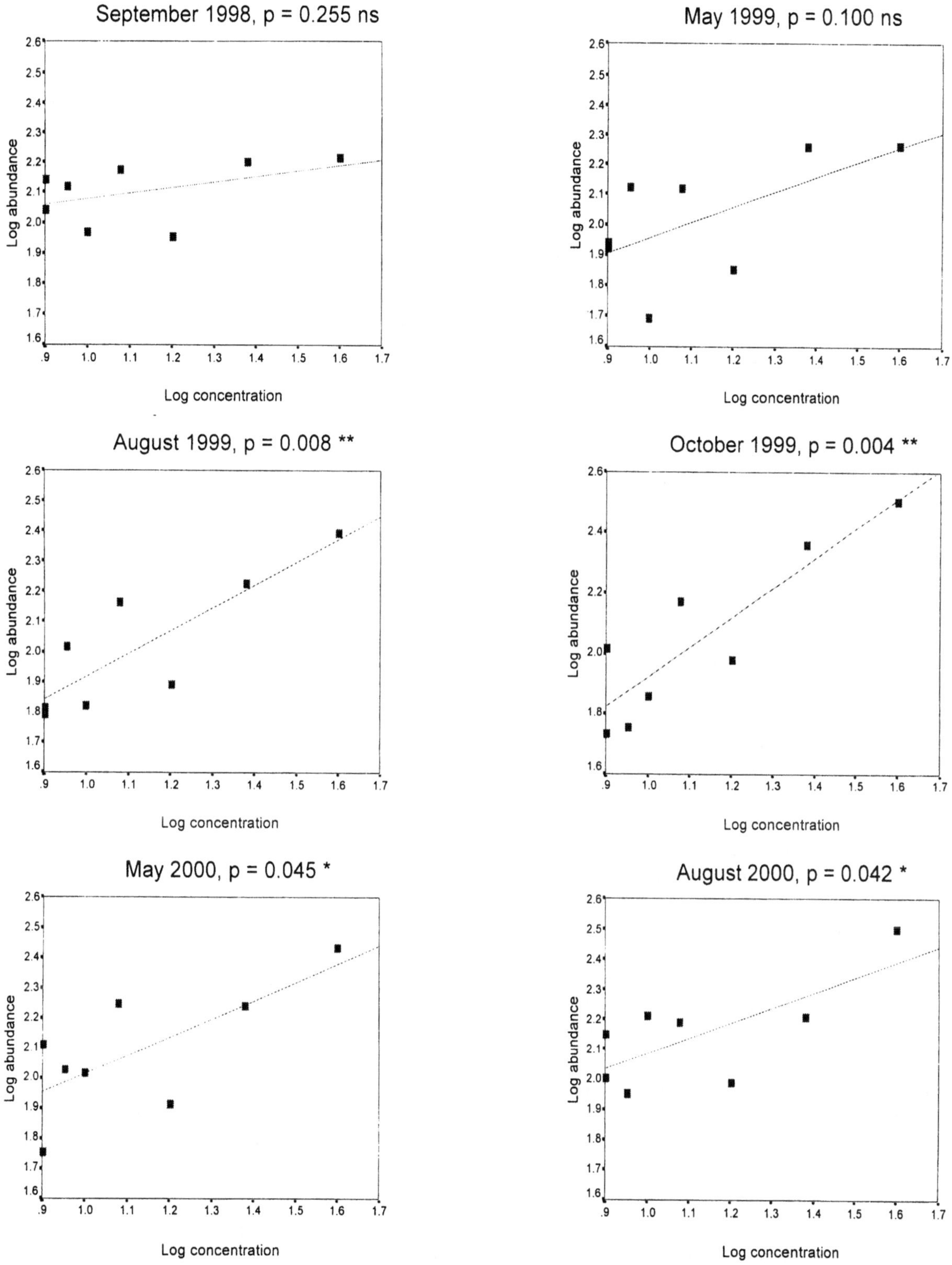

Figure 3. Regression curves over time for *Littorina littorea* abundance values against mean annual nitrogen concentrations in Solbergstrand mesocosms (an average summer background level of 8 μM N added). Both axes were logarithmed, $\log_{10}(x+1)$.

Table 2. BACIP one-way ANOVA on *Littorina littorea* abundance in Solbergstrand mesocosms using the seven measurements until July 1998 as Before values and the six measurements from September 1998 until August 2000 as After values. Original values were transformed by $\log_{10}(x + 100)$ in order to deal with unequal variances. Both uncorrected significance values and values corrected by a sequential Bonferroni are given: $^{*}0.01 < P \leq 0.05$;** $0.001 < P \leq 0.01$;*** $P \leq 0.001$

Concentration	df	F	Uncorrected sign. level	Corrected sign. level
01	1, 11	0.040	0.8459 ns	1.0000 ns
02	1, 11	0.926	0.3565 ns	1.0000 ns
04	1, 11	14.695	0.0028**	0.0112*
08	1, 11	0.054	0.8205 ns	1.0000 ns
16	1, 11	29.301	0.0002***	0.0010***
32	1, 11	59.578	9×10^{-6} ***	0.0001***

Table 3. BACIP one-way ANOVA on changes in percentage cover for filamentous algae (above) and *Mytilus edulis* (middle) as well as *Littorina littorea* abundance (below) at the two Hopavågen sites compared to pooled Verrafjorden reference sites. Values have not been corrected for multiple comparisons

Variable	Df	F	Significance level
Filamentous algae			
Hopavågen I	1, 5	1.440	0.284 ns
Hopavågen II	1, 5	14.016	0.013 *
Mytilus edulis			
Hopavågen I	1, 5	5.247	0.071 ns
Hopavågen II	1, 5	8.091	0.036 *
Littorina littorea			
Hopavågen I	1, 5	0.081	0.788 ns
Hopavågen II	1, 5	6.647	0.050 *

number of comparisons by a sequential Bonferroni, only concentration 02 is still significant, if α is set at 0.05 (Table 1). The regression analysis showed a significantly increased number of *Jaera* in May (uncorrected $P = 0.029$) and August (uncorrected $P = 0.016$) 1999 along the nutrient gradient.

Regarding *Littorina littorea*, six different regression analyses were carried out, one for each of the six sampling occasions in the After period, i.e., September 1998, as well as May, August and October 1999 and May and August 2000 (Fig. 3). In August 1999 the slope became significantly different from zero (uncorrected $P = 0.008$) and this trend was further strengthened in October 1999 (uncorrected $P = 0.004$). However, this development slowed down in May and August 2000 (uncorrected $P = 0.045$ and 0.042, respectively), mainly due to a high mortality rate during the winter of 2000 and poor recruitment the subsequent summer in concentration 16 (visual observation). After correction for five multiple comparisons by Holm's sequential Bonferroni (Legendre & Legendre, 1998), the slopes for the regression lines 1999 are still significantly different from zero ($P = 0.040$ for August 1999 and $P = 0.024$ for October 1999), but not the slopes for May and August 2000. Note that the regression analyses were performed with the x-axis representing the actual (not the nominal) concentrations (i.e., mean annual background levels of 8 μM + the experimental addition). When analysed by a BACIP ANOVA, concentrations 04, 16 and 32 μM are all significantly different from the pooled controls, presenting uncorrected P values of 0.003 (concentration 04), 0.0002 (concentration 16) and 9×10^{-6} (concentration 32). All these P values are still significant after correction for six multiple comparisons by a sequential Bonferroni (Table 2). Note, however, that concentration 08 was not significantly different from the controls, although concentration 04 was. On the contrary the former mesocosm seems to have the most stable *L. littorea* population of all the mesocosms (Fig. 4).

Hopavågen lagoon

The littoral rocky shores of the Hopavågen area proved to be rather species rich environments. In Hopavågen 93 taxa of littoral macrofauna were sampled (66 in Hopavågen I and 76 in Hopavågen II), whereas the total number of sampled littoral taxa was a bit lower, 50, for Verrafjorden (45 in Verrafjorden I and 35 in Verrafjorden II) (Appendix 1). Of the Hopavågen taxonomic groups, 40% (37 taxa) were crustaceans, whereas 34% (32 taxa) were molluscs and 10% (nine taxa) bristle worms. Among the other groups there were four taxa of echinoderms, two taxa each of ascidians, sea anemones and fish, and one taxon each of flatworms, oligochaets, sea spiders, arachnids and insects. Of the taxonomic groups at Verrafjorden, 42% were molluscs (21 taxa), 36% were crustaceans (18 taxa) and 8% bristle worms (four taxa). Sea anemones, flatworms, oligochaets, arachnids, insects, echinoderms and fish were represented by one taxon each. The macroalgal communities of Hopavågen and Verrafjorden were not investigated exhaustively. Just the species identified by counting in frames are included in Appendix 1.

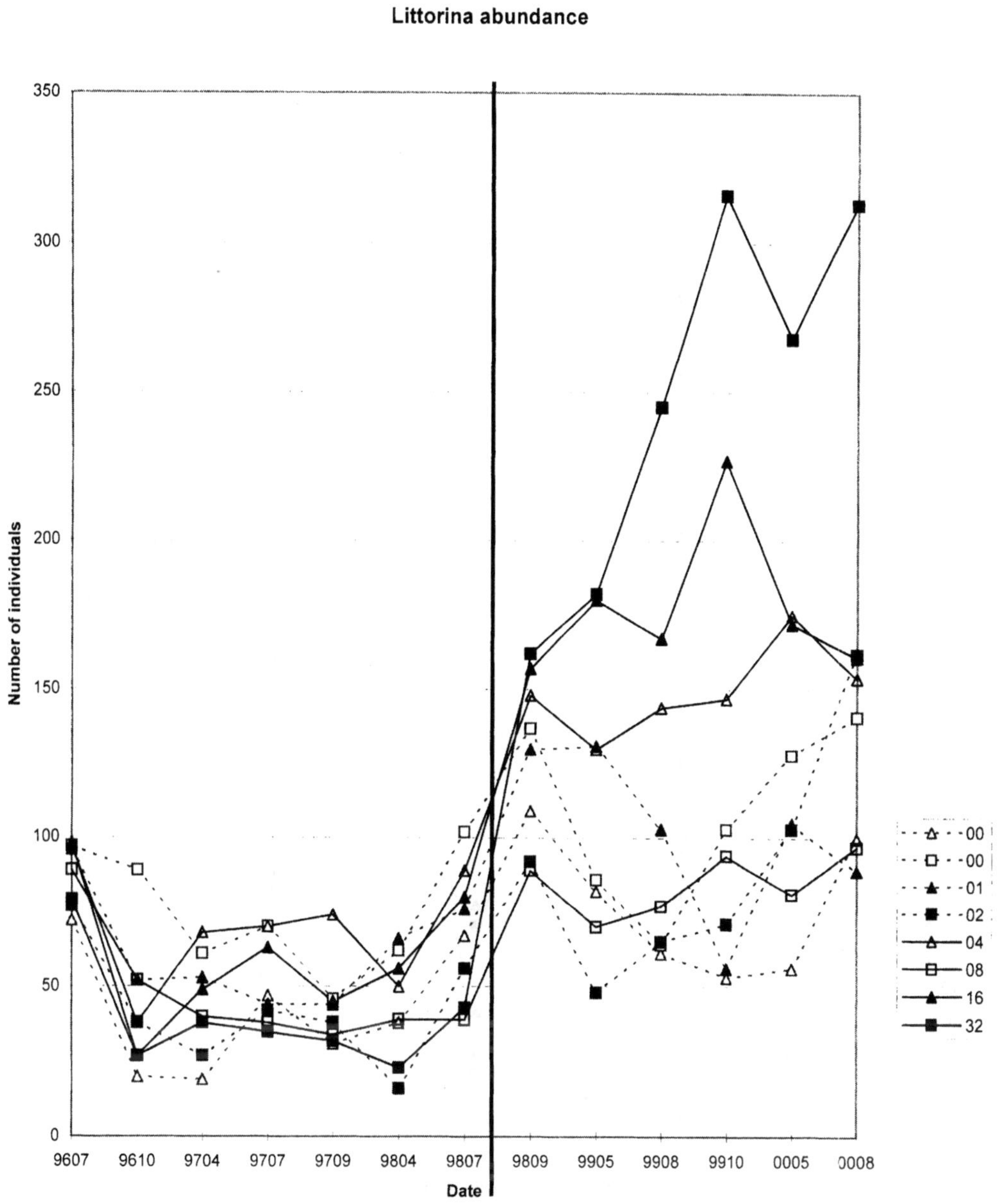

Figure 4. Development of *Littorina littorea* abundance 1996–2000 in Solbergstrand mesocosms. The bold line separates Before measurements from After measurements.

No major responses to the nutrient addition occurred in the degree of cover of Hopavågen macroalgae. Throughout the study period, large brown algae (Fucales) had higher percentage coverage in Verrafjorden (30–60%) than in Hopavågen (10–40%) at the two investigated levels (upper/mid littoral and low littoral/upper sublittoral zone). These differences in coverage between localities did not change over time. Regarding filamentous macroalgae, some stimulation is indicated in Hopavågen after start of the nutrient addition (Fig. 5). A BACIP analysis ends up with a significant increase in degree of cover in Hopavågen II (uncorrected $P = 0.013$), but not in Hopavågen I ($P = 0.284$) (Table 3). The taxa responsible for this increase were the brown alga *Ectocarpus* and the green alga *Cladophora.* It must be noted, however, that the percentage cover of filamentous algae was very low throughout the studied area (always less than 10%), which means that this algal group was of minor importance.

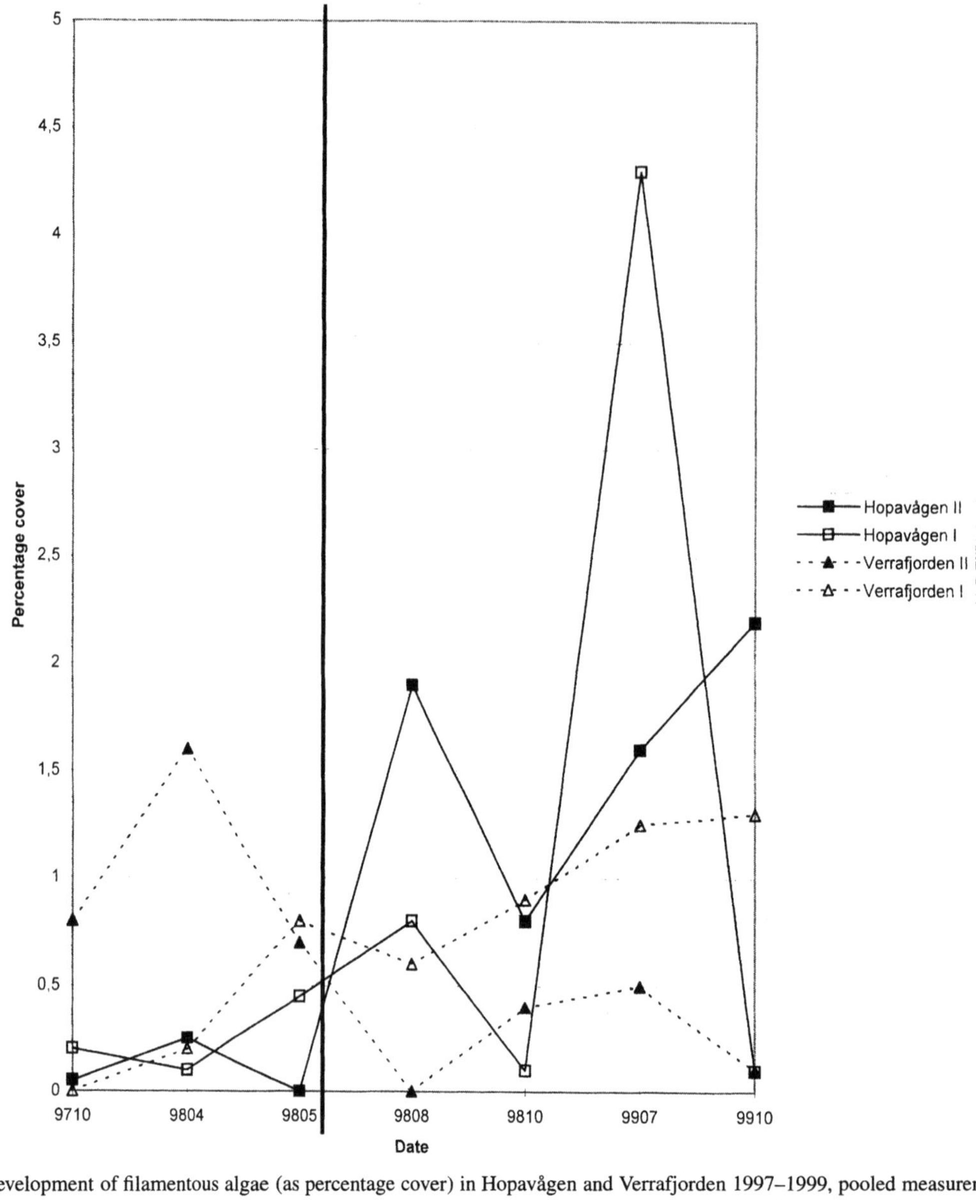

Figure 5. Development of filamentous algae (as percentage cover) in Hopavågen and Verrafjorden 1997–1999, pooled measurements from the both vertical levels. The bold vertical line marks start of nutrient addition in Hopavågen in May 1998 and separates Before measurements from After measurements.

It was found over 65 taxa of mobile macrofauna in the algae samples. The number of taxa was generally a bit higher in Hopavågen than in Verrafjorden, which partly may be explained by the sampling of *Halidrys* at Hopavågen II. Gastropods and crustaceans had generally the highest numbers of individuals. There are differences in abundance values between the localities, as well as differences between years and seasons, although there are no differences that could be related to the nutrient addition. An NMDS ordination on macrofauna abundance data reveals that the animal community structure of *Halidrys* at Hopavågen II is more different from Hopavågen II *Fucus* than *Fucus* in general (Hopavågen compared to Verrafjorden) (Fig. 6). From the NMDS ordination it can also be interpreted that the fauna structure has been changing during 1999, since all later samples cluster out at one side of the plot (the top). Since the same trend is visible in Verrafjorden this cannot be due to the nutrient addition, however. A two-way crossed ANOSIM (using pseudoreplicates) reveals that the community structures of all localities and at all sampling occasions

Table 4. Two-way crossed ANOSIM on Hopavågen and Verrafjorden mobile macrofauna data. For Hopavågen II *Halidrys* data has been used instead of *Fucus vesiculosus*. All P values are still significant (if α is set at 0.05) after correction for multiple comparisons

	Statistical value	Significance value
Differences between localities		
Global ρ	0.812	$P < 0.001$***
Pairwise		
Hopavågen I–Hopavågen II	0.948	$P < 0.001$***
Hopavågen I–Verrafjorden I	0.659	$P < 0.001$***
Hopavågen I–Verrafjorden II	0.830	$P < 0.001$***
Hopavågen II–Verrafjorden I	1.000	$P < 0.001$***
Hopavågen II–Verrafjorden II	1.000	$P < 0.001$***
Verrafjorden I–Verrafjorden II	0.704	$P < 0.001$***
Differences between sampling dates		
Global ρ	0.822	$P < 0.001$***
Pairwise		
October 1997–August 1998	0.815	$P < 0.001$***
October 1997–October 1998	0.907	$P < 0.001$***
October 1997–July 1999	1.000	$P < 0.001$***
October 1997–October 1999	0.861	$P < 0.001$***
August 1998–October 1998	0.537	$P < 0.001$***
August 1998–July 1999	0.898	$P < 0.001$***
August 1998–October 1999	0.741	$P < 0.001$***
October 1998–July 1999	1.000	$P < 0.001$***
October 1998–October 1999	0.898	$P < 0.001$***
July 1999–October 1999	0.824	$P < 0.001$***

are significantly different from each other (Table 4), but due to the experimental set-up with no real replicates it is not possible to relate these differences to the nutrient addition.

The animals, recorded by frame analyses, were large sessile or slowly moving organisms. This community is relatively poor with just a few dominating species. About 10 littoral taxa were registered as common at all investigated localities. Most typical were *Semibalanus balanoides* (L.), *Littorina littorea* (but also *L. obtusata* (L.) and *L. fabalis* Turton), *Nucella lapillus* (L.), *Patella vulgata* L., *Mytilus edulis*, *Pagurus* sp. and sea-urchins. Most of these animal species did not demonstrate any detectable responses to the increased availability of nutrients. The percentage coverage of *M. edulis* (Fig. 7), however, increased significantly at Hopavågen II when analysed by a BACIP one-way ANOVA (uncorrected $P = 0.036$) (Table 3). This is most probably a response to increased nutrient levels (increased amounts of organic particles), since it was concurrent with the nutrient addition. Note, however, that no increase in *M. edulis* coverage occurred at Hopavågen I. On the contrary, the degree of coverage of *M. edulis* at Hopavågen I was not far from being significantly lower than in Verrafjorden ($P = 0.071$), if α is set at 0.05. Regarding *L. littorea*, the Solbergstrand species that responded most readily to the nutrient addition by a stimulated abundance, nothing similar could be observed in Hopavågen. On the contrary, throughout the whole study area, the number of *L. littorea* dropped in the After period compared to the Before period, most evidently in Hopavågen II (even significantly, $P = 0.050$*, compared to pooled Verrafjorden sites) (Table 3, Fig. 8). Note also that Hopavågen I hosted almost five times as many *L. littorea* per m^2 as Hopavågen II in the After period.

Discussion

After 1.5 years of nutrient addition to Hopavågen and 2.5 years of nutrient addition to Solbergstrand mesocosms, only minor responses were visible in the rocky shore communities (both macroalgae and fauna). The

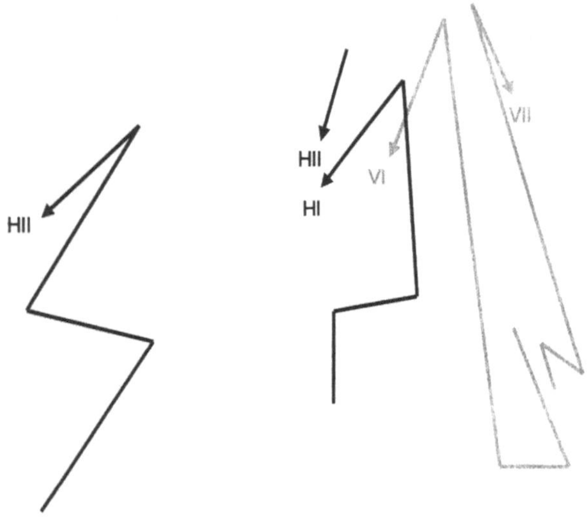

Figure 6. NMDS–ordination of Hopavågen whole ecosystem experiment: macrofauna abundance in *Fucus vesiculosus* in Hopavågen compared to adjacent control sites (2-root transformation, stress = 0.12) from October 1997 (arrow tail, except HII in the middle, Hopavågen II *Fucus* that just was sampled in July and October 1999) to October 1999 (arrow head). HI – Hopavågen I; HII – Hopavågen II *Fucus* (in the middle) or *Halidrys* samples (left); VI – Verrafjorden I; VII – Verrafjorden II.

lack of animal (secondary) responses is perhaps not very surprising when there were so few effects on the plant communities (primary responses). One explanation to the lack of community responses might be that longer time scales are needed in order for clear-cut changes to occur (Warwick, 1993; Bokn et al., 2002). Observations also indicate that wave exposure largely may modulate the impact of increased nutrient load on macroalgal community structure (e.g., Pihl et al., 1999). Hopavågen and also the Solbergstrand mesocosms with their wave machines can be considered as moderately wave exposed systems. Another explanation is the high degree of stability among littoral macroalgae-dominated communities with internal biological regulation factors counteracting the effects (e.g., Connell, 1985). In other words, an increased benthic plant production will soon be grazed down by herbivores that in turn will be consumed by fish or other predators.

Although the causal agents are largely unknown, the lack of general responses indicates that tidal communities have a high resistance against disturbance caused by an excessive nutrient availability as long as the communities are not seriously stressed or perturbed by other processes (chemical, physical or biological) (Connell, 1985). Regarding effects of strong physical forces, preliminary results from EULIT show that on bare substrates (new areas for colonisation), different plant communities will develop along the nutrient gradient, with green algae dominating over red and brown algae at higher nutrient concentrations (Bokn et al., 2003; Karez et al., submitted). Within established macroalgal communities with algal fronds continuously sweeping the surfaces, on the other hand, settlement of more delicate algae and sessile animals is largely prevented (Dayton, 1971; Kiirikki, 1996). Regarding predation, this process was probably of much more importance in Hopavågen than in the well-protected enclosed Solbergstrand mesocosms. In Hopavågen, juvenile fish belonging to the family Gadidae were observed feeding in the macroalgal beds during late summer and autumn. At Solbergstrand, the mesocosm outlets instead served as important structuring factors for the zoobenthic production by continuously exporting free-swimming crustaceans. Up to 8000 specimens of amphipods and 1500 specimens of isopods per basin were exported each summer day or a proportion of 1–2% of the estimated daily standing stocks (Christie & Kraufvelin, 2003). Due to the intensive export through the mesocosm outlets, it is therefore still possible that no significant changes would have occurred in the Solbergstrand mobile macrofauna even in any presence of clear-cut plant responses to the added nutrients (e.g., habitat changes and improved food quality for the animals).

By the increased number of *Littorina littorea* and some indications of increased abundance of *Jaera* spp., also significant changes in Solbergstrand mesocosm fauna appeared due to the nutrient addition. It is interesting that these two possibly stimulated animal groups have two things in common: (1) they are not flushed out of the system through the mesocosm outlets, (2) they feed on microalgae and green algae in the upper littoral zone, i.e., algal groups that also have shown the clearest response to the nutrient addition this far (Bokn et al., 2002).

The responses that were observed in Hopavågen in connection with increased nutrient availability are more complex, however. The markedly increased recruitment of *Mytilus edulis* at Hopavågen II has probably been due to the nutrient addition or more specifically to the increased amount of digestible particles (increased phytoplankton production) caused by increased nutrient availability (see other Hopavågen papers of this issue). In this respect, the lack of response in *M. edulis* coverage at Hopavågen I may be a bit problematic, although this might just have been due to this locality being inherently unsuitable for *M. edulis* settlement/colonisation in the first place

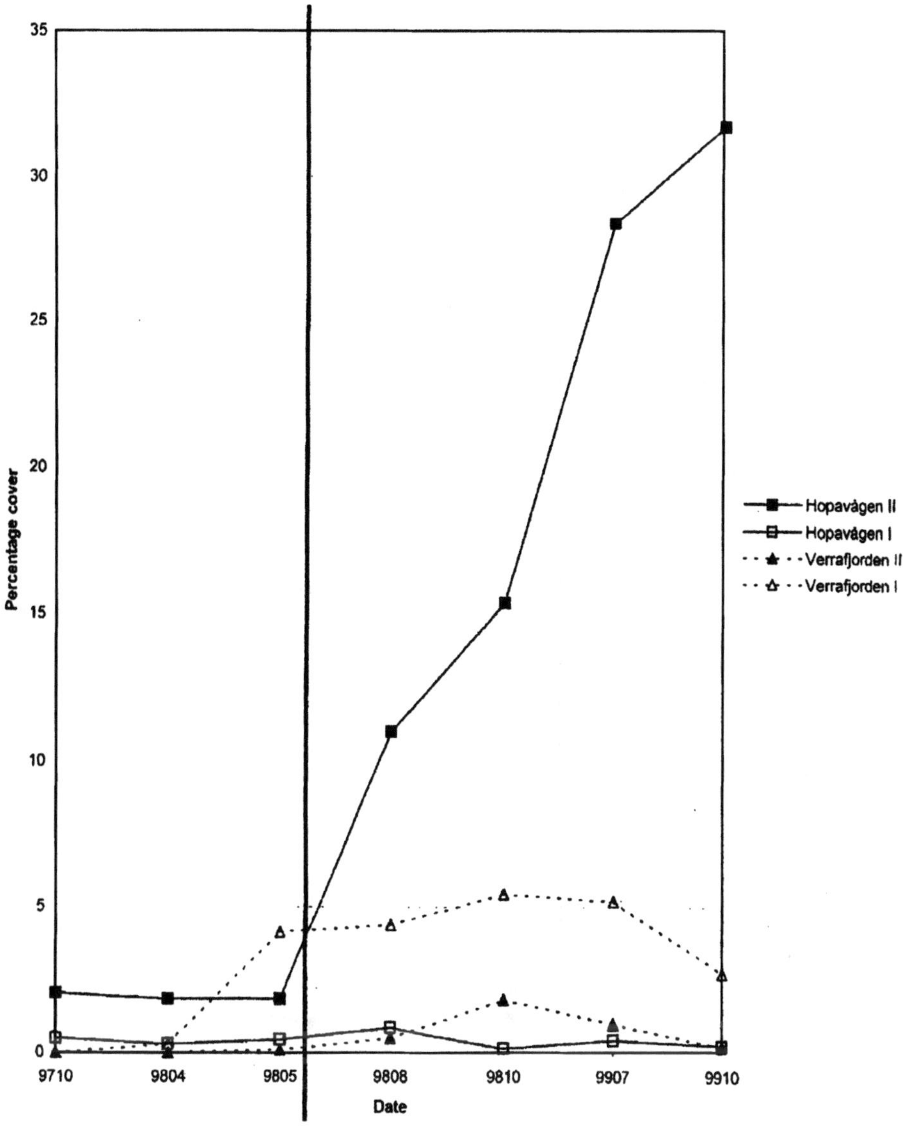

Figure 7. Change in percentage cover for *Mytilus edulis* in Hopavågen and Verrafjorden. The bold vertical line marks start of nutrient addition into Hopavågen in May 1998 and separates Before measurements from After measurements.

(referring to the 0% coverage in the Before period, Fig. 7). The results for *M. edulis* and the fact that the amount of filamentous macroalgae also was just stimulated at Hopavågen II, but not at Hopavågen I, suggests the existence of major community structuring differences between the two sites. The observed difference in abundance of *L. littorea* at Hopavågen I and II is interesting in this respect. It is, however, very hard to tell if the decreased abundance of *L. littorea* at Hopavågen II is a cause or a consequence. Has the difference in *L. littorea* abundance between the Hopavågen sites been responsible for some of the differences in degree of cover of *M. edulis* in the After period or was it rather the difference in *M. edulis* coverage that did suppress the number of *L. littorea*? The high *L. littorea* abundance at Hopavågen I may have been sufficient to prevent settlement and development of a dense *M. edulis* population. The increase in coverage of *M. edulis* at Hopavågen II, which is congruent with the clearly decreased abundance of *L. littorea* (Fig. 8), supports this conclusion. A possible course of events may have been that *L. littorea* abundance occasionally dropped at Hopavågen II in the summer of 1998, which was enough for *M. edulis* to take over and subsequently prevent further dominance of *L. littorea*, despite favourable nutritive conditions for the snail.

A comparison of eutrophication responses across spatial scales, i.e., a comparison between Sol-

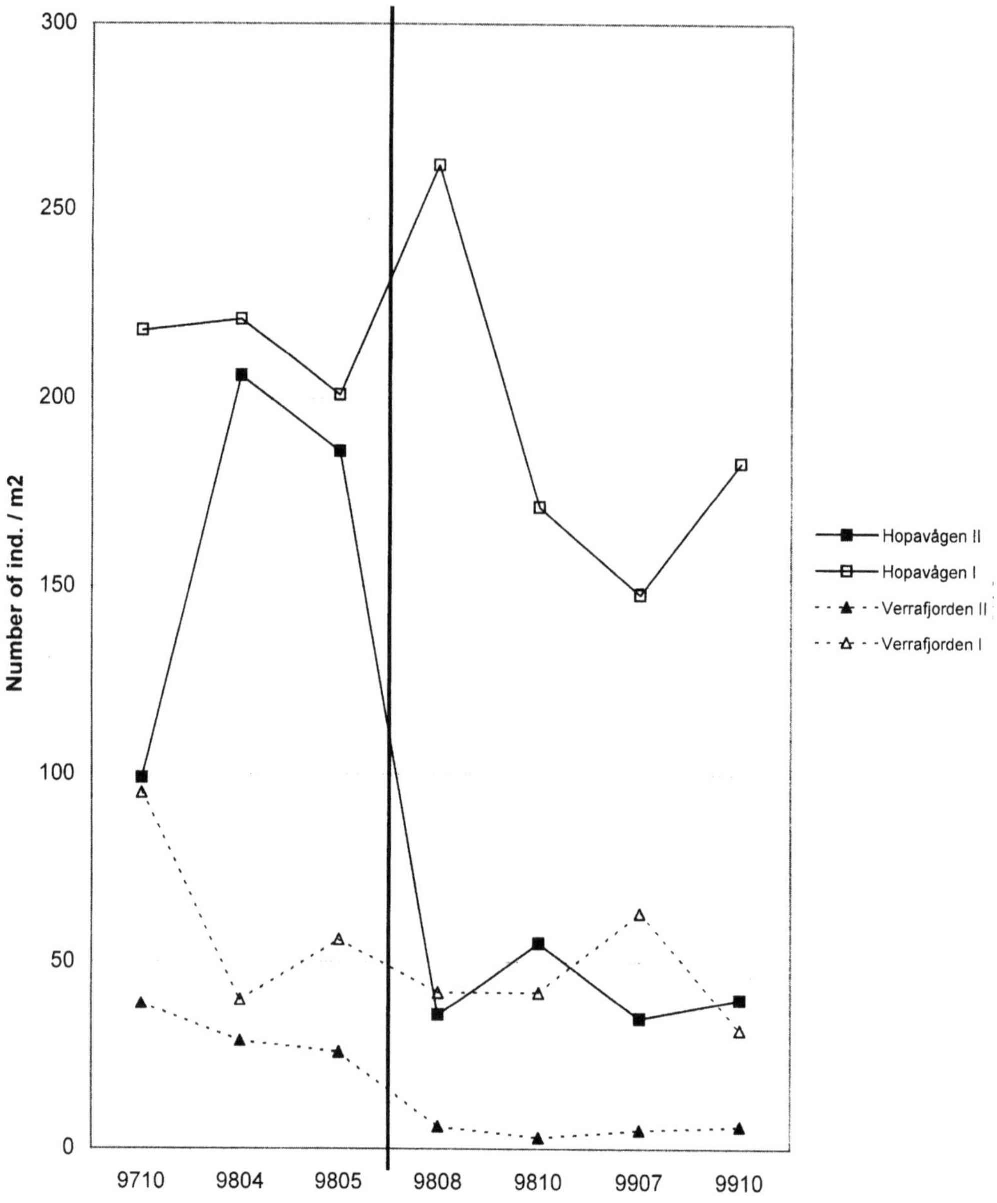

Figure 8. Change in number of individuals of *Littorina littorea* in Hopavågen and Verrafjorden. The bold vertical line marks start of nutrient addition into Hopavågen in May 1998 and separates Before measurements from After measurements

bergstrand mesocosms and Hopavågen whole-ecosystem, was one of the primary objectives of this study. The reduced scale affects the mesocosms in two major ways: species inclusion or exclusion, and potential artifacts associated with containerisation (Perez, 1995). For this paper, this is most evident in: (1) the lack of direct contact between Solbergstrand mesocosms and other eutrophicated ecosystems which meant a continuous import of larvae and spores to the mesocosms from relatively unaffected areas (via the incoming fjord water), (2) the export of certain mobile organisms from the mesocosms while the migration of other species was prevented and (3) the restricted predation impact in the mesocosms (Kraufvelin, 1999). When comparing Solbergstrand and Hopavågen, one must also take into consideration the long distance between the two systems (with the possible occurrence of crucial regional differences with regard to hydrology, meteorology and biogeography) and keep in mind that the nutrient addition continued over one more year in Solbergstrand mesocosms than in Hopavågen, the nutrient concentrations throughout were much higher in the mesocosms and that the nutrient dosing to Hopavågen was turned off during the winter 1998–99, whereas the dosing to the mesocosms was continuous. In terms of background levels, however, both sites included a doubling of the background nutrient level (Hopavågen 1998, Solbergstrand concentration 08, if the background is averaged at 8 μmol l^{-1} N) and a 3-fold increase (Hopavågen 1999, Solbergstrand concentration 16). In addition Solbergstrand also came up with a 5-fold increase (concentration 32). Nevertheless, despite a doubling and a 3-fold increase compared to the mean background levels in both systems, there was a lack of overall responses at the community level, thus indicating that the two studies are able to support each other.

It must anyway be remembered that the few animals that did respond were not the same. There is no agreement between the stimulation of *Mytilus edulis* at Hopavågen II and the results for *M. edulis* in Solbergstrand mesocosms, although this may be due to the minor importance of the phytoplankton component in the mesocosms because of the fast water exchange rate. Neither is there any agreement between the observed responses in *Jaera* spp. and especially *Littorina littorea* in Solbergstrand mesocosms and the responses for these taxa in Hopavågen. This could have been due to an increase in the amount of microalgae at Solbergstrand, that did not take place in Hopavågen, but numerous other explanations are possible. The stimulation of *M. edulis* at Hopavågen II (a more natural 'field site') could be considered as a more realistic response than no response at all (as in the less realistic Solbergstrand mesocosms), just because it took place in a real and not an artificial ecosystem. At the same time we had no control over the mechanisms behind this increase in settlement rate and ultimately degree of cover (see above!). Did the increase, for example, mainly depend on increased food availability or was it just due to an absence of predators or lower numbers of *L. littorea* at Hopavågen II (or perhaps due to presence of predators/grazers at other sites)? In the case of Hopavågen whole-ecosystem study, we suffer from a lack of experimental control. The increased control in a mesocosm investigation, means at least that we may be sure that the stimulated abundances of *Jaera* spp. and *L. littorea* did not depend on an unbalanced lack of predation at higher concentrations, regardless of the realism of these responses in natural ecosystem terms.

The observed high densities of benthic grazers such as snails and sea-urchins, that are controlling the macroalgal communities in Hopavågen, constitute an interesting ecological question in eutrophication context. Especially heavy grazing by sea-urchins is common world-wide (e.g., Lawrence, 1975; Dean et al., 1988) and can often be seen as a problem, where both hiding places and feeding habitats for fish and macroinvertebrates will be reduced. It could be questioned whether an increased nutrient addition could cause a change in dominance between algae and grazers and transfer over-grazed systems into more luxuriant and productive systems. Another possible scenario could be that a change in growth rate of algae or change in algal community structure could alter the competitive relationships between grazers, which could cause a development into totally different animal communities, e.g., initiate a switch from sea-urchin dominated communities to communities dominated by crustaceans. Well-developed macroalgal dominated ecosystems have proved to be especially rich with regard to biodiversity issues (e.g., Hagerman, 1966; Dommasnes, 1969). Knowledge of this could in turn be of interest in management context, e.g., if a more efficient utilisation of biological resources is of primary interest. A separate question is the possibility that nutrient enrichment and increased pelagic production could cause negative changes in the littoral zone such as an increase in benthic diatoms and green algae along the shores, which for example could harm their recreational values or cause an increased export of drift algae to adjacent ecosystems, where for example oxygen deficiency could occur (Norkko & Bonsdorff, 1996; Raffaelli et al., 1998; Vahteri et al., 2000). This far, our results provide little support for such responses being directly caused by increased nutrient availability in rocky shore 'Fucacean communities'.

It is not easy to foretell what ultimately may happen in natural rocky shore ecosystems in connection with large-scale controlled nutrient addition or related management measures. On the basis of these results one could be tempted to predict only small changes. On the other hand, we do not know if we over longer time scales rather could expect an increased animal production, ready to be exploited, than merely detrimental effects. Neither do we know how, where and when we could expect possible effects to show up. What we do know is that there are evident needs for more controlled large-scale experiments examining how rocky intertidal communities respond to increased nutrient availability as well as more targeted information about the role of rocky shore ecosystems in the eutrophication process (Valiela, 1995; Vidal et al., 1999).

Appendix 1. List of species/taxa sampled in the mesocosms (Solbergstrand) and in the whole-system experiment (Hopavågen and Verrafjorden). In the mesocosm experiment, only animals were identified by us and thus algae from Solbergstrand are not given here

Species/taxa	Locality				
	Solberg	Hopa I	Hopa II	Verra I	Verra II
Halichondria panicea(Pallas)	+				
Actinia sp.	+		+		
Metridium senile(L.)	+	+	+		
Anthozoa indetermined	+			+	
Turbellaria indetermined	+	+	+	+	+
Nemertea indetermined	+				
Polynoinae indetermined	+	+	+		+
Arenicolidae indetermined	+				
Arenicola marina(L.)		+			
Capitellidae indetermined				+	
Hesionidae indetermined			+		
Nereidae indetermined			+		
Nereis pelagica L	+				
Phyllodocidae indetermined	+		+		
Spionidae indetermined				+	
Syllidae indetermined	+	+	+		
Typosyllis sp.			+		
Terebellidae indetermined			+		
Pomatoceros sp.	+	+	+		
Polychaeta indetermined		+	+	+	+
Enchytreidae indetermined		+	+	+	+
Oligochaeta indetermined	+				
Semibalanus balanoides (L).	+	+	+	+	+
Praunus inermis (Rathke)	+	+	+		
Praunus flexuosus (Müller)		+	+	+	
Mysidae indetermined			+		
Decapoda indetermined			+		
Idotea baltica (Pallas)	+	+			
Idotea granulosa Rathke	+	+	+	+	+
Idotea pelagica Leach	+	+	+	+	+
Idotea sp.		+		+	+
Arcturidae indetermined	+				
Jaera albifrons Leach	+	+	+	+	+
Jaera sp.	+	+	+	+	+
Lysiannassidae indetermined	+				
Amphilochidae indetermined		+			
Gitanopsis inermis G. O. Sars		+	+		
Stenothoidae indetermined	+	+	+		
Orchestia sp.			+		
Hyale nilssoni (Rathke)	+	+		+	+
Gammarus locusta (L).	+	+		+	
Gammarus sp.	+	+	+	+	+
Apherusa bispinosa (Bate)		+	+		
Apherusa jurinei (Milne Edwards)		+			+
Apherusa sp.	+		+		
Calliopius laeviusculus (Krøyer)	+	+	+	+	+
Gammarellus angulosus (Rathke)	+				+
Atylus swammerdami (Milne Edwards)				+	
Dexamine thea Boeck	+	+	+		
Aoridae indetermined	+	+			
Microdeutopus gryllotalpa Costa	+				
Corophiidae indetermined	+		+		
Corophium bonelli (Sars)		+	+		
Ischyroceridae indetermined	+				

Appendix 1. contd.

Species/taxa	Locality				
	Solberg	Hopa I	Hopa II	Verra I	Verra II
Jassa sp.	+				
Parajassa pelagica (Leach)	+				
Hyperiidea indetermined	+				
Amphipoda indetermined		+	+	+	+
Caprella sp.	+				
Natantia indetermined	+		+		
Palaemon adspersus Rathke	+		+		
Athanas nitescens (Leach)	+				
Hippolyte varians (Leach)		+	+	+	
Eualus gaimardii (Milne Edwards)		+			
Eualus pusiolus (Krøyer)			+		
Processa sp.	+				
Pagarus sp.	+	+	+		
Galathea sp.	+		+		
Hyas araneus (L).			+		
Carcinus maenas (L.)	+	+	+	+	
Macropipus sp.	+	+			
Brachyura indetermined				+	
Halacaridae indetermined	+	+	+	+	+
Chironomidae indetermined	+	+	+	+	+
Nymphon sp.	+				
Polyplacophora indetermined		+	+		
Tectura virginea (Müller)			+		+
Tectura tessulata (Müller)			+		
Acmaeidae indetermined	+	+	+		
Patellidae indetermined	+	+			
Patella vulgata L.		+	+	+	+
Margarites helicinus (Fabricius)	+	+			
Gibbula sp.		+	+		
Lacuna vincta (Montagu)	+	+	+	+	+
Lacuna parva (da Costa)	+			+	
Littorina saxatilis (Olivi)				+	+
Littorina littorea (L.)	+	+	+	+	+
Littorina obtusata (L.)	+	+	+	+	+
Littorina fabalis Turton	+	+	+	+	+
Onoba semicostrata (Montagu)		+	+	+	
Alvania sp.	+		+	+	+
Rissoa parva (da Costa)		+	+	+	
Rissoa sp.	+	+	+	+	+
Skeneopsis planorbis (Fabricius)	+			+	+
Omalogyra atomus (Philippi)			+	+	+
Bittium reticulatum (da Costa)	+				
Nucella lapillus (L.)		+	+	+	+
Neptunea antiqua (L.)		+			
Buccinum undatum L.		+	+	+	
Hinia reticulata (L.)			+		
Prosobranchia indetermined	+	+	+	+	+
Aplysia punctata (Cuvier)			+		
Onchidorididae indetermined		+	+	+	+
Nudibranchia indetermined	+	+	+	+	
Coryphella sp.	+				
Anomia ephippium L.			+		

Continued on p. 165

Appendix 1. contd.

Species/taxa	Locality				
	Solberg	Hopa I	Hopa II	Verra I	Verra II
Monia patelliformis (L.)	+				
Modiolus modiolus (L.)		+			
Mytilus edulis L.	+	+	+	+	+
Musculus discors (L.)	+				
Ostrea edulis L.	+				
Pecten sp.			+		
Cardiidae indetermined	+				
Turtonia minuta	+		+		
Mya sp.		+			
Hiatella arctica (L.)	+	+	+		+
Membranipora membranacea (L.)	+				
Electra crustulenta (Pallas)	+				
Asterias rubens L.	+	+	+		
Marthasterias glacialis (L.)		+			
Psammechinus miliaris (P. L. S. Müller)	+				
Echinus sp.		+	+	+	
Strongylocentrotus droebachiensis (O. F. Müller)		+	+		
Ciona intestinalis (L.)	+		+		
Ascidia sp.	+	+			
Gadidae indetermined		+	+		
Cottidae indetermined	+				
Labridae indetermined	+			+	+
Gobiidae indetermined	+				
Gobiusculus flavescens (Fabricus)			+		
Total animal taxa sampled	80	66	76	45	35
Lithothamnion glaciale Kjellm.			+		
Corallina officinalis L.		+			
Polysiphonia lanosa (Tandy)				+	+
Ceramium rubrum (Hudson) C. *Agardh*				+	+
Chondrus crispus Stackhouse				+	
Dumontia contorta (S. G. Gmel.) Rupr.				+	
Halidrys siliquosa (L.) Lyngbye		+	+		
Pelvetia canaliculata (L.) Decaisne & Thuret				+	
Ascophyllum nodosum (L.) LeJol				+	+
Fucus serratus L.				+	+
Fucus vesiculosus L.		+	+	+	+
Fucus spiralis L.				+	+
Desmarestia sp.		+			
Scytosiphon lomentaria (Lyngbye) Link		+	+		
Ectocarpus siliculosus (Dillwyn) Lyngbye		+	+		
Elachista fucicola (Velley) Areschoug		+	+		
Enteromorpha intestinalis (L.) Nees				+	+
Cladophora rupestris (L.) Kützing		+	+	+	+
Codium fragile (Suringar) Har		+	+		
Total plant taxa sampled		9	8	11	8

Acknowledgements

This work has been financed by the European Commission through MAST III programme MAS3–CT97–0153 and by contributions from MARICULT and Norsk Hydro 1998 and 1999. Financial support has also been received from Carl Cedercreutz stipendiefond at Svenska Kulturfonden (for the first author). We thank the co-ordinator Tor Bokn and the entire EU-LIT group for a fruitful collaboration, Henning Steen for field assistance at Hopavågen as well as Johnny Berglund, Åsa Hägg, Sonja Salovius and Barbro Sinisalo for field assistance at Solbergstrand. The paper is contribution number 42 from Marine Research Station Solbergstrand.

References

Baden, S. P., L.-O. Loo, L. Pihl & R. Rosenberg, 1990. Effects of eutrophication on benthic communities including fish: Swedish west coast. Ambio 19: 113–122.

Bakke, T., 1990. Benthic mesocosms: II. Basic research in hard-bottom benthic mesocosms. In Lalli, C. M. (ed.), Enclosed Experimental Marine Ecosystems: A Review and Recommendations. Coastal and Estuarine Studies, Vol. 37, Springer, Heidelberg: 188–196.

Bokn, T. L., S. N. Murray, F. E. Moy & J. B. Magnusson, 1992. Changes in fucoid distributions and abundances in the inner Oslofjord, Norway: 1974–80 versus 1988–90. Acta phytogeogr. suec. 78: 117–124.

Bokn, T. L., H. Christie, C. M. Duarte, O. Geertz-Hansen, E. E. Hoell, K. Kersting, P. Kraufvelin, C. Lindblad, M. F. Pedersen, U. Sommer & F. Moy, 1998. Effects of eutrophicated seawater on rocky shore ecosystems studied in large littoral mesocosms – EULIT. In Barthel, K.-G., H. Barth, M. Bohle-Carbonell, C. Fragakis, E. Lipiatou, P. Martin, G. Ollier & M. Weydert (eds), Project Synopses from Third European Marine Science and Technology Conference. Lisbon, 23–27 May 1998, Vol. II, Strategic Marine Research: 871–876.

Bokn, T. L., E. E. Hoell, K. Kersting, F. E. Moy & K. Sørensen, 2001. Methods applied in the large littoral mesocosms study of nutrient enrichment in rocky shore ecosystems – EULIT. Continental Shelf Res. 21: 1925–1936.

Bokn, T. L., F. E. Moy, H. Christie, S. Engelbert, R. Karez, K. Kersting, P. Kraufvelin, C. Lindblad, N. Marba, M. F. Pedersen & K. Sørensen, 2002. Are rocky shore ecosystems affected by nutrient-enriched seawater? Some preliminary results from a mesocosm experiment. Hydrobiologia 484/Dev. Hydrobiol. 167: 167–175.

Bokn, T. L., C. M. Duarte, M. F. Pedersen, N. Marba, F. E. Moy, C. Barron, B. Bjerkeng, J. Borum, H. Christie, S. Engelbert, F. L. Fotel, E. E. Hoell, R. Karez, K. Kersting, P. Kraufvelin, C. Lindblad, M. Olsen, K. A. Sanderud, U. Sommer & K. Sørensen, 2003. The response of experimental rocky shore communities to nutrient additions. Ecosystems (in press).

Bonsdorff, E., E. M. Blomqvist, J. Mattila & A. Norkko, 1997a. Long-term changes and coastal eutrophication. Examples from the Åland islands and the Archipelago Sea, Northern Baltic Sea. Oceanol. Acta 20: 319–329.

Bonsdorff, E., E. M. Blomqvist, J. Mattila & A. Norkko, 1997b. Coastal eutrophication: Causes consequences and perspectives in the Archipelago areas of the Northern Baltic Sea. Estuar. coast. Shelf Sci. 44: 63–72.

Carpenter, S. R., 1999. Microcosm experiments have limited relevance for community and ecosystem ecology: reply. Ecology 80: 1085–1088.

Carr, M. R., 1996. PRIMER User Manual, Ver 4.0, Plymouth Marine Laboratory, UK: 1–42.

Christie, H. & P. Kraufvelin, 2003. Mechanisms regulating amphipod population density within macroalgal communities with low predator impact. Sci. Mar. (in press).

Clark, J. R. & C. R. Cripe, 1993. Marine and estuarine multi-species test systems. In Calow, P. (ed.), Handbook of Ecotoxicology. Vol. I. Blackwell Scientific Publications, Oxford: 227–247.

Clarke, K. R. 1993. Non-parametric multivariate analyses of changes in community structure. Aust. J. Ecol. 18: 117–143.

Clarke, K. R. & R. M. Warwick, 1994. Change in Marine Communities: An Approach to Statistical Analysis and Interpretation. Plymouth Marine Laboratory, Plymouth: 1–144.

Connell, J. H., 1985. Variation and persistence of rocky shore populations. In Moore, P. G. & R. Seed (eds), The Ecology of Rocky Coasts. Hodder and Stoughton Educational Press, Kent, England: 57–69.

Dayton, P. K., 1971. Competition, disturbance and community organization: the provision and subsequent utilization of space in a rocky intertidal community. Ecol. Monogr. 41: 351–389.

Dean, T. A., F. R. Jacobsen, K. Thies & S. L. Lagos, 1988. Differential effects of grazing by white sea urchin on recruitment of brown algae. Mar. Ecol. Prog. Ser. 48: 99–102.

Dommasnes, A., 1969. On the fauna of *Corralina officinalis* L. In Western Norway. Sarsia 34: 117–124.

Drenner, R. W. & A. Mazunder, 1999. Microcosm experiments have limited relevance for community and ecosystem ecology: comment. Ecology 80: 1081–1085.

Duarte, C. M., 1995. Submerged aquatic vegetation in relation to different nutrient regimes. Ophelia 41: 87–112.

Duggins, D. O. & J. E. Eckman, 1994. The role of detritus in the growth of benthic suspension feeders in an understory kelp forest. J. exp. mar. Biol. Ecol. 176: 53–68.

Edgar, G. J., 1990. The influence of plant structure on the species richness, biomass and secondary production of macroalgal assemblages associated with Western Australian seagrass beds. J. exp. mar. Ecol. Biol. 137: 215–240.

Gearing, J. N., 1989. The role of aquatic microcosms in ecotoxicologic research as illustrated by large marine systems. In Levin, S. A., M. A. Harwell, J. R. Kelly & K. D. Kimball (eds), Ecotoxicology: Problems and Approaches. Springer, Heidelberg: 411–470.

Hagerman, L. 1966. The macro and microfauna associated with *Fucus serratus* L., with some ecological remarks. Ophelia 3: 1–43.

Huston, H. A., 1999. Microcosm experiments have limited relevance for community and ecosystem ecology: synthesis of comments. Ecology 80: 1088–1089.

Kiirikki, M., 1996. Experimental evidence that *Fucus vesiculosus* (Phaeophyta) controls filamentous algae by means of the whiplash effect. Eur. J. Phycol. 31: 61–66.

Kraufvelin, P., 1998. Model ecosystem replicability challenged by the 'soft' reality of a hard bottom mesocosm. J. exp. mar. Biol. Ecol. 222: 247–267.

Kraufvelin, P., 1999. Baltic hard bottom mesocosms unplugged: Replicability, repeatability and ecological realism examined by non-parametric multivariate techniques. J. exp. mar. Biol. Ecol. 240: 229–258.

Lawrence, J. M., 1975. On the relationships between marine plants and sea urchins. Oceanogr. mar. Biol. ann. Rev. 13: 213–286.

Lawton, J. H., 1995. Ecological experiments with model systems. Science 269: 328–331.

Legendre, P. & L. Legendre, 1998. Numerical Ecology. 2nd Engl. edn. Elsevier, Amsterdam: 1–853.

Lundgren, A., 1985. Model ecosystems as a tool in freshwater and marine research. Arch. Hydrobiol. 70: 157–196.

Moore, P. G. & Seed, R., 1985. The Ecology of Rocky Coasts. Hodder & Stoughton, London: 453 pp.

Norkko, A. & E. Bonsdorff, 1996. Rapid zoobenthic community responses to accumulations of drifting algae. Mar. Ecol. Prog. Ser. 131: 143–157.

Pedersen, M. F. & J. Borum, 1996. Nutrient control of algal growth in estuarine waters. Nutrient limitation and the importance of nitrogen requirements and nitrogen storage among phytoplankton and species of macroalgae. Mar. Ecol. Prog. Ser. 142: 261–272.

Perez, K. A., 1995. Role and significance of scale to ecotoxicology. In Cairns, J. & B.R. Niederlehner (eds), Ecological Toxicity Testing – Scale, Complexity and Relevance. Lewis Publ., CRC Press, Boca Raton, FL: 49–72.

Pihl, L., A. Svenson, P. O. Moksnes & H. Wennhage, 1999. Distribution of green algal mats throughout shallow soft bottoms of the Swedish Skagerrak archipelago in relation to nutrient sources and wave exposure. J. Sea Res. 41: 281–295.

Raffaelli, D. G., J. A. Raven & L. J. Poole, 1998. Ecological impacts of green macroalgal blooms. Oceanogr. mar. Biol. ann. Rev. 36: 97–125.

Robertson, A. I. & J. S. Lucas, 1983. Food choice, feeding rates, and turnover of macrophyte biomass by a surf-zone inhabiting amphipod. J. exp. mar. Biol. Ecol. 72: 99–124.

Seed, R. J. & O'Connor, R. J., 1981. Community organization in marine algal epifaunas. Ann. Rev. Ecol. Syst. 124: 49–74.

Sokal, R. R. & F. J. Rohlf, 1995. Biometry. 3rd edn., W.H. Freeman, New York: 1–887.

Stewart-Oaten, A., W. W. Murdoch & K. R. Parker, 1986. Environmental impact assessment: 'pseudoreplication' in time? Ecology 67: 929–940.

Stewart-Oaten, A., J. R. Bence & C. W. Osenberg, 1992. Assessing effects of unreplicated perturbations: no simple solutions. Ecology 73: 1396–1404.

Vahteri, P., A. Mäkinen, S. Salovius & I. Vuorinen, 2000. Are drifting algal mats conquering the bottom of the Archipelago Sea, SW Finland? Ambio 29: 338–343.

Valiela, I. 1995. Marine Ecological Processes. Springer, New York: 686 pp.

Vidal, M., C. M. Duarte & M. C. Sanches, 1999. Coastal eutrophication research in Europe: Progress and imbalances. Mar. Poll. Bull. 38: 851–854.

Warwick, R. M., 1993. Environmental impact studies on marine communities: Pragmatical considerations. Aust. J. Ecol. 18: 63–80.

Hydrobiologia **484:** 167–175, 2002.
O. Vadstein & Y. Olsen (eds), Sustainable Increase of Marine Harvesting: Fundamental Mechanisms and New Concepts.

Are rocky shore ecosystems affected by nutrient-enriched seawater? Some preliminary results from a mesocosm experiment

Tor L. Bokn[1], Frithjof E. Moy[1], Hartvig Christie[2], Silke Engelbert[3], Rolf Karez[3], Kees Kersting[4], Patrik Kraufvelin[5], Cecilia Lindblad[6], Nuria Marba[7], Morten F. Pedersen[8] & Kai Sørensen[1]

[1]*Norwegian Institute for Water Research (NIVA) Oslo, Norway*
E-mail: tor.bokn@niva.no
[2]*Norwegian Institute for Nature Research (NINA), Oslo, Norway*
[3]*Institut für Meereskunde an der Universität Kiel, Kiel, Germany*
[4]*ALTERRA, AD Den Burg, The Netherlands*
[5]*Åbo Akademi University, Environmental and Marine Biology, Åbo, Finland*
[6]*Department of Botany, Stockholm University, Stockholm, Sweden*
[7]*Instituto Mediterraneo de Estudios Avanzados, (CSIC-UIB), Esporles, Spain*
[8]*Department of Life Sciences and Chemistry, 17.2 Roskilde University, Roskilde, Denmark*

Key words: macroalgae, benthic fauna, ecological experiments, eutrophication, system metabolism, concrete mesocosms

Abstract

The response of rocky shore ecosystems to increased nutrient availability was examined in eight land-based mesocosms designed for hard-bottom littoral communities built at Marine Research Station Solbergstrand (Norway). The average seawater volume in each basin was 9 m^3 with an average water residence time of about 2 h. A tidal regime resembling that in the fjord was maintained in the basins, and waves were generated regularly. NH_4NO_3 and H_3PO_4, at a constant molar NP ratio of 16:1, was added into 6 basins at concentrations 1, 2, 4, 8, 16, 32 μM DIN above the background DIN concentration during $1\frac{1}{2}$ years. Two mesocosms were kept as control treatment. Marine communities were introduced into the basins two years prior to the start of nutrient dosage. The effects of nutrient enrichment were few and only marginal during the first year of nutrient addition, while some effects became more obvious during the second year. The growth rate of the periphyton and fast-growing macroalgae communities was stimulated by nutrient enrichment, while the response was less evident among the perennial fucoids. The structure of the macroalgal communities, however, did not change during 16 months' measurements. In contrast, growth on artificial rock substrates during the same period of time revealed intensive growth of the fast-growing *Ulva lactuca* in high-dosed basins compared with low-dosed and control basins, which were dominated by the fucoid *Fucus serratus.* The fauna communities exhibited only a minor response to nutrient treatment. The common periwinkle *Littorina littorea*, however, appeared with increased abundance in the high-dosed basins. The total system metabolism tended to increase slightly, but not significantly, with increased nutrient loading.

Introduction

The present article is aimed to give an overview of results from a comprehensive EU-project and will be restricted to present available information from some of the subtasks.

Increased nutrient inputs from land have led to eutrophication of many European coastal areas, which have caused widespread environmental deterioration (e.g. Duarte, 1995; Richardson & Jørgensen, 1996; Borum, 1996). Well-known effects of eutrophication include loss of seagrass cover, increased dominance of ephemeral macroalgae and noxious blooms of *Phaeocystis* and other toxic algae (e.g. Lancelot, 1990; Wennberg, 1992; Duarte, op. cit., Richardson & Jørgensen, op. cit.). In the Baltic Sea and the Outer

Oslofjord reduced light penetration has led to decreased depth limits for *Fucus vesiculosus* and for several other seaweed species, respectively (Kautsky et al., 1986; Rueness & Fredriksen, 1991; Eriksson et al., 1998).

The effects of increased nutrient loading are rather well described for phytoplankton and seagrass dominated ecosystems (e.g. Ryther & Dunstan, 1971; Schindler, 1978; Graneli & Sündback, 1985; Hecky & Kilham, 1988; Sand-Jensen & Borum, 1991; Duarte op. cit.; Borum op. cit.; Valiela et al., 1997). In contrast, the present knowledge of the effects of increased nutrient inputs to rocky shore ecosystems is largely qualitative (Vidal et al., 1999). Rocky shores are important components of the coastal ecosystems along the Atlantic coasts of Europe and in the Baltic. These ecosystems are dominated by macroalgal communities, which, together with their epiphytes, provide substantial amounts of organic matter, a production calculated to a level of 1000 g C per m^2 per yr, which is similar to the production of a rain forest (Mann, 1982; Bird & Benson, 1987). The high primary production of rocky shore ecosystems supports a complex and highly productive animal community, consisting of both sessile and motile invertebrates that live on the produced algal matter (Hagerman, 1966; Hayward, 1988; Duggins et al., 1989). In addition to mussels, the majority of the fauna is made up by periwinkles, amphipods and isopods, and their community structure is expected to depend on the composition of the algal community (Seed & O'Connor, 1981; Edgar, 1990).

Increased anthropogenic nutrient input to nutrient poor (or pristine) coastal ecosystems is expected to stimulate the growth rate of most or all photosynthetic organisms (e.g Fujita, 1985; Moncreiff et al., 1992), on rocky shores and, therefore, also the total productivity of the ecosystem. Increased production does, however, not occur solely by the increased growth rate of the existing vegetation. Changes in productivity are also hypothesized to occur as a consequence of structural changes in the macroalgal communities. Slow-growing species are expected to be replaced by faster-growing due to lower competition for nutrients and increased competition for light (Bokn et al., 1992; Duarte, 1995; Borum & Sand-Jensen, 1996; Pedersen & Borum, 1997; Valiela et al., 1997).

The critical nutrient loading needed to exceed the buffering capacity of a rocky shore ecosystem and, thus, lead to undesirable effects is presently unknown. Quantitative knowledge is, therefore, needed in order to be able to predict the responses of rocky shore ecosystems to increased nutrient inputs. Experimental ecosystems (i.e. mesocosms) are regarded as essential tools when studying marine pollution because they bridge the gap between single-species tests (laboratory) and the natural environment (field experiments) (Lalli, 1990). Mesocosm experiments have some clear logistical advantages. Experimental installations such as settling panels, cages, electrodes, etc. are better protected against accidental or deliberate damage, and sampling is very well controlled. Further, some environmental variables (e.g. physical disturbance, water retention time etc.) can be controlled to a large extent.

The aim of this study was to predict the response of rocky shore ecosystems to increased nutrient availability. This overall goal was approached through the following specific objectives:

- to investigate structural and functional changes of rocky shore ecosystems as a result of nutrient enrichment.
- to assess the response of community metabolism to increased nutrient supply through measurements of changes in oxygen, nitrate, ammonia and phosphate in enclosed sub-systems.
- to test if increased nutrient load stimulates the production of individual algal populations and communities (e.g. periphyton, ephemeral and perennial algae) and total system productivity.
- to examine whether the composition of the faunal community changes in response to structural changes of the algal assemblage (the fish community was negligible).

Materials and methods

The research project was conducted in eight land-based mesocosms, designed for hard-bottom littoral communities at the Marine Research Station, Solbergstrand (Oslofjord, Norway). Each mesocosm is a concrete basin, 4.75 m long, 3.65 m wide and 1.35 m deep. In the eastern part a series of five steps are constructed to mimic different tidal levels on the rocky shore. The seawater volume in each mesocosm was between 6 m^3 and 12 m^3 (low and high tide) with tidal amplitude of 36 cm similar to the outside fjord. The mesocosms were fed with water from 1 m depth in the fjord, flowing through the basins at 5 m^3 h^{-1} giving an average water residence time of about 2 h, dependent of the tide level. Waves were generated regularly with 17 strokes/min. A detailed, specific description of the mesocosms and the associated facilities is presented in Bokn et al. (2001). Littoral communities were initiated by transferring small boulders with associated macroalgae and fauna from the Oslofjord into the mesocosms two years ahead of the start of the project. Import and settlement of algal spores and animal larvae ensured the continuous development during the pre-project period and have resulted in a luxuriant flora and fauna resembling that of the fjord outside, except for the fish community.

NH_4NO_3 and H_3PO, at a constant molar N:P ratio of 16:1, were added into 6 basins at concentrations 1, 2, 4, 8, 16, 32 μM DIN above the variable background DIN concentration during the year (i.e. surface water from the Oslofjord). Two mesocosms were kept as control treatment, i.e. no extra nutrients were added. Nutrient additions started in spring (May) 1998, have been maintained for 16 months and will continue for another year. Nutrient levels in water entering from the fjord and in outlets from the eight basins have been measured on weekly pooled samples, normally consisting of 5 samples taken each workday. Automatic registration equipment ensured continuous measurements of insolation, temperature and salinity in all mesocosms.

The community structure (species composition, abundance, biomass), nutrient content in plant tissues of periphyton and macroalgae and ecosystem processes (succession, growth of periphyton and macroalgae, export of biological matter, metabolism) were examined during intensive field campaigns (spring, summer and autumn) each year.

Cover of macroalgae and sessile macrofauna (% cover/m^2) and the amount of large motile animals (number/m^2), were estimated using quadrate analyses within a fixed grid system (frame) established in each basin (repeated measurements) and by random samples. The biomass was calculated by combining the cover estimates with measurements of species specific biomass per unit of cover ratios, obtained through destructive sampling. Destructive sampling were taken three times a year, outside the fixed repeated measurement area. These samples also provided the material for the determination of carbon, nitrogen and phosphorus content in macroalgal species from each basin.

Recruitment and succession of periphyton and macroalgal communities during early stages of colonisation were investigated using clean artificial substrates (5 $\times$ 5 cm ceramic and 10 $\times$ 10 cm granite chips) mounted in the mesocosms at different time scales. Recruitment of periphyton and macroalgae was measured repeatedly (6$\times$) over a time schedule of 10 days and one month, respectively. Macroalgae succession on granite chips mounted at start of the experiment in four of the basins (treatment: +0, +2, +8 and +32 μM N) was measured on three replicate chips randomly selected every second months in the summer season. The chips were analysed for biomass (DW and AFDW), taxa composition, cover degree of sessile organisms (per 100 cm^2) and number of mobile fauna (per 100 cm^2). Growth and production of the periphyton community were measured as increase in chlorophyll *a* and C-, N-, P-content, determined following the methods of Lorenzen (1967) and Hillebrand & Sommer (1997), respectively.

Biomass and species composition of macroscopic epiphytes on fucoids, were investigated by scraping of all epiphytes on 8–10 randomly selected specimens of *Fucus serratus* and *F. vesiculosus* collected in each mesocosm and by harvesting clean (without epiphytes) *Fucus vesiculosus* specimens introduced into the mesocosms in summer 1999.

Growth and production of individual key species of macroalgae (*Ulva lactuca, Ceramium rubrum* (group),*Fucus serratus* and *F.vesiculosus*) were measured using species specific methods (Pedersen & Borum, 1996).

The amount (number/m^2) of smaller mobile animals (macrofauna and meiofauna) were estimated from artificial substrates mounted at different depths (8 replicates/basin at each field campaign) and by direct sampling from the dominating macroalgae associations (Kraufvelin et al., 2002). The total abundance of each fauna group was estimated from knowledge of the total proportion of different substrates and algal groups in each mesocosm. Biomass was estimated from standardised weight analyses of the different species.

System metabolism was calculated from continuous measurement of oxygen concentrations in each mesocosm as described in Bokn et al. (2001) and Kersting & Lindblad (2001). For each mesocosm the metabolism of dominating macroalgal, taxonomic groups (red, green and brown algal species) was measured in transparent enclosures with a volume of 20 L. The enclosures were incubated in the mesocosms during 36 h in all mesocosm campaigns, i.e. six times during April–September, 1998–1999.

The amount of exported coarse organic matter (i.e. floating macroalgae and fauna) have been determined monthly (more intensively during campaigns) by trap-

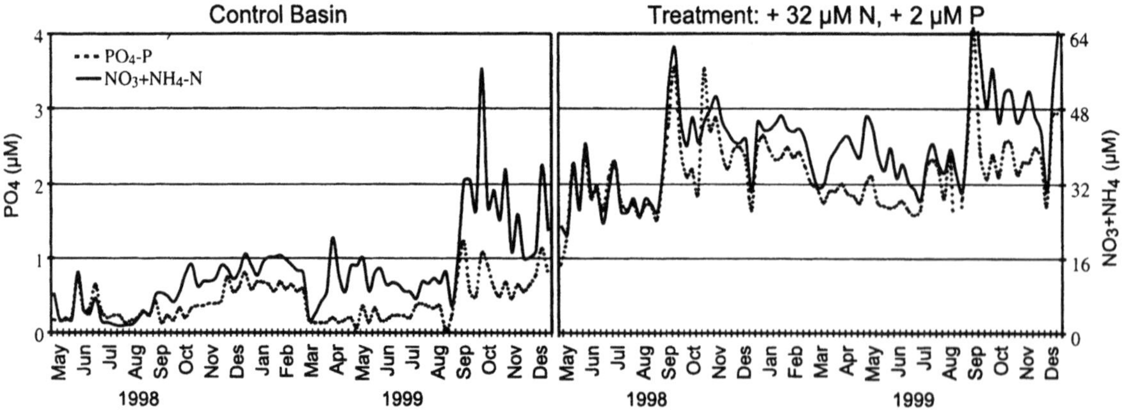

Figure 1. Nitrate, ammonium and phosphate concentrations measured in one of the control basins and the highest continuously dosed basin, May 1998–Dec. 1999.

ping all material larger than 5 mm from the outlet of the mesocosms for 24 h periods.

Results

The nutrient level in the surface water of the Oslofjord water is normally high during winter (0.8–1 μM DIP, 12–14 μM DIN) and low during summer (0.1 μM DIP, 2–4 μM DIN), as was recorded the first experimental year (1998). The second year was characterised by high precipitation and flooding of neighbouring rivers causing high nutrient concentrations in the surface layer. The nutrient level in 1998 and 1999 of the inlet fjord water (surface water) and in the basin treated with 32 μM N and 2 μM P, is shown in Fig. 1.

The effects of nutrient enrichment were few and only marginal during the first year of nutrient addition, while some effects became more obvious during the second year. But contrary to expected effects, no obvious eutrophication response to nutrient treatment was observed.

Right from the beginning of the nutrient enrichment, the growth rate of the periphyton communities were stimulated by nutrient enrichment and they increased their biomass with increasing nutrient richness, particularly during July–August 1999 (Fig. 2a) (Bokn et al., 2001). The growth response of the periphyton community was fast since their C:N and C:P ratios remained with few exceptions constant (but with large variations) across nutrient treatments (Fig. 3a), indicating that excess nutrients were converted into production of new biomass very rapidly.

Growth rate in fast-growing macroalgae such as *Ulva lactuca* was stimulated by nutrient enrichment,

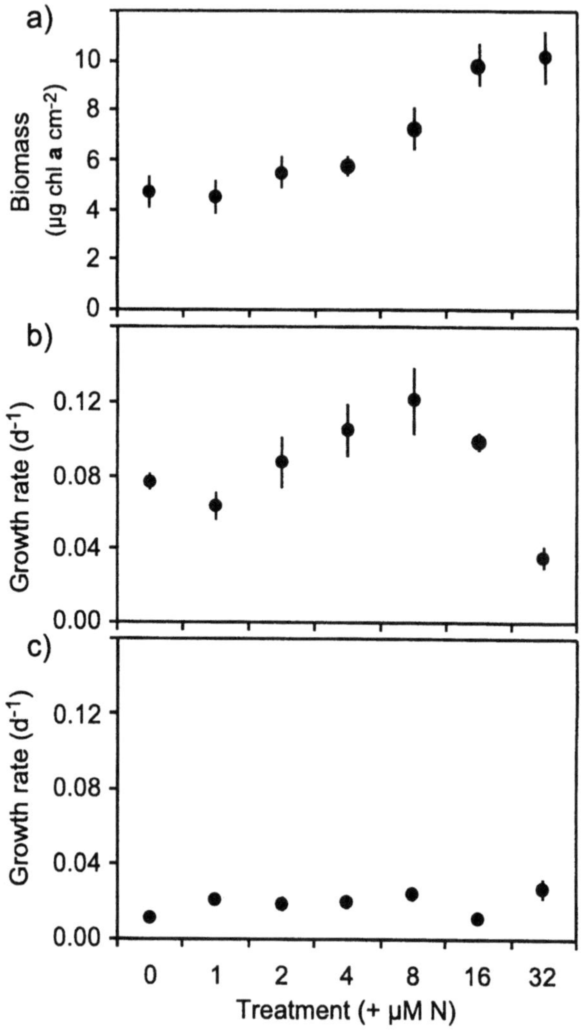

Figure 2. Growth during July/August 1999 of: (a) accumulated biomass of periphyton (μg chl *a* cm^{-2}, $n = 4$), (b) *Ulva lactuca* ($n = 6$) and (c) *Fucus serratus* (n=6). Values are means ±SE.

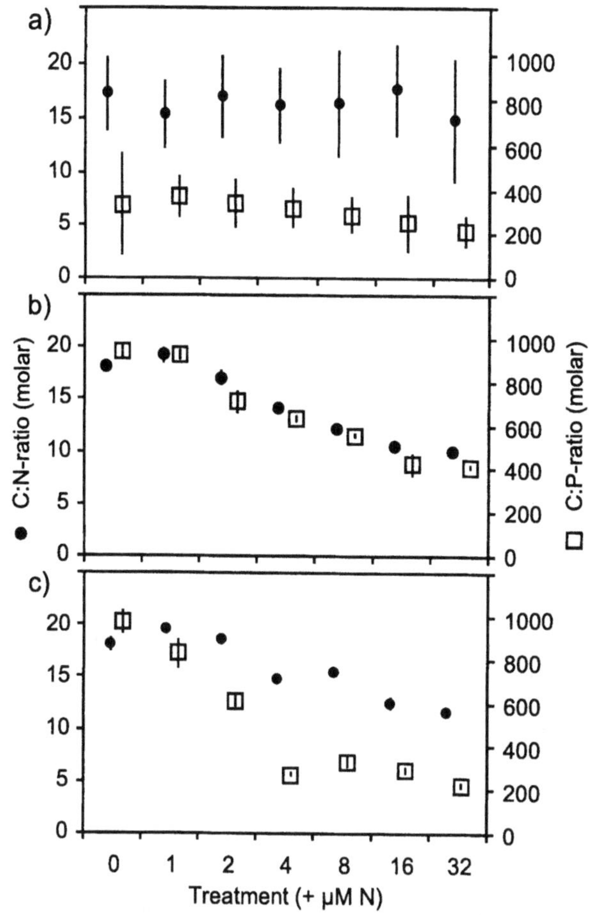

Figure 3. C:N and C:P ratios (molar basis) along the nutrient gradient (July/August 1999) from: (a) early successional periphyton communities (n = 4), (b) green alga *Ulva lactuca* (n = 6) and (c) brown alga *Fucus serratus* (n = 6). Values are mean ±SE. (increased N and P values show decreased values in the figure).

but culminated at 16 μM N, while the response was less evident among the fucoids such as *Fucus serratus* (Figs 2b and c). *Ulva lactuca* and *Fucus serratus* both responded to nutrient enrichment by increasing their tissue nutrient (N and P) concentrations (Figs 3b and c). The response as accumulation of nutrients and as growth rate, differed with algal types (fast-growing green algae/slow-growing fuciods). As result of stimulated growth, the abundance of fast-growing green algae increased with time in the enriched communities (Fig. 4a) while the abundance of fucoids remained stable over time (Fig. 4b). Since the start of the project, these long living brown algae species have accounted for, on average, 83% of the total macroalgae biomass. Thus the structure of these communities (species composition and abundance), did not change significantly as a result of nutrient enrichment during the first 16 months of nutrient addition. Moreover, the macroalgal species composition did not vary in response to nutrient loading.

Macroalgae recruitment and succession on artificial substrates measured over short time scales (one month) have so far only revealed small variations with different nutrient levels indicating that the recruitment potential was equal among treatments. In contrast, recruitment and succession on artificial substrates measured over longer time scales (mounted at start) indicated differences in community development of the algae. The fucoids tended to dominate the communities at low nutrient enrichment, also resulting in quite high biomass, while fast-growing algae like *Ulva lactuca* became almost entirely dominating at high nutrient enrichment (Fig. 5).

The biomass of macroscopic epiphytes on *Fucus* spp. randomly sampled in the basins, did not increase with increased nutrient enrichment. Introduced clean specimens of *F. vesiculosus* developed, with considerable individual variations, a heavy epiphyte load from summer to autumn. However, the biomass of red epiphytes were high in the control mesocosms, while the biomass of green epiphytes proportionally tended to increase with nutrient loading (Fig. 6).

The fauna communities have only slightly changed in response to nutrient treatment (Kraufvelin et al., this volume). But a significant increase in the abundance (number/m^2) of the common periwinkle, *Littorina littorea*, to nutrient loading, were stated, as tested by regression analysis (Fig. 7).

Total system metabolism (i.e. system productivity measured from oxygen balances) tended to increase slightly with increased nutrient loading, although not significantly (Fig. 8). Only in spring and early summer 1999 (April–June) there was a weak indication of a repeated pattern with somewhat higher gross production in the four highest dosed mesocosms compared to gross production in the controls and the low dosed mesocosms. This difference between the four high dosed mesocosms and the four low dosed mesocosms was significant. Whole basin respiration rates were much lower than the production rates and did not differ in response to nutrient treatment either.

Monitoring of the export of matter from the mesocosms through the pipeline outlets showed that the total export of macroalgae was small and relatively insignificant (Fig. 9) compared to the total biomass and productivity of algae. The average export ranged from 20 to 45 g FW d^{-1} corresponding to less than 0.1% of the total biomass or between 5 and 10% of

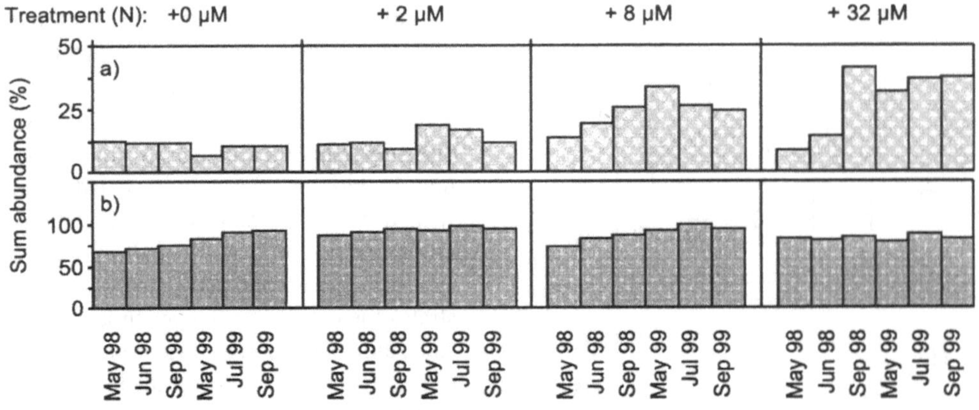

Figure 4. Summarised abundance (% cover) of: (a) fast-growing green algae and (b) fucoids measured by quadrate analyses. Results shown from 4 of 8 mesocosms.

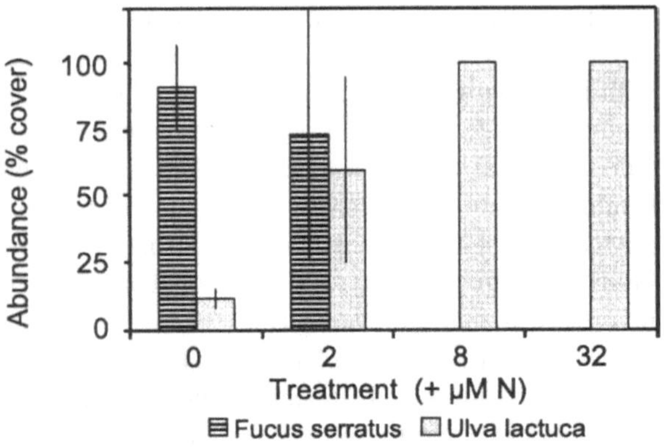

Figure 5. Abundance (% cover) of the brown alga *Fucus serratus* and the green alga *Ulva lactuca* in a successional community after 16 months growth on artificial substrates. (Mean values ±SD).

the daily production. The total export of macroalgae did not vary systematically with nutrient loading.

Discussion and conclusions

Increased nutrient loading enhanced nutrient uptake and subsequently the tissue nutrient concentration in all macroalgal types investigated in this study (but not for periphyton). Accumulation of periphyton biomass on empty substrates and community growth rates were greater at high nutrient loading, indicating that the periphyton communities were nutrient limited during mid-summer under ambient conditions. Fast-growing macroalgae, such as *Ulva lactuca*, showed the same response to increased nutrient availability (except to the highest level), whereas slow-growing, perennial macroalgae, i.e. species of *Fucus* and *Laminaria*, only responded marginally to nutrient enrichment. Thus, increased nutrient loading stimulated the growth and productivity of fast-growing algal types (i.e. periphyton and ephemeral macroalgae) while the productivity of more slow-growing macroalgae remained unaffected by increased nutrient availability, at least for the 16 months reported here.

Tissue nutrient concentrations, CN- and CP-ratios suggested that phosphorus was the main limiting nutrient in ambient seawater from the Oslofjord during summer for both the periphyton community and the macroalgae. Hence, increased P-loading is expected to stimulate growth and productivity of these algal types as observed in these experiments. Work during several decades in the inner Oslofjord has shown phosphorus limitation (Paasche & Erga, 1988) and during 1981–1983 the limitation is measured for 70% of the year (Källqvist, 1988).

The recruitment and succession experiments have shown that the recruitment potential is similar along the nutrient gradient, and that post-recruitment processes (i.e. stimulated growth of certain species, etc.) must be responsible for the observed dominance of ephemeral algae at high nutrient loading in these primary successional communities outcompete ousted.

Although the growth of fast-growing algae were stimulated by increased nutrient loading and the fucoids were ousted on denuded substrates, however, no significant structural changes in the established macroalgal community, have been observed over the 16 months of nutrient enrichment. First of all, the established communities are heavily dominated by large perennial species such as *Fucus* spp. and *Ascophyllum nodosum*. These species are only expected to disappear with increased nutrient loading if shading from

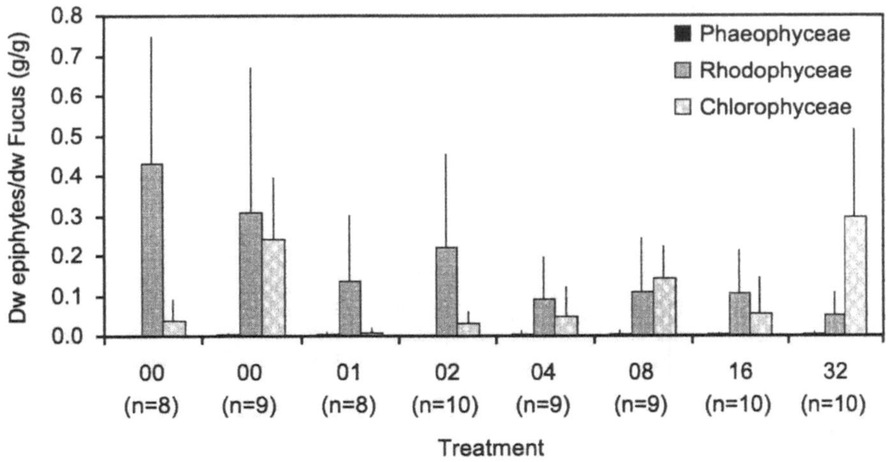

Figure 6. Mean macro-epiphyte biomass (gDW/g *Fucus*) on young *Fucus vesiculosus* thalli exposed from July to September 1999. On the x-axis the nutrient treatment (μM N) and the number of surviving thalli are indicated. (Mean values ±SD).

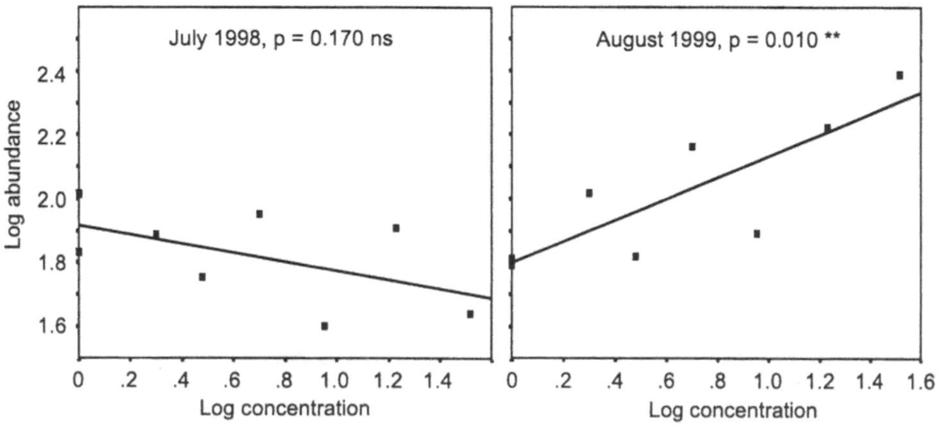

Figure 7. Regression curves for *Littorina littorea* abundance values against nutrient concentrations in July 1998 (1st summer, not significant) and August 1999 (2ns summer, significant). Both axes were logarithmed, $\log_{10}(x+1)$.

epiphytes becomes severe. Our results showed that epiphytic load (biomass) were more or less unaffected by increased nutrient loading. Secondly, the growth rate of ephemeral species was indeed stimulated by increased nutrient availability, but why then, do they not become more dominating with time? One obvious reason could be that most herbivores prefer ephemeral algae due to higher nutrient content and lower content of phenolic compounds etc (Karez et al., 2000). However, our preliminary results (not presented in this article) showed that apparent grazing was very low and certainly insignificant relative to the production of new algal biomass, although a significant increase in the abundance of the common periwinkle with nutrient load was stated.

Instead, physical disturbance created by the wave bar in the mesocosms may be more important for the structuring of the algal communities than hypothesised (e.g. Pihl et al. 1999). Thin, delicate ephemeral macroalgae (i.e. *Ulva, Enteromorpha, Ceramium*) are detached from their substrate and exported from the basins more easily than the perennial species (i.e. *Fucus, Ascophyllum, Laminaria*).

In summary, we can conclude from these preliminary results that increased nutrient loading to hard-bottom littoral communities dominated by perennial algae will stimulate the growth of ephemeral algae as expected. The stimulated growth may, however, not result in significant changes of the community structure in areas with wave exposure, because ephemeral algae probably will be removed by physical disturbance before they will cause increased competition for the large perennial algae. If the substrate is completely cleared from perennial algae (by storms, ice-scouring,

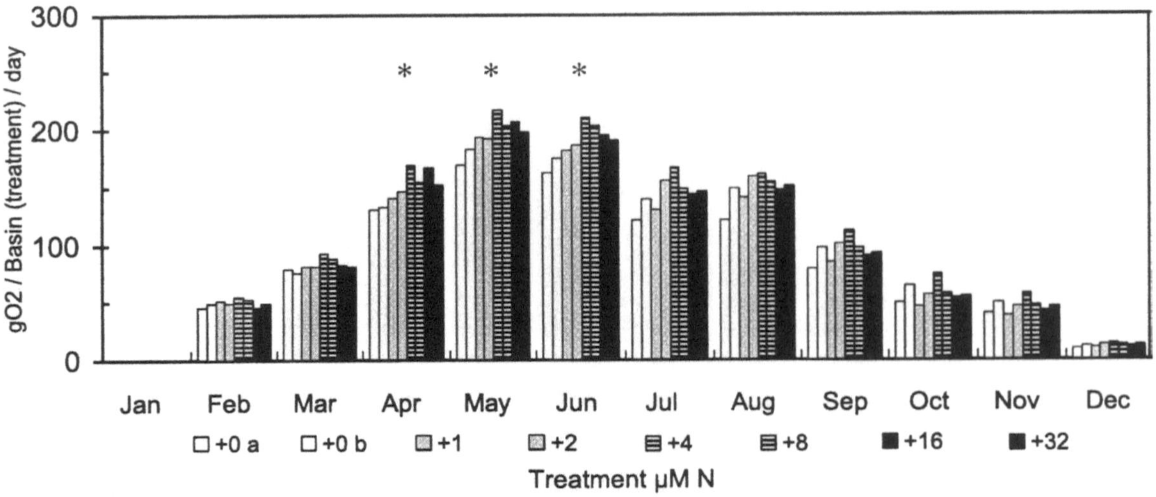

Figure 8. Basin Gross Primary Production based on in situ oxygen measurements during 1999. * Indicates significant difference between low (four basins incl. controls) and high doses (four basins).

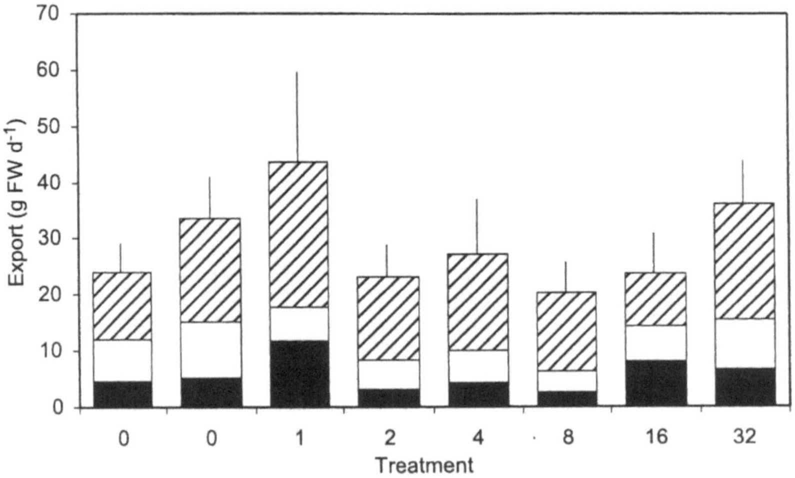

Figure 9. The average daily export of macroalgae measured during May, July and September 1999. Black: Phaeophyceae. White: Rhodophyceae. Hatch: Chlorophyceae. Values are means ±SE.

outbreak of grazers, etc.) ephemeral algae may be able to colonise and become dominant if the nutrient loading is high enough, cf. recruitment and succession on artificial substrates, this work. However, our results indicate that well-established communities of perennial algae seem to be able to persist under circumstances of increased nutrient loading. Because of their ability to store nutrients, they, moreover, may play an important role as nutrient sinks in nitrogen and phosphorus rich environment.

Acknowledgements

This is contribution no. 282/11 of the ELOISE Programme under the Research DG in the European Commission, of which the EULIT project carried out under Contract MAS3-CT97-0153 was part. The project was also partly funded by the MARICULT/Norsk Hydro Programme. We like to thank Torgeir Bakke for valuable advice and the technical staff at the Marine Research Station for running the research facilities so properly. This paper represents Contribution No. 41 from the Marine Research Station, Solbergstrand, Norway.

References

Bird, K. T. & P. H. Benson, 1987. Seaweed Cultivation for Renewable Resources. Elsevier Science Publishers, Amsterdam: 382 pp.

Bokn, T. L., S. N. Murray, F. E. Moy & J. B. Magnusson, 1992. Changes in fucoid distributions and abundances in the inner Oslofjord, Norway: 1974–80 versus 1988–90. Acta Phytogeogr. Suec. 78: 117–124.

Bokn, T. L., E. E. Hoell, K. Kersting, F. E. Moy & K. Sørensen, 2001. Methods applied in the large littoral mesocosms study of nutrient enrichment in rocky shore ecosystems – EULIT. Continental Shelf Res. 21: 1925–1936.

Borum, J., 1996. Shallow waters and land/sea boundaries. In Jørgensen, B. B. & K. Richardson (eds), Eutrophication in Coastal Marine Ecosystems. Chapter 9, American Geophysical Union, Washington DC, USA.

Borum, J. & K. Sand-Jensen, 1996. Is total primary production in shallow coastal marine waters stimulated by nitrogen loading? OIKOS 76: 406–410.

Duarte, C. M., 1995. Submerged aquatic vegetation in relation to different nutrient regimes. Ophelia 41: 87–112.

Duggins, D. O., C. A. Simenstad & J. A. Estes, 1989. Magnification of secondary production by kelp detritus in coastal marine ecosystems. Science NY. 245: 170–173.

Edgar, G. J., 1990. The influence of plant structure on the species richness, biomass and secondary production of macroalgal assemblages associated with Western Australian seagrass beds. J. exp. mar. Ecol. Biol. 137: 215–240.

Eriksson, B. K., G. Johansson & P. Snoeijs, 1998. Long-term changes in the sublittoral zonation of brown algae in the Southern Bothnian Sea. Eur. J. Phycol. 33: 241–249.

Fujita, R. M., 1985. The role of nitrogen status in regulating transient ammonium uptake and nitrogen storage by macroalgae. J. exp. mar. Biol. Ecol. 92: 283–301.

Graneli, E. & K. Sündback, 1985. The response of planktonic and microbenthic algal assemblages to nutrient enrichment in shallow coastal waters, southwest Sweden. J. exp. mar. Biol. Ecol. 85: 253–268.

Hagerman, L., 1966. The macro and microfauna associated with *Fucus serratus* L., with some ecological remarks. Ophelia 3: 1–43.

Hayward, P. J., 1988. Animals on Seaweed. Naturalists Handbooks 9. Richmond Publishing Co. Ltd.

Hecky, R. E. & P. Kilham, 1988. Nutrient limitation of phytoplankton in freshwater and marine environments: a review of recent evidence on the effects of enrichment. Limnol. Oceanogr. 33: 796–822.

Hillebrand, H. & U. Sommer, 1997. Response of epilithic microphytobenthos of the western Baltic Sea to in situ experiments with nutrient enrichment. Mar. Ecol. Prog. Ser. 160: 35–46.

Karez, R., S. Engelbert & U. Sommer, 2000. 'Co-consumption' and 'protective coating': two new proposed effects of epiphytes on their macroalgal hosts in mesograzer-epiphyte-host interactions. Mar. Ecol. Prog. Ser. 205: 85–93.

Kautsky, N., H. Kautsky, U. Kautsky & M. Waern, 1986. Decreased depth penetration of *Fucus vesiculosus* since the 1940's indicates eutrophication of the Baltic Sea. Mar. Ecol. Prog. Ser. 28: 1–8.

Kersting, K. & C. Lindblad, 2001. Nutrient loading and metabolism in hard-bottom littoral mesocosms. Continental Shelf Res.21: 2117–2125.

Kraufvelin, P., H. Christie & M. Olsen, 2002. Littoral macrofauna (secondary) responses to experimental nutrient addition to rocky shore mesocosms and a coastal lagoon. Hydrobiologia 484/Dev. Hydrobiol. 167: 149–166.

Källqvist, T., 1988. Nitrogen or Phosphorus – what is Limiting Nutrient in coastal areas? Examples from Norwegian fjords. Vatten 44: 11–18.

Lalli, C. M. (ed.), 1990. Enclosed Experimental Marine Ecosystems: A Review and Recommendations. Springer Verlag, Berlin, (Coastal and Estuarine Studies): 218 pp.

Lancelot, C., 1990. *Phaeocystis* blooms in the continental coastal area of the Channel and the North Sea. In: Lancelot C., G. Billen & H. Barth (eds). Eutrophication and Algal Blooms in North Sea Coastal Zones, The Baltic and Adjacent Areas: Prediction and Assessment of Preventive Actions. Wat. Poll. Res. Rep. 12: 27–54. CEC, Brussels.

Mann, K. H., 1982. Ecology of Coastal Waters. A system approach. Blackwell, Oxford: 322 pp.

Moncreiff, C. A., M. J. Sulivan & A. E. Daehnick, 1992. Primary production dynamics in seagrass beds of Mississippi Sound: the contribution of seagrass, epiphytic algae, sand microflora, and phytoplankton. Mar. Ecol. Prog. Ser. 87: 161–171.

Paasche, E. & S. R. Erga, 1988. Phosphorus and nitrogen limitation of phytoplankton in the inner Oslofjord. Sarsia 73: 229–243.

Pedersen, M. F. & J. Borum, 1996. Nutrient control of algal growth in estuarine waters. Nutrient limitation and the importance of nitrogen requirements and nitrogen storage among phytoplankton and species of macroalgae. Mar. Ecol. Prog. Ser. 142: 261–272.

Pedersen, M. F. & J. Borum, 1997. Nutrient control of estuarine macroalgae: growth strategy and the balance between nitrogen requirements and uptake. Mar. Ecol. Prog. Ser. 161: 155–163.

Pihl, L., A. Svenson, P. O. Moksnes & H. Wennhage, 1999. Distribution of green algal mats throughout shallow soft bottoms of the Swedish Skagerrak archipelago in relation to nutrient sources and wave exposure. J. Sea Res. 41: 281–295.

Richardson, K. & B. B. Jørgensen, 1996. Eutrophication: definition, history and effects. In Jørgensen, B. B. & Richardson K. (eds), Eutrophication in Coastal Marine Ecosystems, Chapter 1. American Geophysical Union, Washington DC, USA.

Rueness, J. & S. Fredriksen, 1991. An assessment of possible pollution effects on the benthic algae of the outer Oslofjord, Norway. Int. J. mar. Biol. Oceanogr. (Suppl. 1): 223–235.

Ryther, J. H. & W. H. Dunstan, 1971. Nitrogen, phosphorus, and eutrophication in the coastal marine environment. Science 171: 1008–1013.

Sand-Jensen, L. & J. Borum, 1991. Interaction among phytoplankton, periphyton, and macrophytes in temperate freshwaters and estuaries. Aquat. Bot. 41: 137–176.

Seed, R. J. & R. J. O'Connor, 1981. Community organization in marine algal epifaunas. Ann. Rev. Ecol. Syst. 124: 49–74.

Schindler, D. W., 1978. Factors regulating phytoplankton production and standing crop in the world's fresh waters. Limnol. Oceanogr. 23: 478–486.

Valiela, I., J. McClelland, J. Hauxwell, P. Behr, D. Hersh & K. Foreman, 1997. Macroalgal blooms in shallow estuaries: controls and ecophysiological and ecosystem consequences. Limnol. Oceanogr. 42: 1105–1118.

Vidal, M., C. M. Duarte & M. C. Sánchez, 1999. Coastal eutrophication research in Europe: progress and imbalances. Mar. pollut. bull. 38: 851–854.

Wennberg, T., 1992. Colonization and succession of macroalgae on a breakwater in Laholm Bay, a eutrophicated brackish water area (SW Sweden). Acta Phytogeogr. Suec. 78: 65–77.

Hydrobiologia **484:** 177–190, 2002.
O. Vadstein & Y. Olsen (eds), Sustainable Increase of Marine Harvesting: Fundamental Mechanisms and New Concepts.

Production enhancement by artificial upwelling: a simulation study

Jarle Berntsen[1], Dag L. Aksnes[2] & Arne Foldvik[3]
[1]*Department of Mathematics, University of Bergen, N-5008 Bergen, Norway*
E-mail: jarle.berntsen@mi.uib.no
[2]*Department of Fisheries and Marine Biology, University of Bergen, N-5008 Bergen, Norway*
E-mail: dag.aksnes@ifm.uib.no
[3]*Geophysical Institute, University of Bergen, N-5008 Bergen, Norway*
E-mail: arne.foldvik@gfi.uib.no

Key words: primary production, upwelling, ecosystem models, hydrodynamics

Abstract

A coupled physical–chemical–biological ocean model was applied to study primary production in an idealized 60-km long and 4-km wide fjord. Three different scenarios were simulated: (a) Without fresh water runoff; (b) A river at the head of the fjord adds 100 m^3 s^{-1} fresh water to the surface layer; (c) The river adds 90 m^3 s^{-1} fresh water to the surface layer and 10 m^3 s^{-1} enters the fjord through a pipe at 50 m depth. The average productions in the inner 40 km of the fjord for the three scenarios are 68, 70 and 233 g C m^{-2} $year^{-1}$ respectively. It is thus shown that there is a considerable potential for increasing the primary production in many fjords and coastal areas by submerging some of the river runoff. Because of the higher silicate content of the deeper water, artificial upwelling tends to stimulate diatom more than flagellate growth. This may be beneficial to mariculture developments. The sensitivity of the simulated primary production to horizontal and vertical resolution, horizontal viscosity, vertical diffusivity and viscosity, discharge depth, wind forcing, sill depth and pulsating the discharge is also studied. A simulation where a large river flux (90 m^3 s^{-1}) was submerged, showed that primary production was significantly enhanced in the outer parts of the fjord as well as along the coastline.

Introduction

In the research program 'Marine Production and Cultivation (Maricult)' (Anon., 1994), it is suggested that future harvestable production in the sea can be increased by use of fertilisers. The world population increase, and nowadays intense use of land, call for alternative food production methods. The pioneering work on fertilisation of Loch Craiglin with artificial fertilisers (Gross et al., 1944) showed, despite some unexpected shortcomings, the great potential of such endeavours. In an analysis of the Loch Craiglin experiment, Cooper & Steven (1948) pointed at several improvements, including substitution of the artificial fertilisers with natural nutrients from deep water outside the loch. This idea was further developed by Shiels & Hood (1970), who suggested utilising the tidal range to draw deep, nutrient-rich water into a shallow bay through a pipe. Expected increase in primary production was estimated and supported in experiments carried out by Nevø et al. (1976) and Paul et al. (1979). Deep water is an obvious source of nutrients, and there are a number of enrichment experiments with deep water (e.g., Brown & Parsons, 1972; Platt et al., 1977; Roels, 1980, 1983; Terry & Caperon, 1982; Aksnes et al., 1985). Some of these studies point at the economic potential of developing artificial upwelling for mariculture.

The Maricult research program suggests that the least expensive fertilisation method is addition of formulated fertiliser to surface waters. An alternative is to create upwelling of nutrient-rich deep waters by submerging freshwater outlets from the surface to greater depths. Due to buoyancy, submerged freshwater in a saline ocean will rise towards the surface. On the way freshwater will mix with ambient nutrient-rich waters. A dam with surface level a few meters above the surface level of the fjord can provide sufficient potential energy to drive freshwater through a pipe with outlet at for instance 50 m depth.

In the present study, we simulate effects of artificial upwelling, driven by such submerged freshwater outlets, on the primary production of a fjord. The numerical simulation model is based on a model system developed in order to study responses of marine systems to natural, as well as anthropogenic nutrients (e.g., Aksnes & Lie, 1990; Aksnes & Egge, 1991; Skogen, 1993; Aksnes et al., 1994, 1995; Skogen et al., 1995; Berntsen et al., 1996). Model sensitivity to horizontal and vertical resolution, horizontal viscosity, vertical diffusivity and viscosity, discharge depth, wind forcing, sill depth and pulsating the discharge is also presented.

The physical–chemical–biological model system

The governing equations for the physical model are the equation of continuity for an incompressible fluid, momentum equations with hydrostatic approximation, conservation equations for salinity and temperature and an equation of state from Wang (1984). Physical variables and governing equations for the mode split σ-coordinate model used in the present study are given in Berntsen (2000). Symbols used are:

$\vec{U} = (U, V)$ Horizontal velocities in x- and y-direction, respectively
$\vec{U_A} = (U_A, V_A)$ Depth averaged values of (U, V)
W = Vertical velocity in z-direction
η = Surface elevation
H = Bottom depth
S = Salinity
T = Temperature
K_M = Vertical eddy viscosity
A_M = Horizontal eddy viscosity
A_{M2D} = 2-D horizontal eddy viscosity
K_H = Vertical eddy diffusivity
f = The Coriolis parameter (1.3×10^{-4} s^{-1})
$\vec{U_b} = (U_b, V_b)$ Horizontal velocities at the bottom

Representation of sub grid scale mixing processes on the larger scale fields is important in our experiments. Effects of these processes are parameterised by terms of the form $\frac{\partial}{\partial z}(K_M \frac{\partial U}{\partial z}) + \frac{\partial}{\partial x}(A_M \frac{\partial U}{\partial x}) + \frac{\partial}{\partial y}(A_M \frac{\partial U}{\partial y})$ in the horizontal momentum equations, and $\frac{\partial}{\partial z}(K_H \frac{\partial F}{\partial z})$ in the conservation equations for salinity and temperature. The horizontal 3-D viscosity, A_M, and the 2-D viscosity, A_{M2D}, are computed according to Smagorinsky (1963)

$$A_M = C_M \Delta x \, \Delta y \, \frac{1}{2}\left[\left(\frac{\partial U}{\partial x}\right)^2 + \frac{1}{2}\left(\frac{\partial V}{\partial x} + \frac{\partial U}{\partial y}\right)^2 + \left(\frac{\partial V}{\partial y}\right)^2\right]^{\frac{1}{2}}, \quad (1)$$

$$A_{M2D} = C_{M2D} \Delta x \, \Delta y \, \frac{1}{2}\left[\left(\frac{\partial U_A}{\partial x}\right)^2 + \frac{1}{2}\left(\frac{\partial V_A}{\partial x} + \frac{\partial U_A}{\partial y}\right)^2 + \left(\frac{\partial V_A}{\partial y}\right)^2\right]^{\frac{1}{2}}, \quad (2)$$

where C_M and C_{M2D} are viscosity parameters. Unless otherwise stated, values used for C_M and C_{M2D} are 10 and 500, respectively. No explicit diffusivity for the scalar fields is applied, but there is enough diffusivity associated with the TVD-advection scheme to avoid over-/undershooting.

The vertical viscosity, K_M, and vertical diffusivity, K_H, are computed according due to Mellor & Yamada (1982) unless otherwise stated. Minimum allowed values of K_M and K_H are 1×10^{-5} and 1×10^{-7} m^2 s^{-1} respectively.

The prognostic variables of the biological–chemical model are nitrate, N; phosphate, P; silicate, Si; detritus, Det; flagellates, Fla; and diatoms, Dia. A detailed specification and discussion of the phytoplankton growth model is found in Aksnes et al. (1995). The main features are: (1) There are two functional groups of primary producers. Both groups require nitrogen, phosphorus as well as light for growth. One, however, also requires silicate and is denoted diatoms. The other group is denoted flagellates. (2) According to Furnas (1990) and Egge & Aksnes (1992), the diatom group is given a higher inherent growth rate than the flagellate groups. Hence, unless silicate limits diatom growth rate, the diatom group of the model will be favoured. (3) Self-shading of the phytoplankton groups are incorporated similar to other state-of-the art simulation models: Irradiance is attenuated through the vertical layers of the model by a light extinction coefficient that depends on the water itself as well as the concentration of diatoms and flagellates within the layers. (4) The currency of the biomass is nitrogen as this element is considered as the main limiting nutrient. Estimates of primary production, however, is presented throughout the paper in terms of carbon by assuming a C:N=6.

Equations for light in the water column are taken from Skartveit & Olseth (1986, 1988).

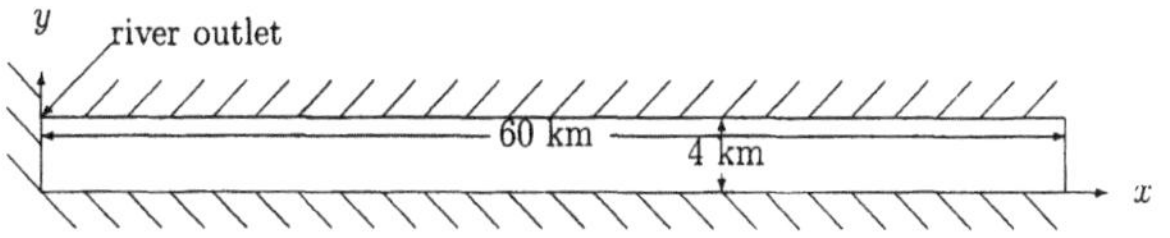

Figure 1. The computational domain.

A three-dimensional σ-coordinate circulation model (Berntsen, 2000) is used to approximate prognostic physical variables and also to advect prognostic chemical–biological variables with the computed currents. Routines for light and sink and source terms for the chemical and biological variables are taken from NORWECOM (Skogen, 1993).

The numerical experiments

The model domain described in Figure 1 is $0 \leq x \leq L_x$, $0 \leq y \leq L_y$ with L_x = 60 km and L_y = 4 km. There are vertical closed boundaries at $y = 0$ and $y = L_y$. There is also a vertical boundary at $x = 0$, except for a river outlet to be specified. At $x = L_x$ there is an open boundary. A constant $H(x, y)$ = 250 m is assumed unless otherwise stated. The model is run for different scenarios for 360 days starting January 1.

Initial and boundary values

Initially, $U = V = W = 0$ m s^{-1} and $\eta = 0$ m. A seven-grid cell flow relaxation zone is added at the outflow of the 60-km long fjord (Martinsen & Engedahl, 1987). At each time step, all velocities in this zone, are updated according to

$$\phi = (1-\alpha)\phi_{\text{int}} + \alpha\phi_{\text{ext}},$$

where ϕ_{int} contains the unrelaxed values computed by the model and ϕ_{ext} is a specified external value. The relaxation parameter α varies from 0 at $x = L_x$ to 1 at the end of the zone.

For each time step approximations to the integral

$$U_{\text{ext}}(y, z, t) = \frac{1}{12\text{ km}} \int_{x=40\text{ km}}^{x=52\text{ km}} U(x, y, z, t)\text{d}x$$

are computed from approximations to $U(x, y, z)$ in the interior model domain and applied as the external boundary values for U. This choice of U_{ext} allows the average interior velocity profile to propagate through the open boundary.

At the free surface no wind stress is assumed, i.e.,

$$K_{\text{M}}\frac{\partial \vec{U}}{\partial z} = 0, \qquad z = \eta,$$

except in the studies of effects of wind forcing. A quadratic law for the bottom stress

$$K_{\text{M}}\frac{\partial \vec{U}}{\partial z} = c_{\text{d}}(U_{\text{b}}^2 + V_{\text{b}}^2)^{1/2}\vec{U}_{\text{b}},$$
$$z = H(x, y),$$

where

$$c_{\text{d}} = \max[0.0025, \frac{\kappa^2}{(\ln(z_{\text{b}}/z_0))^2}]$$

and z_b is the distance of the nearest grid point to the bottom, is applied. The von Karman constant κ is 0.4, and the bottom roughness parameter z_0 is 0.01 m. For the closed horizontal boundaries no normal flux is assumed.

On inflow the salinity at the open boundary is specified as a linear combination between a specified winter profile

$$S_1(z) = 0.5 \times (S_{\max} + S_{\text{minw}}) - \Delta S_1 \times \tanh\left(\frac{z - Z_{\text{s}}}{\Delta Z_{\text{s}}}\right),$$

and a typical summer stratified profile

$$S_2(z) = 0.5 \times (S_{\max} + S_{\text{mins}}) - \Delta S_2 \times \tanh\left(\frac{z - Z_{\text{s}}}{\Delta Z_{\text{s}}}\right),$$

where $S_{\max}$ = 35.0, S_{minw} = 32.0, S_{mins} = 25.0, $\Delta S_1 = (S_{\max} - S_{\text{minw}})/2$, $\Delta S_2 = (S_{\max} - S_{\text{mins}})/2$, $Z_{\text{s}} = -10$ m and ΔZ_{s} = 5 m. We assume a sinusoidal variation over the year and define

$$\text{Frac}(t) = 0.5 + 0.5 \times \cos(2\pi(t - 30)/360), \quad (3)$$

where t is time in days since the start of simulation (January 1). On inflow, the salinity at the outlet of the fjord is specified according to

$$S(L_x + L_{\text{FRS}}, y, z, t) = \text{Frac}(t) \times S_1(z) + (1 - \text{Frac}(t)) \times S_2(z)\forall y. \quad (4)$$

On outflow ($U_{\text{ext}} > 0$), a Neumann condition is used for S and all other scalar variables. This allows coastal water masses to enter the fjord on inflow, and interior water masses to leave the fjord on outflow. We compute boundary conditions for temperature and chemical–biological variables with same technique.

Initial values are

$$S(x, y, z, 0) = S(L_x + L_{\text{FRS}}, y, z, 0) \quad \forall x, y.$$

No precipitation or evaporation is assumed. The initial and boundary conditions for temperature are given by either an assumed fjord profile $T_{\text{fjord}}(z,t)$ or an assumed coastal profile $T_{\text{coast}}(z,t)$. Let a typical winter value of temperature in a fjord be

$$T_1(z) = 0.5 \times (T_{\min} + T_{\text{aver}}) - \Delta T_1 \times \tanh\left(\frac{z - Z_t}{\Delta Z_t}\right),$$

and a typical summer stratified profile be

$$T_2(z) = 0.5 \times (T_{\max} + T_{\text{aver}}) + \Delta T_2 \times \tanh\left(\frac{z - Z_t}{\Delta Z_t}\right),$$

where $T_{\max} = 18.0$, $T_{\min} = 0.0$, $T_{\text{aver}} = 8.0$, $\Delta T_1 = (T_{\text{aver}} - T_{\min})/2$, $\Delta T_2 = (T_{\max} - T_{\text{aver}})/2$, $Z_t = -10$ m and $\Delta Z_t = 5$ m. Let Frac(t) be given by (3). The fjord profile is then given by

$$T_{\text{fjord}}(z,t) = \text{Frac}(t) \times T_1(z) + (1 - \text{Frac}(t)) \times T_2(z).$$

To compute $T_{\text{coast}}(z,t)$, we apply the same formula with $T_{\max} = 16.0$, $T_{\min} = 4.0$, $T_{\text{aver}} = 10.0$, $Z_t = -20$ m and $\Delta Z_t = 10$ m.

Initial values are

$$T(x,y,z,0) = T_{\text{fjord}}(z,0) \qquad \forall x, y.$$

In lack of information on surface heat fluxes, we relax the sea surface temperature towards typical fjord surface temperatures (Cox & Bryan, 1984). The surface flux of heat is specified by

$$K_H \frac{\partial T(x,y,0,t)}{\partial z} = \gamma(T_{\text{fjord}}(0,t) - T(x,y,0,t)),$$

where K_H is the vertical diffusivity in the top layer and γ a time constant selected to be 1.735×10^{-5} m s^{-1}. This means that during weak forcing, the temperature in the upper 15 m obtains the fjord surface value on a time scale of 10 days.

On inflow, $T(L_x + L_{FRS}, y, z, t) = T_{\text{coast}}(z,t)$ is used as a Dirichlet boundary condition for all y.

At the open boundary external values for inorganic nitrogen, phosphate and silicate boundary are on inflow specified as a linear combination between winter unstratified profiles ($N_1(z) = N_{\max}$, $P_1(z) = P_{\max}$ and $Si_1(z) = Si_{\max}$), and typical summer stratified profiles

$$N_2(z) = 0.5 \times (N_{\max} + N_{\min}) - \Delta N \times \tanh\left(\frac{z - Z_n}{\Delta Z_n}\right),$$

$$P_2(z) = 0.5 \times (P_{\max} + P_{\min}) - \Delta P \times \tanh\left(\frac{z - Z_n}{\Delta Z_n}\right),$$

$$Si_2(z) = 0.5 \times (Si_{\max} + Si_{\min}) - \Delta Si \times \tanh\left(\frac{z - Z_n}{\Delta Z_n}\right).$$

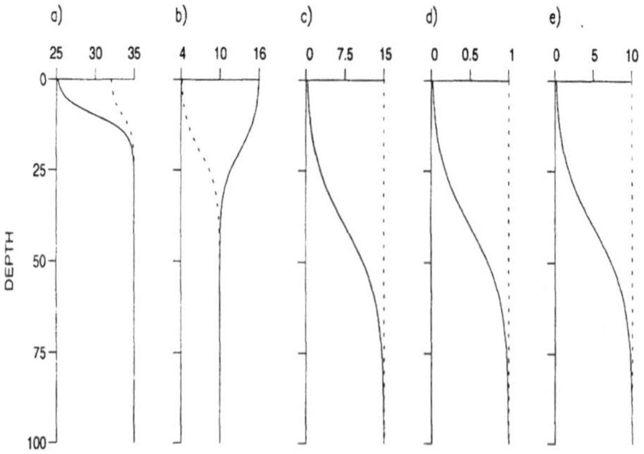

Figure 2. Vertical profiles of (a) salinity, (b) temperature of the coastal water in °C, (c) nitrate in μM, (d) phosphate in μM, (e) silicate in μM. Solid lines indicate the summer profiles and dashed lines the winter profiles.

Table 1. Parameters of nutrient boundary value parameters

Symbol	Value/unit	Explanation
N_{max}	15.0 (μM)	Maximum value of inorganic nitrogen
N_{min}	0.0 (μM)	Minimum value of inorganic nitrogen
P_{max}	1.0 (μM)	Maximum value of inorganic phosphate
P_{min}	0.0 (μM)	Minimum value of inorganic phosphate
Si_{max}	10.0 (μM)	Maximum value of inorganic silicate
Si_{min}	0.0 (μM)	Minimum value of inorganic silicate
ΔN	$0.5 \times (N_{max} - N_{min})$	
ΔP	$0.5 \times (P_{max} - P_{min})$	
ΔSi	$0.5 \times (Si_{max} - Si_{min})$	
Z_n	40 (m)	Depth of nutricline
ΔZ_n	20 (m)	Thickness of nutricline
Dia_{min}	2.75 (mg N m^{-3})	Minimum value of diatoms
Fla_{min}	2.75 (mg N m^{-3})	Minimum value of flagellates

A sinusoidal variation over the year for these variables is achieved by using the interpolation technique given by Equations 3 and 4. As initial values, we apply

$$N(x,y,z,0) = N(L_x + L_{FRS}, y, z, 0)$$
$$P(x,y,z,0) = P(L_x + L_{FRS}, y, z, 0)$$
$$Si(x,y,z,0) = Si(L_x + L_{FRS}, y, z, 0)$$

For diatoms and flagellates the initial values are Dia_{min} and Fla_{min} respectively (Table 1). On inflow, these values are also used as Dirichlet boundary values. During summer the boundary values for diatoms and flagellates are probably too small. However, the focus is on new production in the fjord, and we want to keep the imported biomass at a minimum level. Detritus is initialised to be zero and we also assume zero concentration of detritus through the open boundary

Table 2. Average concentrations of the scalar field C as a function of z and spatial resolution

DX (km)	KB	$\bar{C}_{10}$	$\bar{C}_{20}$	$\bar{C}_{30}$	$\bar{C}_{40}$
1	21	0.0384	0.3986	0.5612	0.6376
1	30	0.0117	0.3479	0.5085	0.6026
1	43	0.0219	0.3442	0.4896	0.5619
1	52	0.0241	0.3478	0.5376	0.6462
1	61	0.0119	0.3444	0.5309	0.6442
1	70	0.0101	0.3392	0.5245	0.6340
2	70	0.0043	0.3321	0.5011	0.5921
4	70	0.0001	0.3496	0.5155	0.6062

on inflow. Table 1 summarises the nutrient parameter values and Figure 2 shows the vertical profiles of salinity, temperature, nitrate, phosphate and silicate at the open boundary.

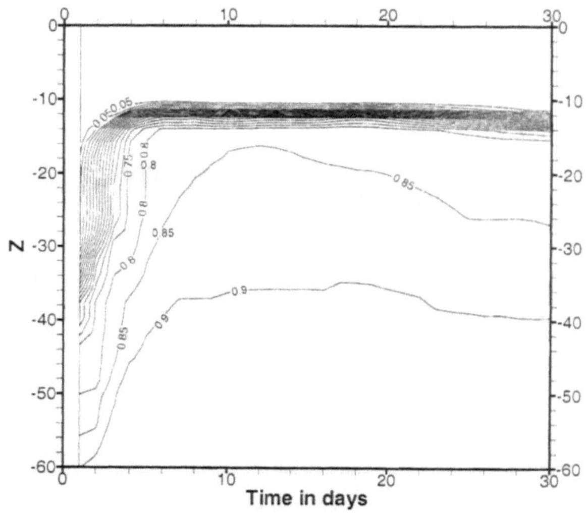

Figure 3. Concentrations of C in the upper 60 m of the fjord over 30 days at (x, y) = (6 km, 2 km) produced with 4 km horizontal resolution and 70 layers in the vertical. The fresh water flux at the surface is 90 $m^3\ s^{-1}$ and the flux at 50 m depth is 10 $m^3\ s^{-1}$. The contour levels are 0(0.05)1.

Results

Sensitivity to the vertical and horizontal resolution

Primary production is governed by supply of nutrients to the upper layers. To run the model for 1 year, with 1 km resolution in the horizontal and 52 vertical layers, requires more than 100 CPU hours on a powerful workstation. Therefore, we studied the significance of resolution on the models ability to supply nutrients from the deeper part of the fjord to the surface layer.

A scalar field, C, is initiated as $C(x, y, z) = 1.$ for $z < -40$ m and $C(x, y, z) = 0.$ for $0 \geq z \geq -40$ m, $\forall x, y$. These values are also kept as Dirichlet boundary conditions at the open boundary. The model is forced with a freshwater flux of 90 $m^3\ s^{-1}$ released at the surface and a freshwater flux of 10 $m^3\ s^{-1}$ released at 50 m depth, both at $x = 0$. The concentration of C in the river runoff is zero. The model is run for 720 h in these experiments. Average values of C over the upper layers are then computed from

$$\bar{C}_z = \frac{1}{z \times 720\,\text{h} \times 8\,\text{km} \times 4\,\text{km}} \int_0^{720\,\text{h}} \int_0^{8\,\text{km}} \int_0^{4\,\text{km}} \int_{-z}^{0} C(x, y, z, t)\text{d}x\text{d}y\text{d}z\text{d}t$$

for z = 10, 20, 30 and 40 m and for different number of layers in the vertical, KB, and different horizontal resolutions, DX (Table 2).

The development of C in time in the upper 60 m of the fjord at the point (x, y) = (6 km, 2 km) is given in Figure 3 for the 4-km resolution case. Other choices of grid resolution gave quite similar results. We always find a sharp gradient at approximately 10 m depth. This gradient is only marginally shallower with finer horizontal resolution. For the case with 100 $m^3\ s^{-1}$ released at the surface the development of the concentration field over the first month is given in Figure 4. The separation between depleted water and enriched water is deeper than 20 m in this case. Since the results from this study were relatively robust to changes in horizontal resolution, the sensitivity studies of physical parameters were performed with 4 km horizontal resolution and 70 layers in the vertical. This allowed good resolution of the photic layer, and gave a CPU time of 2 h for 1 year simulations on a workstation.

Processes on scales smaller than 4 km are important. Resolution at length scales of the buoyant plume, approximately 1 m, would significantly reduce the errors due to spatial resolution. This, however, is unachievable with state-of-the-art models and computers. At such small length scales, a non-hydrostatic model would be necessary, but in the present study we expect to find only minor non-hydrostatic effects (Cushman-Roisin, 1994).

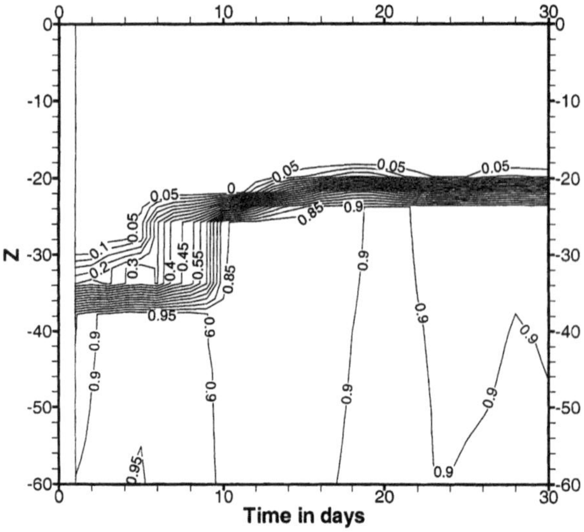

Figure 4. Concentrations of C in the upper 60 m of the fjord over 30 days at $(x, y) = (6$ km, 2 km$)$ produced with 4 km horizontal resolution and 70 layers in the vertical. The fresh water flux at the surface is 100 $m^3\ s^{-1}$ and no fresh water is submerged. The contour levels are 0(0.05)1.

Three basic scenarios

In the first three experiments, the horizontal resolution is 4 km, 70 layers are used in the vertical, the 3-D time steps are 360 s and the 2-D time steps 10 s. The forcing is:

(a) No freshwater flux. The primary production is only driven by light, nutrients and temperature.

(b) Freshwater with constant flux (100 $m^3\ s^{-1}$) is added to the upper 8 layers (2 m depth) at $x = 0$ m.

(c) The same total freshwater flux as in (b), but where 90 $m^3\ s^{-1}$ is added into the upper eight layers at $x = 0$ m. We assume that the remaining 10 $m^3\ s^{-1}$ is led through a pipe down to 50 m depth at $x = 0$ m.

The introduced freshwater at 50 m is subject to strong buoyant forces and rises. On the way towards the surface, the ambient water masses will be mixed with the freshwater.

Circulation and hydrography

In Figures 5 and 6, the currents in the upper 50 m at day 135 are shown. Over the last 11 months the average transports out of the fjord through a cross section at $x = 54$ km are for scenario (b) 644 $m^3\ s^{-1}$ when integrated over the upper 10 m, and 2499 $m^3\ s^{-1}$ when integrated over the upper 50 m. For scenario (c) the corresponding numbers are 425 $m^3\ s^{-1}$ and 3790

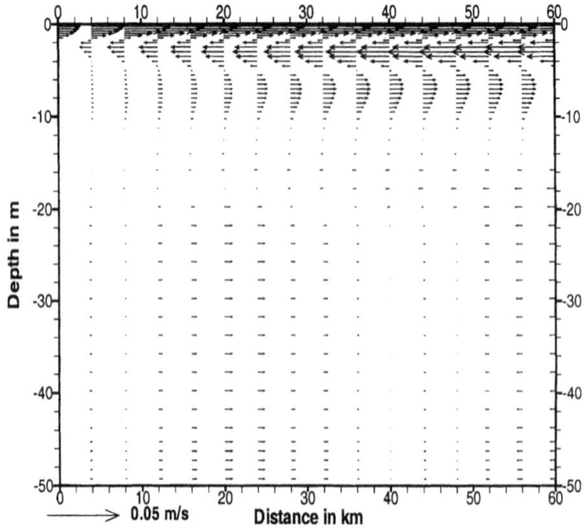

Figure 5. Along fjord currents for scenario (b) after 135 days in the upper 50 m of the fjord.

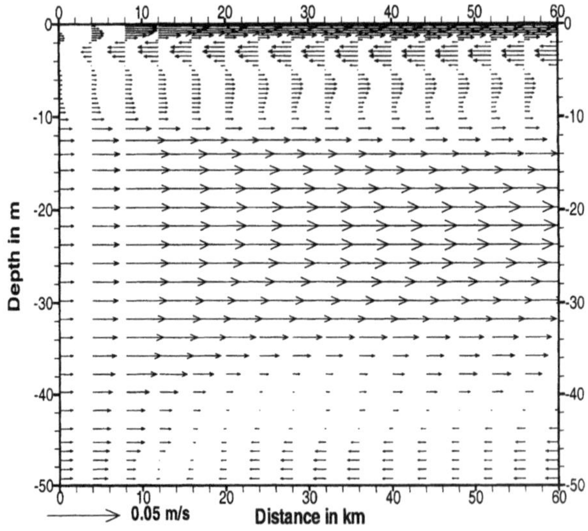

Figure 6. Along fjord currents for scenario (c) after 135 days in the upper 50 m of the fjord.

Table 3. Primary productions of flagellates ($Prod_{Fla}$), diatoms ($Prod_{Dia}$) and total production ($Prod_{Tot}$) averaged over the inner 40 km of the fjord in the three basic experiments. All values are in g C m^{-2} $year^{-1}$

Exp.	$Prod_{Fla}$	$Prod_{Dia}$	$Prod_{Tot}$
(a)	28	40	68
(b)	28	46	74
(c)	26	207	233

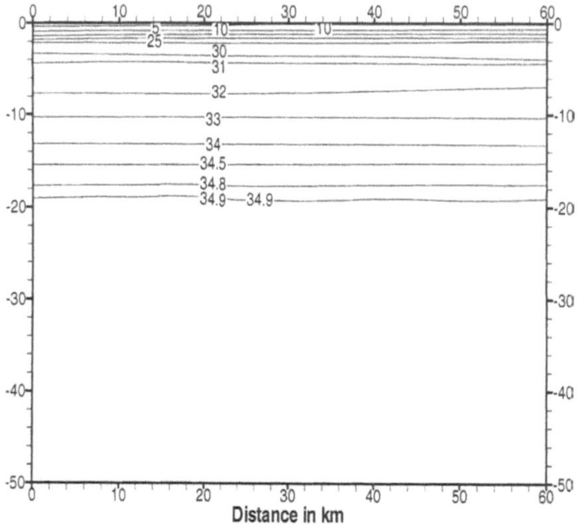

Figure 7. Along fjord salinities for scenario (b) after 135 days in the upper 50 m of the fjord.

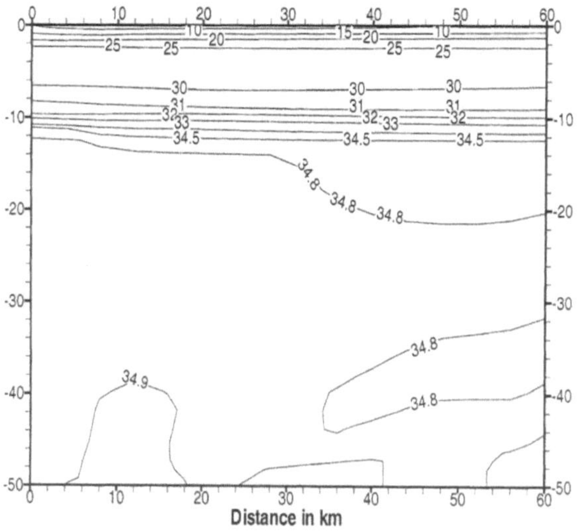

Figure 8. Along fjord salinities for scenario (c) after 135 days in the upper 50 m of the fjord.

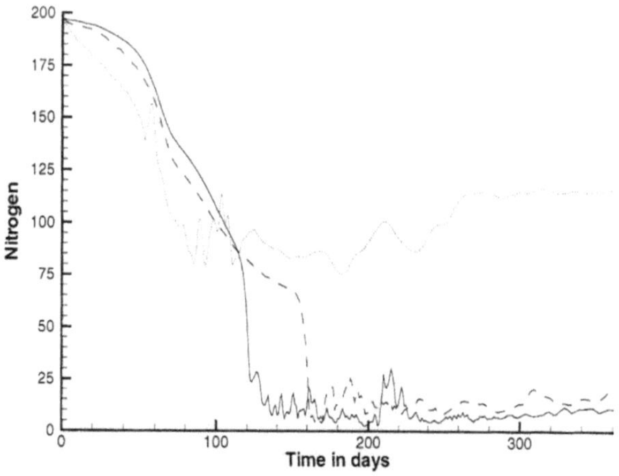

Figure 9. Time series of average concentrations of inorganic nitrogen in mg m^{-3} for scenario (a) (solid line), scenario (b) (dashed line) and scenario (c) (dotted line). The averages are computed over the upper 30 m of the inner 40 km of the fjord.

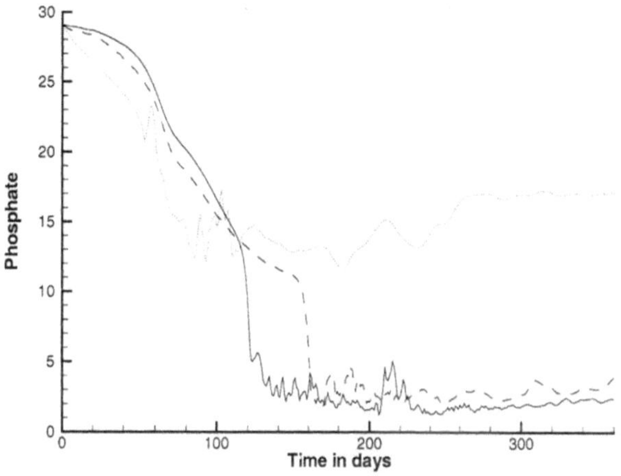

Figure 10. Time series of average concentrations of inorganic phosphate in mg m^{-3} for scenario (a) (solid line), scenario (b) (dashed line) and scenario (c) (dotted line). The averages are computed over the upper 30 m of the inner 40 km of the fjord.

m^3 s^{-1}. In Figures 7 and 8 the salinity fields corresponding to the currents shown in Figures 5 and 6 are given.

By comparing Figures 5 and 7 with Figures 6 and 8, we notice the effects of the submerged fresh water supply. In scenario (b), we have a thin layer ($\sim$ 2 m) of outflowing water above a layer, approximately 2–5 m, of inflowing water. Between 5 and 10 m there is also a layer of outflowing water building up in strength as one goes out the fjord. For scenario (c) the situation is very similar above 10 m depth, but between 10 and 40 m we get a very strong outflow. Connected to this outflow we find a return flow of nutrient-rich water at depths greater than 40 m.

Phytoplankton dynamics

In Figures 9–13 time series of inorganic nitrogen, phosphate, silicate, diatoms and flagellates are given. Figure 14 shows the concentration of diatoms in the upper 50 m after 135 days. Total production averaged over the inner 40 km of flagellates and diatoms are summarised in Table 3. Yearly production as a function of the distance from the head of the fjord are given in Figure 15.

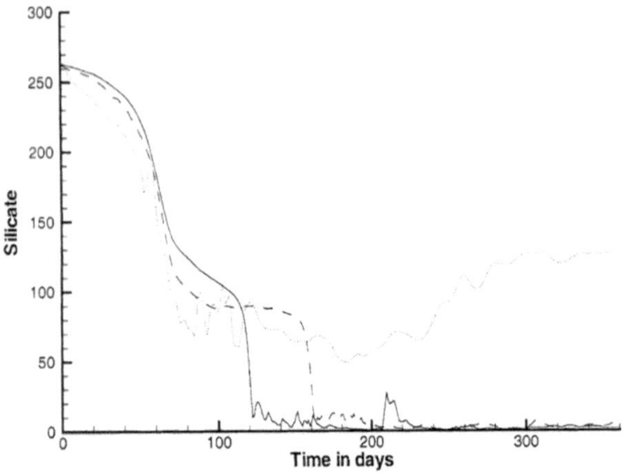

Figure 11. Time series of average concentrations of inorganic silicate in mg m^{-3} for scenario (a) (solid line), scenario (b) (dashed line) and scenario (c) (dotted line). The averages are computed over the upper 30 m of the inner 40 km of the fjord.

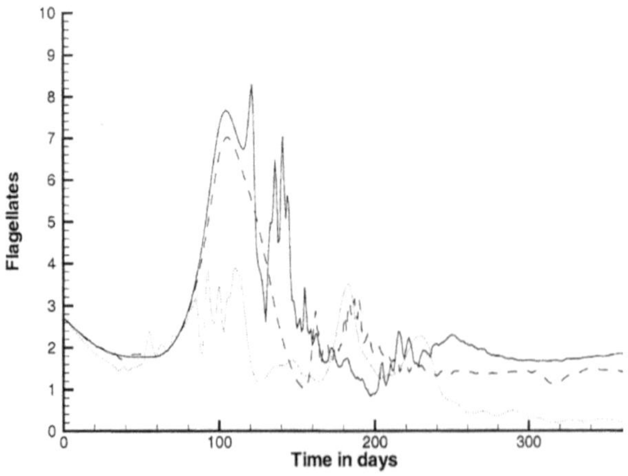

Figure 13. Time series of average concentrations of flagellates in mg N m^{-3} for scenario (a) (solid line), scenario (b) (dashed line) and scenario (c) (dotted line). The averages are computed over the upper 30 m of the inner 40 km of the fjord.

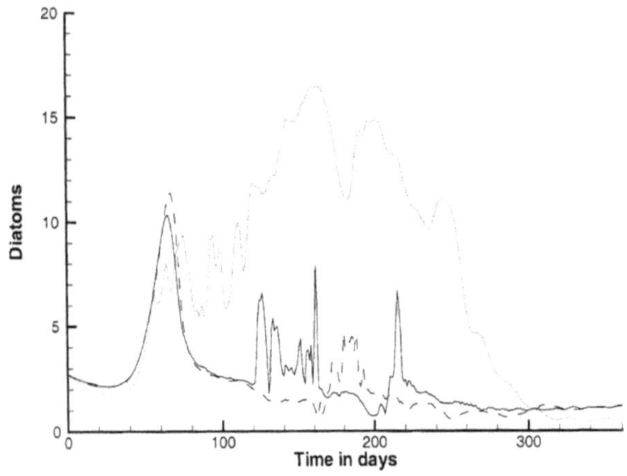

Figure 12. Time series of average concentrations of diatoms in mg N m^{-3} for scenario (a) (solid line), scenario (b) (dashed line) and scenario (c) (dotted line). The averages are computed over the upper 30 m of the inner 40 km of the fjord.

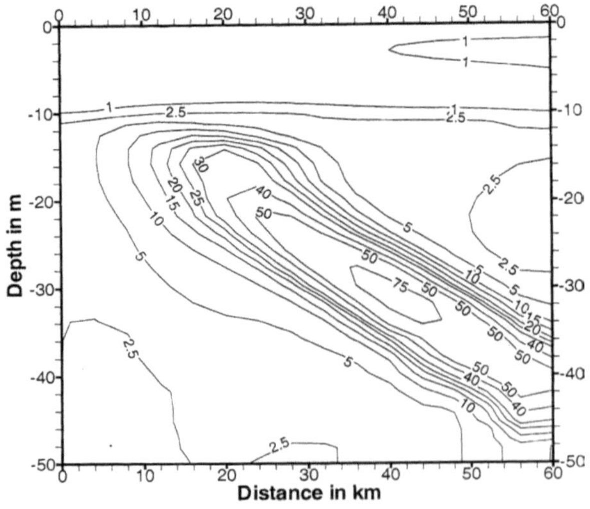

Figure 14. Along fjord concentrations of diatoms in mg N m^{-3} after 135 days in the upper 50 m of the fjord for scenario (c).

From the simulation results (Figs 9–15, Table 3) we conclude:

(i) The total production is almost three times higher for the submerged case than for the normal case.

(ii) Diatom production clearly dominated over flagellates in scenario (c) (Table 3 and Fig. 15). Diatoms are fed continously with silicate from the upwelled water. As long as diatom growth is not silicate limited, flagellates compete poorly with diatoms due to the lower inherent growth rate of the flagellates (Egge & Aksnes, 1992; Aksnes et al., 1995).

(iii) Despite the strong increase in primary production (scenario (c)), the production is still nutrient limited in the surface layer because the buoyant plume only reaches up to 10 m (Fig. 14). This means that the primary production can be enhanced even more with higher supply of nutrients to the surface layer either at the head of the fjord and/or throughout the fjord.

Sensitivity to parametrisation of vertical mixing

For the submerged discharge, the dilution and resulting primary production is to a large extent governed by values of the diffusivities and viscosities produced by the turbulence scheme when the water masses are unstable.

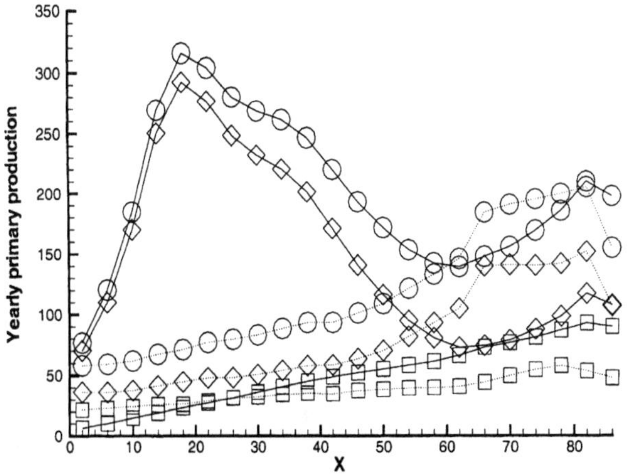

Figure 15. Yearly primary production of flagellates (squares), diatoms (diamonds) and total production (circles) for scenario (b) (dotted line) and scenario (c) (solid line). The values are in g C m^{-2} $year^{-1}$. On the horizontal axis the distance out the fjord in km is given. The production through the boundary zone is included.

In Berntsen et al. (1997) the algorithm due to Munk & Anderson (1948) was applied to compute the diffusivities and viscosities. Vertical viscosity, K_M, and vertical diffusivity, K_H, are computed as functions of the Richardson number according to the formula

$$K_H = 10^{-3}\ m^2\ s^{-1} \times (1 + 3.33\ Ri)^{-1.5};$$
$$K_M = 10^{-3}\ m^2\ s^{-1} \times (1 + 10\ Ri)^{-0.5};$$

where Ri is the Richardson number.

These formula apply when Ri is positive (stable water masses), and other choices must be made when Ri is negative (unstable water masses). In such cases we apply $K_M = K_H = K_{unstable}$. Table 4 gives sensitivity of the primary production to various values of $K_{unstable}$ for the Munk & Anderson formulation and for the Mellor & Yamada scheme. For larger values of $K_{unstable}$, primary production is smaller than for the Mellor & Yamada (1982) scheme which is a more high level and widely used closure scheme. For small values of $K_{unstable}$ the productions are similar and also greater than the values reported in Berntsen et al. (1997).

The smaller values of $K_{unstable}$ create a dilution that occurs on a longer time scale. This allows the buoyant plume to reach all the way up to the surface. The circulation is then dominated by a continuous outflow in the upper 40 m, and an inflowing counter current below as we found in Berntsen et al. (1997). In a real experiment there are also degrees of freedom, as discussed by Fischer et al. (1979), that may be utilised to optimise the dilution, and thereby the production.

Table 4. Sensitivity of the primary production averaged over the inner 40 km of the fjord to the choice of turbulence closure scheme for the submerged discharge case. M &Y indicate that the Mellor &Yamada (1982) scheme is applied and M &A that the Munk & Anderson (1948) scheme is applied. Values of $K_{unstable}$ are given in the parenthesis behind M & A. All productions are in g C m^{-2} $year^{-1}$

Turbulence closure	$Prod_{Fla}$	$Prod_{Dia}$	$Prod_{Tot}$
M & Y	26	207	233
M &A (0.5 ($m^2\ s^{-1}$)	51	101	153
M &A (0.2 ($m^2\ s^{-1}$)	38	135	172
M &A (0.1 ($m^2\ s^{-1}$)	35	155	190
M &A (0.05 ($m^2\ s^{-1}$)	38	148	185
M &A (0.02 ($m^2\ s^{-1}$)	34	216	250
M &A (0.01 ($m^2\ s^{-1}$)	27	213	240
M &A (0.005 ($m^2\ s^{-1}$)	56	310	366
M &A (0.002 ($m^2\ s^{-1}$)	38	356	394
M &A (0.001 ($m^2\ s^{-1}$)	34	387	422

Table 5. Sensitivity of the primary production averaged over the inner 40 km of the fjord to the horizontal viscosity parameters C_M and C_{M2D} for the submerged discharge case. All productions are in g C m^{-2} $year^{-1}$. Averages of A_M and A_{M2D} over the whole year and the whole fjord $\bar{A}_M$ and $\bar{A}_{M2D}$, respectively, are given in $m^2\ s^{-1}$

C_M	C_{M2D}	$Prod_{Fla}$	$Prod_{Dia}$	$Prod_{Tot}$	$\bar{A}_M$	$\bar{A}_{M2D}$
50	2500	44	132	176	366	254
20	1000	30	195	221	200	108
10	500	26	207	233	121	51
2	100	27	71	99	43	12
1	50	23	103	125	37	15

In all other studies reported here the Mellor & Yamada (1982) scheme has been applied.

Sensitivity to parametrisation of horizontal viscosity

Primary production is sensitive to the horizontal viscosities. The viscosities, A_M and A_{M2D}, are computed from Equations (1) and (2), respectively. They are determined by the $2\Delta x$ scale shear in the velocity fields and the parameters C_M and C_{M2D}, respectively. The sensitivity of primary production to C_M and C_{M2D} is reported in Table 5. Average values of A_M and A_{M2D} over year and fjord are also given.

The depth integrated velocities are determined by the river fluxes and are in average equal to 0.0001 ms^{-1} (100 $m^3\ s^{-1}$/(4000× 250 m)). The correspond-

Table 6. Sensitivity of the primary production averaged over the inner 40 km of the fjord to the depth of the submerged discharge. All productions are in g C m^{-2} year^{-1}

Depth (m)	$Prod_{Fla}$	$Prod_{Dia}$	$Prod_{Tot}$
10	25	44	69
20	38	59	97
30	36	81	117
40	33	228	260
50	26	207	233
60	25	211	236
70	29	258	287
80	23	111	134
100	23	146	170

Table 7. Sensitivity of the primary production averaged over the inner 40 km of the fjord to the wind for the all discharge at the surface case. All productions are in g C m^{-2} year^{-1}

W_{max} (m s^{-1})	T (h)	$Prod_{Fla}$	$Prod_{Dia}$	$Prod_{Tot}$
0		28	46	74
2	24	25	44	69
5	24	22	44	66
8	24	25	56	81
8	120	74	131	205
11	24	28	40	68
11	120	85	146	231

ing shears tend to be small. To produce values of A_{M2D} that are of the same order as A_M and large enough to filter away $2\Delta x$ noise in the 2D-fields, C_{M2D} must be chosen much larger than C_M.

Except for the sensitivity analysis the values of C_M and C_{M2D} are 10. and 500., respectively. Even for these large values, we notice from Figure 15 that there is an enhanced production at the open boundary which is mainly due to artificial divergence/convergence close to the open boundary. For smaller values of the coefficients this gets worse, and eventually the noise propagate into the interior domain and destroys the general circulation pattern that supplies the upper layer with nutrients in the submerged case. On the other hand, too large values also filter away modes that may be well represented on the given grid, weakens the flow pattern and reduces the production. To decide on horizontal viscosities is therefore a trade off between the need to filter out $2\Delta x$ noise and to maintain well represented features.

No explicit horizontal diffusivity is applied to the scalar fields S and T, or to the chemical and biological variables in our experiments.

Sensitivity to subcharge depth

Primary production is a function of the depth of the discharge (Table 6). Production is similar for discharge depths between 40 and 70 m with apparent maxima at 40 and 70 m. With shallower discharge, the water masses brought to the surface do not contain much nutrients and production is gradually decreased towards the normal case with river flow only at surface. With large discharge depths the buoyant plume does not reach far enough into the photic zone and production is reduced.

Sensitivity to wind forcing

For the scenario with all river outflow at surface (case (b)), the flow is also forced with wind. For most fjords wind direction is typically in or out of the fjord because of surrounding mountains. To illustrate the response of production to wind, we forced with wind oscillating periodically according to the equation

$$W_x = W_{max} \sin(2\pi t/T),$$

where W_x is wind speed at time t in the x-direction, W_{max} the amplitude of the oscillation and T the period. There is no wind across the fjord. From this, the wind stress is computed according to

$$\tau_x = 1.2 \times 10^{-3} \rho_a |W_x| W_x$$

where ρ_a is the density of air which is set to 1.3 kg m^{-3}. The production for various values of W_{max} and T is given in Table 7.

In summer T is typically equal to the period of the daily sea breeze which is 24 h. Table 7 indicates that this period is too short to create upwelling that enhances production. To illustrate possible effects of longer T, a period of 120 h was used. In this case production may be enhanced similarly to the submerged discharge case (Table 7). However, a noticeable difference is that the production of flagellates is much stronger when the enhancement is created by wind forcing instead of submerged freshwater.

Sensitivity to sill depth

Norwegian fjords are formed during the last glacial period. Most, but not all, have a sill at the fjord mouth.

Table 8. Sensitivity of the primary production averaged over the inner 40 km of the fjord to the sill depth for the submerged discharge case. All productions are in g C m^{-2} year^{-1}

Discharge depth (m)	Sill depth (m)	$Prod_{Fla}$	$Prod_{Dia}$	$Prod_{Tot}$
50	250	26	207	233
50	200	51	213	264
50	150	52	136	188
50	100	40	73	114
70	100	34	133	167
100	100	21	80	101
50	50	33	63	96
70	50	30	123	153
100	50	25	109	134

Table 9. Sensitivity of the primary production averaged over the inner 40 km of the fjord to the period in hours of the discharge puls. All productions are in g C m^{-2} year^{-1}. The asterix indicate that the production values are for the case with constant flow

Period	$Prod_{Fla}$	$Prod_{Dia}$	$Prod_{Tot}$
*	26	207	233
0.5	25	207	232
1.0	24	200	225
2.0	22	180	202
4.0	20	160	180
5.1	23	195	218
8.0	23	193	217
24.0	21	159	180
48.0	23	189	212
94.3	23	188	211
112.0	29	208	237
120.0	22	175	197
240.0	28	182	210

Typical sill depths are 50–120 m and there might be more than one sill. However, often the exchange between the fjord and the ocean is controlled by wide shallow banks outside the sill which then determine the effective barrier for water exchange.

We studied the effects of sill depth on primary production (Table 8). We defined depth equal to 250 m in the inner 60 km and depth equal to H_{Sill} outside 64 km. The depth varies linearly in between.

Production is considerably reduced as flow is gradually constrained by reducing the sill depth (Table 8). Shallow sill depth leads to recirculation of depleted, outward flowing water masses. This water flows back to the submerged discharge, and submerged freshwater mixes with fjord water already depleted in nutrients. However, below this recirculation we find, even for the case with 50 m sill, an inflow of nutrient-rich water. Table 8 indicates that with increased discharge depths, this water is brought into the photic zone and still enhances the production.

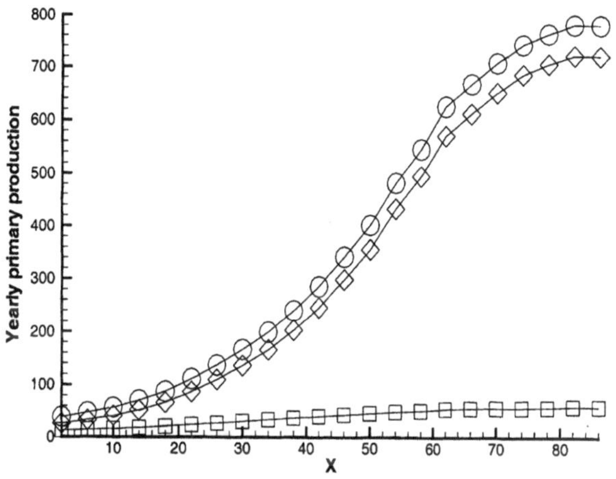

Figure 16. Yearly primary production of flagellates(squares), diatoms (diamonds) and total production (circles) for the case with 90 m^3 s^{-1} submerged to 200 m depth. The values are in g C m^{-2} year^{-1}. On the horizontal axis the distance out the fjord in km is given. The production through the boundary zone is included.

Effects of pulsating the freshwater discharge

Pulsating the submerged water may enhance the mixing and thereby primary production. Instead of assuming constant discharge (10 m^3 s^{-1}), we assume

$$F = 10\,\mathrm{m}^3\,\mathrm{s}^{-1} + 10\,\mathrm{m}^3\,\mathrm{s}^{-1}\sin(2\pi t/T),$$

where F is flux at time t and T the period. A number of periods were simulated, including those appearing in the time series of temperature and that are believed to be connected to internal waves of fjords. The major result of this study is that primary production is not further enhanced if the submerged water enter the fjord in pulses (Table 9). Even if we apply periods observed in the system, 5.1 h, 94.3 h or 112.0 h, the total production becomes similar to the production we get with a constant discharge. The primary production seems to benefit from a steady supply of nutrients and steady behaviour of the flow field, while pulsating of the discharge may have a negative effect on production.

Effects of increasing the submerged freshwater discharge

In the experiments so far only 10% of the total river runoff has been submerged. We also studied the effects of leading 90 m^3 s^{-1} through a pipe and submerging it at 200 m depth, and leaving only 10 m^3 s^{-1} as surface outlet. The total production (inner 40 km average) for this experiment is only 116 g C m^{-2} $year^{-1}$. The production pattern throughout the fjord, however, is completely different (Fig. 16). The entrainment and the circulation in the fjord become so strong that phytoplankton growth is moved away from the head of the fjord. At the mouth of the fjord the maximum production is 781 g C m^{-2} $year^{-1}$. It is possible that peak production in a real experiment will be moved further out than our 60-km model fjord. Notice also the clear dominance of diatoms over flagellates in Fig. 16. The results from this experiment suggest that by submerging major fractions of river fluxes, one may enhance the primary production considerably over large geographical scales.

Discussion

We have shown that there may be a considerable potential for enhancing the primary production in fjords and coastal areas by utilising freshwater to create upwelling of nutrient water. The forcing and topography of real fjords will be more complex than our simplified case. The model should be run with more realistic forcing and topography in order to calculate the potential for increased primary production for a specific fjord. The interpretation of the results for a more realistically specified fjord, however, would have been complicated by a number of processes acting more or less simultaneously. Our idealised fjord makes it much easier to interpret results.

Fischer et al. (1979) describe the dilution in buoyant plumes from submerged discharges and discuss factors governing the dilution such as stratification, discharge flux, and the length and spacing of diffusers. Dilution of submerged freshwater is the key to enhanced primary production in our experiments. We want to maximise the dilution in order to bring nutrient-rich water into the buoyant plume. At the same time, however, the plume should rise as high as possible into the photic zone. With no dilution, no nutrients are brought to the surface. With too high dilution, the buoyant plume will not reach the photic zone. An optimum in between will maximise primary production.

The buoyant plume in scenario (c) rises to approximately 10 m depth. Budget calculations indicate a the dilution factor in the range from 17 to 174 which is in the range given by Fischer et al. (1979). Our simulations (Figs 3 and 8) indicate that the buoyant plume reaches up to approximately 10 m depth. This was sufficient to create a considerable enhancement of the production.

The grid spacing applied is far from sufficient to resolve the buoyant plume. The results therefore depend on the quality of the parametrisations of processes at sub grid scale. Further studies, especially with enhanced resolution around the buoyant plume, are necessary to improve the quality of the numerical results. The quality is also improved if the fjord model is connected more realistically to a model system for the coast.

The simulated production of 70–80 g C m^{-2} $year^{-1}$ in the run without any freshwater supply seems plausible. Although measured primary production of Norwegian fjords is likely to exceed this estimate (Aksnes & Lie, 1990), we must take into account that the present simulation assumes no supply of nutrient from land run-off, and furthermore that the model tends to underestimate the remineralized production due to the missing food web structure of the model. The artificial upwelling experiments of Aksnes et al. (1985) demonstrated that an artificial upwelling rate corresponding to approximately 1 m day^{-1} resulted in an 8–10-fold increase in the primary production during mid-summer. The upwelling rate in the present simulation corresponds to 0.22 m day^{-1}, assuming that the submerged water upwells in the nearest model cell. This gave rise to a yearly production enhancement factor of about 3 for the whole fjord. The actual enhancement near the outlet, however, is approximately 5 (Fig. 15). Results from the experiment with 90 m^3 s^{-1} freshwater submerged at 200 m created production enhancements comparable to those reported in Aksnes et al. (1985).

In accordance with the physical experiments of Aksnes et al. (1985) our simulated production enhancement was due to the growth of diatoms (Fig. 15 and Table 3). In contrast to commercial fertilisers, and to anthropogenic nutrient inputs in general, the deep water contains high levels of silicate. In contrast to flagellates, the fast-growing diatoms require silicate for their growth, and hence deep water is a suitable growth medium for this phytoplankton group. During summer, coastal water becomes stratified and devoid of silicate. Consequently, diatoms become out-competed by flagellates. This is of particular significance to shell farming. Fjords and coastal areas of Sweden and Norway are frequently unsuitable for mussel farming due to the presence of toxic flagellates. This is effectively hindering a development of shell farming in Scandinavia. Haamer (1995), however, showed that the toxicity of mussels was related to the phytoplankton and nutrient compositions in Swedish coastal water. Low concentrations of toxin were found in areas with high silicate supply, while high concentrations were found in silicate depleted areas. Haamer (1995) hypothesised that presence of silicate promotes growth of diatoms which are non-toxic, whereas absence of silicate favours dinoflagellates including DSP-producing Dinophysis species. Mesocosms experiments carried out by Egge & Aksnes (1992) and Strand (1993) also indicate that high silicate supply may ensure diatom dominance during summer. This points to the possible 2-fold benefits of using artificial upwelling in connection with shell farming. Firstly, shell will feed directly upon the enhanced primary production, and secondly the problem of algal toxins may be greatly reduced or even eliminated.

In our sensitivity studies it is shown that the primary production is highly sensitive to the vertical mixing that takes place above the submerged discharge. In a real experiment it is possible to optimise the dilution and thereby production, and higher production levels than we have indicated may be obtained. The production was relatively robust to discharge depths in the range from 40 to 70 m. Discharges placed above or below this range reduced the productions. It was furthermore shown that daily see breeze, typical for summer, did not enhance the production substantially. Wind events lasting for days may enhance the production to levels found for our simulated submerged case. Primarily flagellates, however, were stimulated in this case. The production was reduced with sill depth. However, with increased discharge depth production may be enhanced considerably even for sill depths of 50 m. It was shown that the simulated primary production seemed to benefit from a steady supply of nutrients, while pulsating the discharge might have a negative effect on production. In 1998 a Norwegian project, FJORDCULT, was established to analyse the potential and possible bottlenecks to enhanced primary production using artificial upwelling and to locate a suitable fjord. The project ended up proposing the Samnangerfjord east of Bergen as a test fjord. Scientists supporting this idea are now working to get the discharge system and the required monitoring program funded, see Aure et al. (2000).

Acknowledgements

The numerical experiments have been run on a Cray ORIGIN 2000 installed at Parallab, University of Bergen, and we thank Parallab for kind support. This research has also received support from The Research Council of Norway (Programme for Supercomputing) through a grant of computing time. An anonymous referee is thanked for valuable comments and suggestions.

References

Aksnes, D. & J. Egge, 1991. A theoretical model for nutrient uptake in phytoplankton. Marine Ecology Progress Series 70: 65–72.

Aksnes, D. & U. Lie, 1990. A coupled Physical-Biological Pelagic Model of a Shallow Sill Fjord. Estuar. Coast. Shelf Sci. 31: 459–486.

Aksnes, D., T. Magnesen & U. Lie, 1985. Nutrient enrichment experiments in plastic cylinders and the implications of enhanced primary production in Lindåspollene, western Norway. Sarsia 70: 45–58.

Aksnes, D., J. Egge, R. Rosland & B. Heimdal, 1994. On the representation of *Emilania huxleyi* in phytoplankton simulation models – A first approach. Sarsia 79: 291–300.

Aksnes, D., K. Ulvestad, B. Balino, J. Berntsen, J. Egge & E. Svendsen, 1995. Ecological modelling in coastal waters: Towards predictive physical-chemical-biological simulation models. OPHELIA 41: 5–36.

ANON, 1994. MARICULT – Programme description, Hydro Research Centre, Porsgrunn, Norway.

Aure, J., L. Asplin & Ø. Strand, 2000, Kan vi gjøre fjordene våre mer produktive? (in Norwegian). In Aure, J. (ed.), Havets miljø 2000. Fisken og Havet 2-2000, Institute of Marine Research, Nordnesgt. 50, N-5817 Bergen, Norway, ISSN 0802 0620: 90–94.

Baliño, B., 1996. Eutrophication of the North Sea, 1980-1990: an evaluation of anthropogenic inputs using a 2D phytoplankton production model. Ph.D. thesis, Department of Fisheries and Marine Biology, University of Bergen, Norway.

Berntsen, J., 2000. USERS GUIDE for a modesplit σ-coordinate numerical Ocean model. Technical Report 135, Dept. of Applied Mathematics, University of Bergen, Johs. Bruns gt.12, N-5008 Bergen, Norway: 48 pp.

Berntsen, J., D. Aksnes & A. Foldvik, 1997. Fresh water driven primary production in a fjord. Technical Report 109, Dept. of Applied Mathematics, University of Bergen.

Brown, P. & T. Parsons, 1972. The effect of simulated upwelling on the maximization of primary productivity and the formation of phytodetritius. Mem. Ist. ital. Idrobiol. Dott 29: 169–183.

Cooper, L. & G. Steven, 1948. An experiment in marine fish cultivation. Nature 161: 631–633.

Cox, M. & K. Bryan, 1984. A numerical model of the ventilated thermocline. J. Phys. Oceanogr. 14: 674–687.
Cushman-Roisin, B., 1994. Introduction to Geophysical Fluid Dynamics. Prentice Hall, Englewood Cliffs, NJ, ISBN-0-13-353301-8.
Egge, J. & D. Aksnes, 1992. Silicate as regulating nutrient in phytoplankton competition. Mar. Ecol. Prog. Ser. 83: 281–289.
Fischer, H., E. List, R. Koh, J. Imberger & N. Brooks, 1979. Mixing in Inland and Coastal Waters. Academic Press, New York. ISBN-0-12-258150-4.
Furnas, M., 1990. In situ growth rates of marine phytoplankton: approaches to measurement, community and species growth rate. J. Plankton Res. 12: 1117–1151.
Gross, F., J. J.E.G. Raymont, M. Marshall & A. Orr, 1944. A fish-farming experiment in a sea loch. Nature 153: 483–489.
Haamer, J., 1995. Presence of the phycotoxin okadic acid in mussels *Mytilus edulis*) in relation to nutrient composition in a Swedish coastal water. J. Shellfish Res. 14: 209–216.
Lynch, D., J. Ip, C. Naimie & F. Werner, 1995. Convergence studies of tidally-rectified circulation on Georges Bank. In Lynch, D. & A. Davies (eds), Quantitative Skill Assessment for Coastal Ocean Models. American Geophysical Union.
Martinsen, E. & H. Engedahl, 1987. Implementation and testing of a lateral boundary scheme as an open boundary condition in a barotropic ocean model. Coast. Eng. 11: 603–627.
Mellor, G. & T. Yamada, 1982. Development of a turbulence closure model for geophysical fluid problems. Rev. Geophys. Space Phys. 20: 851–875.
Munk, W. & E. Anderson, 1948. Notes on a theory of the thermocline. J. Mar. Res. 7: 276–295.
Nevø, A., R. Clasby, J. Goering & D. Hood, 1976. Enhancement of primary production by artificial upwelling. Mar. Sci. Commun. 2: 109–124.
Paul, A., J. Paul & P. Shoemaker, 1979. Artificial upwelling and phytoplankton production in Alaska. Mar. Sci. Commun. 5: 78–89.
Platt, T., D. Subba & K. Denman, 1977. Quantitative stimulation of phytoplankton productivity by deep water mixture in a coastal inlet. Estuar. Coast. mar. Sci. 5: 567–573.
Roels, O., 1980. From the deep sea: food, energy and fresh water. Mech. Eng. 102: 37–43.
Roels, O., 1983. Potential for mariculture development in the Bahamas. Proc. Gulf Carribean Fish. Inst. 35: 17–18.
Shiels, W. & D. Hood, 1970. Artificial upwelling in Alaskan fjord estuaries. Northern Eng. 2: 41–45.
Skartveit, A. & J. Olseth, 1986. Modelling slope irradiance at high latitudes. Solar Energy 36: 333–344.
Skartveit, A. & J. Olseth, 1988. Varighetstabellar for timevis belysning mot 5 flater på 16 Norske stasjonar (In Norwegian). Technical Report Rapport 7-1988, Geophysical Institute, Meteorological Division, Allegt. 70, N-5007 Bergen, Norway.
Skogen, M., 1993. A User's Guide to NORWECOM. The NORWegian ECOlogical Model system. Technical Report 6, Institute of Marine Research, Norway.
Skogen, M., E. Svendsen, J. Berntsen, D. Aksnes & K. Ulvestad, 1995. Modelling the primary production in the North Sea using a coupled three-dimensional physical- chemical-biological ocean model. Estuar. Coast. Shelf Sci. 41: 545–565.
Smagorinsky, J., 1963. General circulation experiments with the primitive equations, I. The basic experiment. Mon. Weather Rev. 91: 99–164.
Strand, Ø., 1993. Enhancement of bivalve production capacity in Norwegian heliothermic polls. Ph.D. thesis, Department of Fisheries and Marine Biology, University of Bergen.
Terry, K. & J. Caperon, 1982. Phytoplankton growth response to deep ocean water. Mar. environ. Res. 7: 211–225.
Wang, D.-P., 1984. Mutual intrusion of a gravity current and density front formation. J. Phys. Oceanogr. 14: 1191–1199.

Hydrobiologia **484:** 191–202, 2002.
O. Vadstein & Y. Olsen (eds), Sustainable Increase of Marine Harvesting: Fundamental Mechanisms and New Concepts.

Controlled artificial upwelling in a fjord using a submerged fresh water discharge: computer and laboratory simulations

T.A. McClimans[1], G. Eidnes[1] & J. Aure[2]
[1] *SINTEF Fisheries and Aquaculture, 7465 Trondheim, Norway*
[2] *Institute of Marine Research, 1870 Nordnes, 5817 Bergen, Norway*

Key words: upwelling, plume, nutrients, primary production, laboratory model

Abstract

To enhance the conditions for producing shellfish in coastal waters, the possibility of employing artificial upwelling of nutrients is explored. The effectiveness of a submerged discharge of fresh water is studied by means of a numerical buoyant plume model, BJET, and laboratory simulations. An optimisation of the entrainment of deeper, nutrient-rich water to the proper intrusion depth is demonstrated. The studies show that a downward directed jet of fresh water below the euphotic zone can lift significant amounts of nutrients to the primary production near the surface. The outlet must be large enough to lift the deeper water through the pycnocline to the desired depth of primary production. The results are applied to a possible discharge arrangement in the Samnangerfjord, to the east of Bergen, Norway, using field data from 1999. With a discharge of up to 8 m^3/s of fresh water at 35 m depth, the entrainment of deeper water into the buoyant plume, up to the 15 m depth, is 12 to 13 times as much. The chosen arrangement could give an expected vertical transport to the euphotic zone of 467 kg d^{-1} N, 46 kg d^{-1} P and 555 kg d^{-1} Si during the summer growth period. This includes periods of coastal downwelling with greatly reduced values of nutrients, but not periods of strong, deep stratification with deeper intrusions. Further optimisation is possible using active controls of the discharge system.

Introduction

Several coastal regions of the world's nations are plagued with eutrophication and the growth of harmful alge. There are, however, major parts of the globe that have little biological activity due to the lack of nutrients. The Norwegian research program MARICULT (MARIne CULTivation) addresses the question of the economy of fertilising the ocean and/or promoting the artificial upwelling of deeper, nutrient-rich seawater in coastal regions. The present project, within the MARICULT program, aims to suggest practical ways to enhance the primary growth in fjords through controlled artificial upwelling (McClimans & Eidnes, 2000; Aure et al., 2001). One scheme is based on our earlier experience with submerged outlets from hydro-power plants.

Due to the development of hydro-power in Norway, much more water is discharged to some fjords during the winter, when natural fresh water sources freeze up. If no measures were taken, this would cause an unnatural amount of ice, making it difficult to traffic the fjords. The purpose of submerging the outlets is therefore to reduce this ice cover. The criterion used is to induce a mixing that raises the salinity of the plume to at least 25 ppt. This promotes natural convection in the upper layer during atmospheric cooling. Two major outlets were built with discharges between 25 and 40 m depth (McClimans, 1981; Gjerp & McClimans, 1990).

These hydro-power schemes have caused upwelling of deeper water, but they are not efficient for primary production. Furthermore, the intrusions of the mixed water are too close to the surface to promote persistent, well-distributed photosynthesis (primary production). The main objective of the present work is therefore to apply the experience and tools from the earlier work to recommend an efficient (optimal) upwelling system using water resources that are not allocated to other use. This optimisation depends not

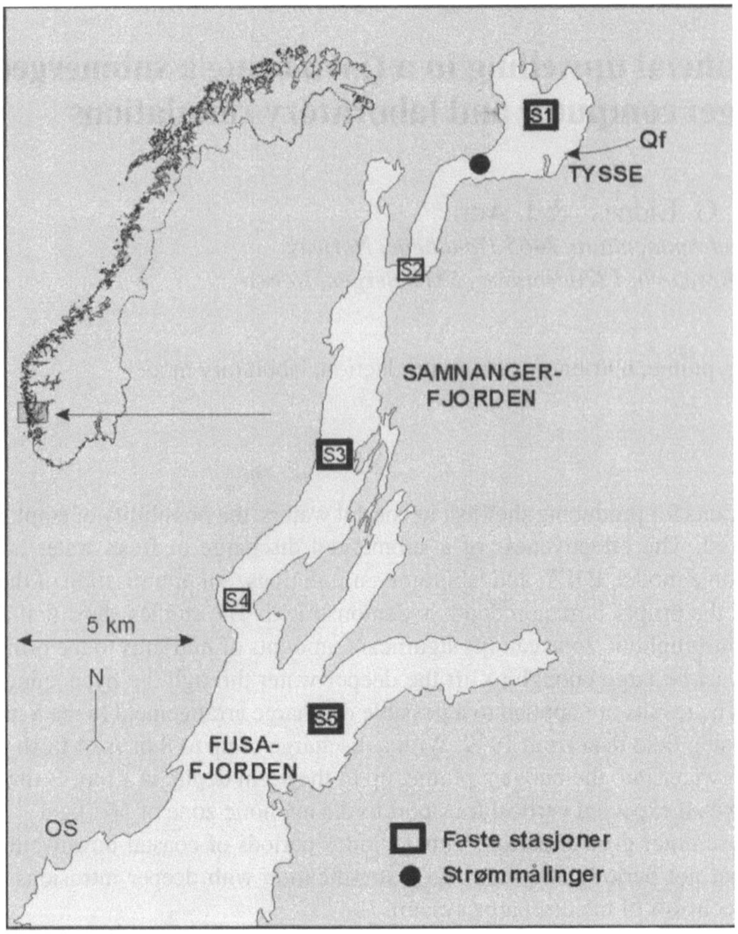

Figure 1. Map of the Samnangerfjord and the proposed discharge location (Q_f).

only on the hydraulics of a submerged, buoyant discharge, but also on the availability of nutrients below the euphotic zone. In this study, we look specifically at the distribution of available nitrogen (N), phosphorous (P) and silicon (Si), which are computed from measurements of nitrates, phosphates and silicates.

The Samnangerfjord (see Fig. 1) to the east of Bergen, Norway is used for a case study of controlled artificial upwelling because it has all of the necessary ingredients to demonstrate the use of unregulated water resources. In the inner part of the fjord, where up to 20 m^3 s^{-1} of fresh water is available for this purpose, there are only weak currents and a reasonably persistent stratification throughout the summer months. The sill depth of the fjord is 85 m. Hydrography, including the important nutrients, was measured from March to October 1999 (Aure et al., 2001). Figure 2 shows typical salinity and nutrient distributions.

The goal of the present application is to lift nutrient-rich water from depths below 30 m to a depth of about 10 m throughout the summer (growth) period. Both density and nutrient distributions are needed for this type of simulation (see Fig. 2). This is a problem of optimisation of the discharge parameters/variables (depth, diameter, angle and volume flux) in the presence of the ambient parameters/variables (stratification, river discharge, nutrients).

Based on our successful experience from earlier hydro-power studies, the synergy of numerical and laboratory methods is exploited. The numerical model BJET (Nilsson, 1991; Sørgård, 1992) is used to evaluate a range of alternatives that are studied in more detail in a scale model of the actual fjord situation. These results are used to validate the model for studying a wider range of conditions.

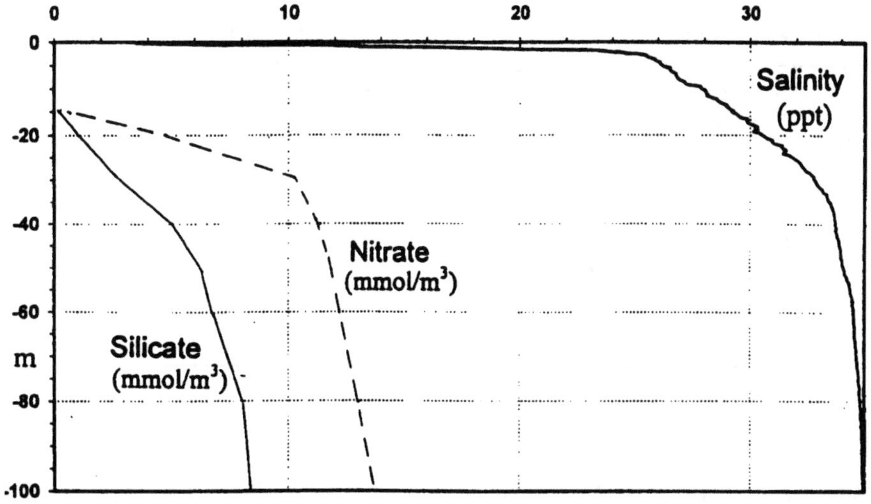

Figure 2. Typical salinity and nutrient profiles in the Samnangerfjord.

After validation, BJET is used first in a parametric study to recommend an outlet diameter that suits the application. The final stage of this work is the application of the recommended design to the available field data from 1999 in order to compute the depths and amounts of upwelling that can be expected. In addition, computations are made for a discharge arrangement suggested by the Bergen Community Energy Works (BKK), taking into account all practical aspects, including an upper limit of 8 m^3 s^{-1} discharge volume.

Methods

Numerical model

The numerical model BJET is based on the equations of conservation of mass, momentum and buoyancy for a submerged, round, buoyant jet in a stratified cross-flow (List, 1979). It is a one-dimensional, trajectory model to compute the 3-D position of the plume from a persistent buoyant discharge in a stratified, flowing recipient. It computes the turbulent entrainment, diameter, dilution and velocity as a function of distance along the centerline of the plume. It uses a radially symmetric Gaussian distribution along the trajectory out to the location where the ambient stratification causes a flattening and intrusion. It does not consider the effects of solid boundaries or the surface, and the laboratory simulations show that the bottom may capture the jet and change both the entrainment and intrusion depth. Likewise, there may be a break-through (overshoot) to the surface. Here, the mixing with the surface water is reduced and the actual plume is expected to sink to a slightly deeper level than computed. The model has been applied extensively to single-port outlets, from small sewage outlets to large hydro-power and thermal-power discharges (e.g. Eidnes, 1999).

Preliminary results from several typical scenarios in the Samnangerfjord (without bottom topography) were computed by using BJET to recommend an optimal way to discharge the water resources available. In this initial study the effects of up to 6 m^3 s^{-1} of freshwater discharge were compared with a 2 kW, vertical, propeller-driven momentum jet below the euphotic zone. The computations show that the propeller alternative has difficulty in producing a high enough intrusion in the stronger stratification. For a 6 m^3 s^{-1} fresh water volume flux, there is no advantage of discharging it through several pipes (diffuser). This gives too deep an intrusion of the nutrient rich-water.

From the preliminary study it was concluded that a downward directed jet can give a better mixing in the deeper water, below the euphotic zone, and could be less sensitive to changes in the stratification. This type of arrangement is also easier to build than other alternatives.

Laboratory simulations

The second phase of the study includes the use of laboratory simulations to check the validity of the computations of BJET. The laboratory procedure follows earlier tests with the laboratory simulation of

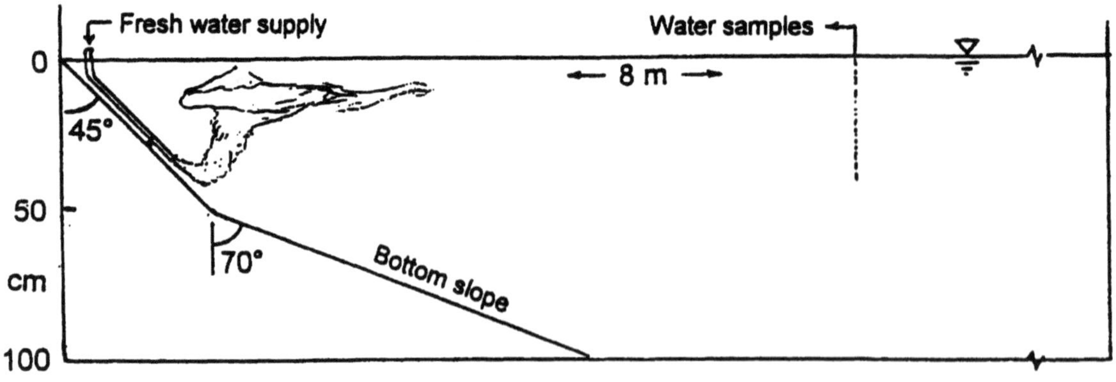

Figure 3. Side view of the laboratory model. The basin is 1 m wide.

submerged hydroelectric discharges (e.g. McClimans, 1980). Although the objective of the earlier studies was to secure an intrusion under the surface, the present application is quite different. To measure the amounts of nutrients that are lifted from the deeper layers, the seawater in the model is seeded with fluorescent Rhodamine tracer below 25 m depth. This allows us to quantify the 'nutrients' and makes it easy to define the intrusions.

The laboratory basin is limited in length (8 m), and the discharges cannot persist longer than the time for the front to reach the end of the basin and cause a reflected signal that changes the local density stratification in the near field. This limits the test runs to 6 min for the largest discharges. During this time, the mixing is considered to be quasi-stationary. Intrusion dynamics for this type of process are given in McClimans, *loc. cit.* for more general conditions.

The measurements comprise video monitoring of dye clouds injected into the source water and the analysis of salinity and fluorescence of water samples at 30 depths down to a simulated depth of 40 m (see Fig. 3). These are taken before and after each mixing event (test run).

Simulations of various discharges are performed in a 1:100 scale, undistorted Froude model of the inner Samnangerfjord. Figure 3 shows a cross-section of the simplified (two-dimensional) topography in the vicinity of the discharge, and Table 1 gives the test conditions. In the table, Q_o is the fresh water discharge volume flux, and the densimetric Froude and Reynolds numbers of the discharges are defined as

$$F = u_o/(g'D)^{1/2}; \quad \mathrm{Re} = u_o D/v, \tag{1}$$

where g' is the specific buoyancy (reduced gravity) of the fresh water in the ambient salinity S_a, u_o is the discharge velocity, D is the diameter of the outlet and v is the kinematic viscosity.

Table 1. Test conditions for the laboratory simulation of the induced upwelling of deeper water to the euphotic zone. (The missing tests (9–11) are propeller simulations)

Test	Diam. (m)	Depth (m)	Angle (°)	Q_o ($m^3\,s^{-1}$)	S_a (ppt)	F	Re
1	1	20	−45	3	31.7	7.6	3800
2	1	20	−45	6	31.9	15.3	7600
3	1	30	−45	3	32.8	7.5	3800
4	1	30	−45	6	32.4	15.3	7600
5	1	20	−35	3	31.7	7.6	3800
6	1	20	−35	6	31.1	15.6	7600
7	1	30	−35	3	32.7	7.5	3800
8	1	30	−35	6	32.1	15.3	7600
12	2	20	−45	3	32.3	1.35	1900
13	2	20	−45	6	31.9	2.70	3800
14	2	30	−45	3	33.6	1.32	1900
15	2	30	−45	6	33.6	2.65	3800

Too much mixing will not give the lift to the desired intrusion depth, which is here interpreted as 10 m depth. This can be remedied by increasing the diameter of the outlet, giving a lower F, as seen in tests 12–15. The different conditions are run in series of four tests (except for the propeller runs). Thus, there are differences in the initial conditions for each run, and this has to be taken into account in the later comparisons.

Model theory

The simulation of the full-scale conditions in nature requires that F is the same in the model (F_m) as it is in nature (F_n), ($F_n = F_m$). For submerged buoyant plumes, the turbulence that is necessary to create the entrainment is achieved a very short distance from the

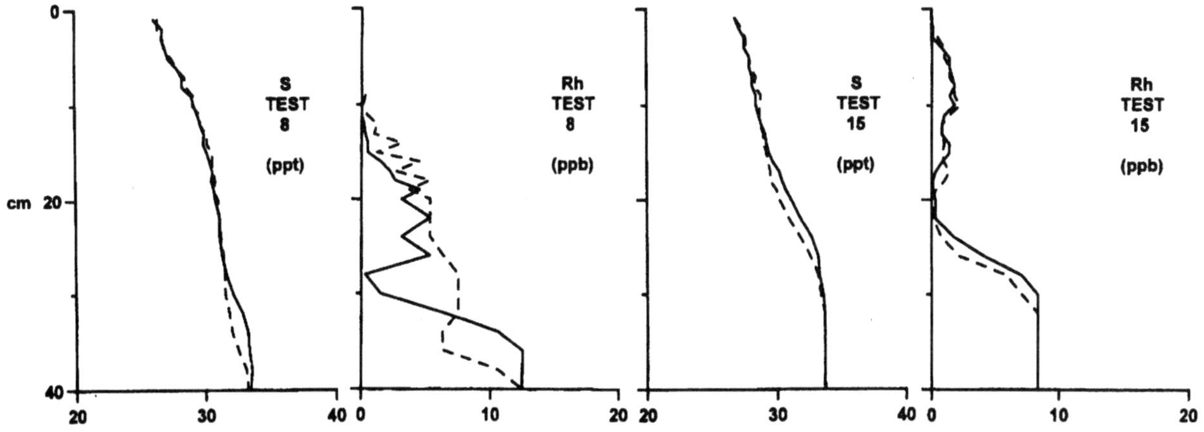

Figure 4. Comparison of salinity and 'nutrient' profiles before (——) and after(- - -) discharges 8 and 15.

Table 2. Gross properties from the laboratory tests

Test =>	1	2	3	4	5	6	7	8	12	13	14	15
Intrusion depth (m)	16	19	15	25	16	20	18	20	13	8	10	11
Entrainment (*E*)	10	17	15	22	9	16	14	20		4		6
Equilibrium depth Z_o (m)	12	17	13	23	11	19	15	24		0		15

outlet (List, 1979, p. 329). For our laboratory simulations, this translates to Reynolds numbers (1) of only a few hundred. All tests in Table 1 fulfil this condition. The scaling is therefore computed from so-called Froude similitude where the ratios $(\,)_r = (\,)_n/(\,)_m$ have to be computed. Since $F_r = 1$, we apply (1) to obtain

$$F_r = 1 = u_{or}/(g'_r D_r)^{1/2}. \quad (2)$$

Since $D_r = 100$ in this reduced (geometrical) scaling, and we use the same reduced gravity ($g'_r = 1$), then $u_{or} = (D_r)^{1/2}$.

The velocity scale is length scale divided by time scale (T_r), from which the time scale is

$$T_r = D_r{}^{1/2} = u_{or} = 10. \quad (3)$$

These scaling laws apply to the interpretation of the laboratory data at full scale. In the following, all reference is made to full scale quantities. A 6-min test run in the model is equivalent to an hour's discharge in the fjord.

Results of the laboratory simulations and comparison with BJET

Vertical profiles of salinity and Rhodamine before and after two discharge tests are shown in Figure 4. The trajectories computed from the video tapes are compared with BJET for all tests in Figure 5. In most cases, there is a good correspondence between the BJET results and the laboratory results. However, it can be seen that the high-speed jets that are directed along the bottom slope (tests 2 and 4) are pulled to larger depths, due to the suction of the proximity of the wall (the so-called Coanda effect). This effect is not part of the BJET dynamics yet. From these comparisons, however, it is recommended that the outlet be directed a few degrees away from the bottom (>7°). The model computations were terminated beyond the first rise as the plume approached the surface.

The concentration data of Figure 4 reveal the vertical changes in the 'nutrients' during the run and the intrusion depths as well, but the absolute values are not correct because the 3-D intruding cloud of Rhodamine does not fill the horizontal plane to the profiler, even after several hours. This could be seen in the model. Adjustments in these values have to be made on the basis of integral constraints of the conservation of Rhodamine. However, the good validation of BJET (Fig. 5) gives us the alternative of using the synergy of the two models and estimate the entrainment E from the numerical model. Here, $E = (Q - Q_o)/Q_o$, where Q is the local volume flux at large distances (near in-

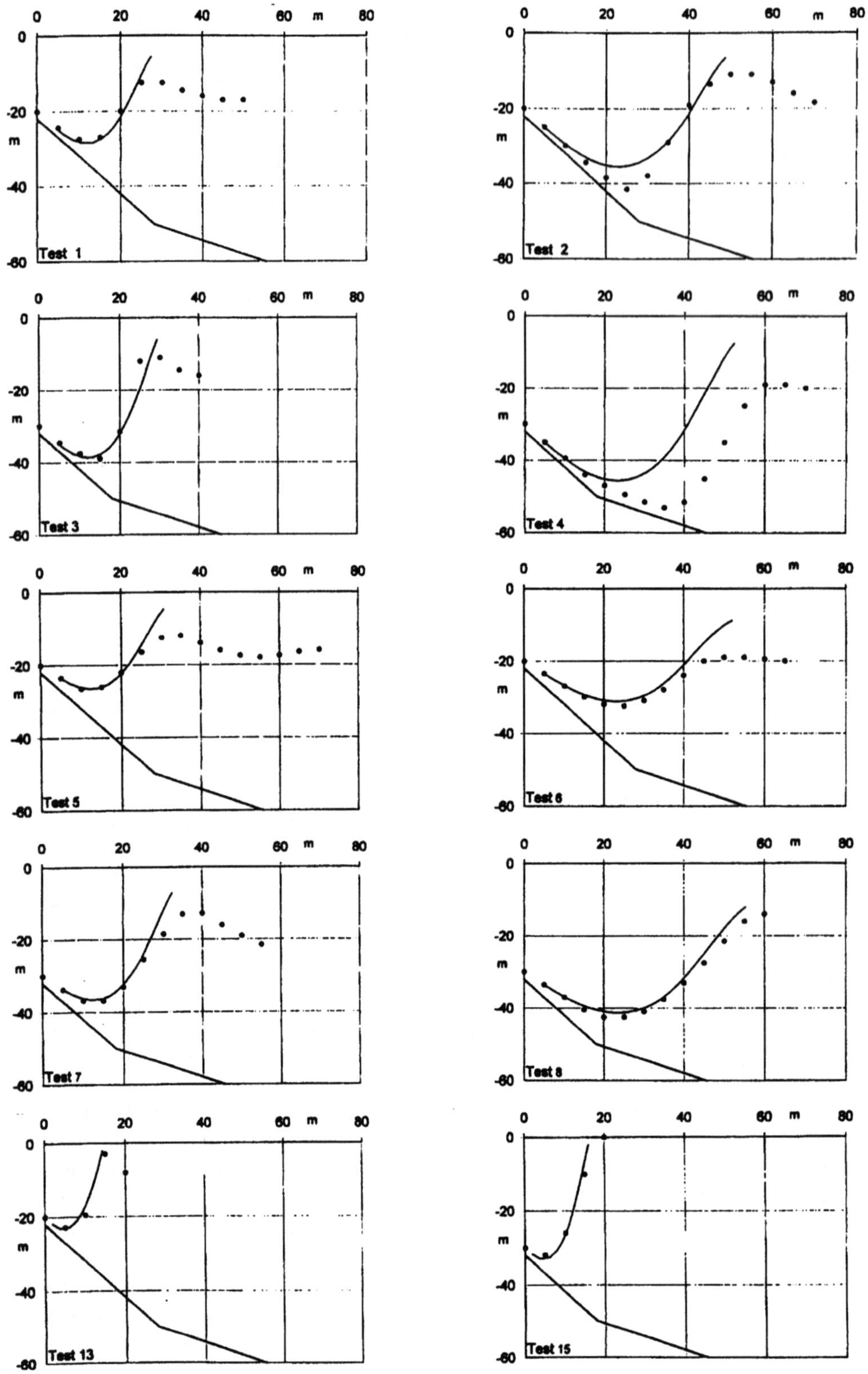

Figure 5. Comparison of laboratory (•) trajectories with BJET (——).

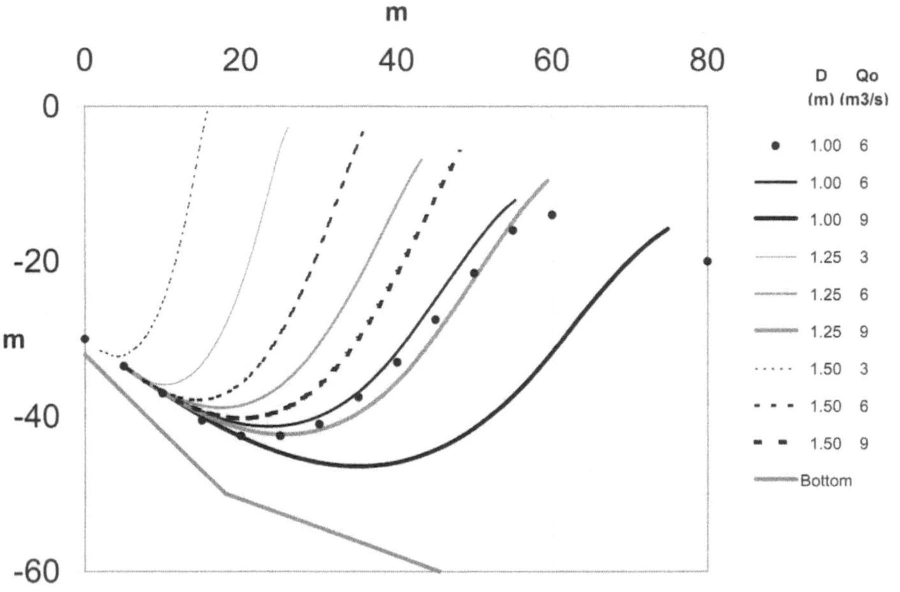

Figure 6. A comparison of trajectories for different diameters and discharges. The thick dots represent the reference laboratory data (see Fig. 5, Test 8).

Table 3. Conditions of a parametric study

Data	Test 8	Test 8A	Test 8B	Test 8C	Test 8D	Test 8E	Test 8F	Test 8G
Discharge ($m^3\ s^{-1}$)	6	9	3	6	9	3	6	9
Pipe diameter (m)	1.0	1.0	1.25	1.25	1.25	1.50	1.50	1.50
Discharge velocity ($m\ s^{-1}$)	7.6	11.5	2.5	4.9	7.3	1.7	3.4	5.1
F	15.7	23.5	4.5	10	15	2.8	5.7	8.5
Results								
QxRh (15) (ppb $m^3\ s^{-1}$)	1139	2441	216	679	1444	121	469	944
E (15)	20.5	27.2	10.8	13.5	17	7.6	10.2	12.1
Z_o (m)	24	28	12.5	15	19.5	8	10.7	14.5

trusion). BJET can also compute the depth Z_o at which the density of the rising plume is equal to the ambient density (equilibrium depth). These computations are given in Table 2 together with the observed intrusion in the far field. For the high F discharges at the smaller (1 m) outlet diameter, BJET functioned well. For low F (tests 13 and 15), the zone of flow establishment had to be reduced. This is done by placing the virtual nozzle 5 D up the outlet pipe. For the near-critical outlet Froude numbers (tests 12 and 14), BJET could not be used to compute E or Z_o.

The simulations show that a large diameter is needed to raise significant amounts of the nutrient-rich seawater to the chosen depth. However, as the densimetric Froude number is reduced to values close to the hydro-power outlets, the entrainment of the deeper water is reduced and the efficiency of the system for artificial upwelling is reduced (tests 13 and 15). For the larger values of discharge that are available in the Samnangerfjord, up to 20 $m^3\ s^{-1}$, the present results imply a diameter in excess of 1 m and a nozzle angle of −35°. We choose Test 8 as a reference for further optimisation.

Table 4. Nutrient fluxes for a 1.25 m diameter outlet at 30 m depth and an angle of −35°. The fresh water discharge is assumed to have concentrations of 14 mmol m^{-3} of N and Si, but no P

Date (1999)	Q_o ($m^3 s^{-1}$)	u_o ($m s^{-1}$)	F	Rho40-15 ($kg m^{-3}$)	Z_o (m) (m)	E (15 m)	N FLUX ($kg d^{-1}$)	P FLUX ($kg d^{-1}$)	Si FLUX ($kg d^{-1}$)
16 March	12	9.78	18.3	1.96	21	20.9	1607	206	1919
15 April	4	3.26	5.9	1.4	8	11.5	319	38	277
29 April	3	2.44	4.5	0.68	6	10.8	254	28	242
28 May	12	9.78	17.8	2.87	25	20.3	1656	210	1746
16 June	8	6.52	12.2	1.63	14	16	248	18	428
7 July	12	9.78	17.6	4.14	24	20.5	2225	303	2187
4 August	12	9.78	17.4	1.52	16	20.5	2575	319	2630
18 August	12	9.78	18	2.45	23	20.6	1199	59	1324
1 September	1.5	1.5	2.7	3.24	12	13.2	84	5	121
22 September	3	2.44	4.7	1.38	0	10.8	71	5	148
20 October	10	8.12	14.8	2.8	22	18.1	129	7128	1195

Results of a parametric study with BJET

With the validated results above it is now possible to make a parametric study of the effects of discharge volume flux and pipe diameter with exactly similar external conditions using BJET. This is done for the conditions noted in Table 3, where Test 8 is the reference test from the laboratory study (1 m diameter outlet at 30 m depth and −35° angle). In particular, the diameter of the outlet is increased to 1.25 and 1.5 m, and discharges of 3, 6 and 9 $m^3 s^{-1}$ are simulated.

The initial salinity (density) profile for Test 8 is used, but to get a better definition of the flux of nutrients, the (passive) Rhodamine profile is taken from Test 5, with 12.5 ppb below 30 m depth. The upward flux of nutrients through the 15 m depth level is therefore given in ppb $m^3 s^{-1}$. In addition to the flux, the results for the entrainment $E(15) = (Q - Q_o)/Q_o$ at 15 m depth, and the depth to the level where the density in the rising plume is equal to the ambient density (Z_o) are given. For these conditions, the intrusion depth is expected to be a few meters above Z_o. For the largest discharges, the larger diameter (1.5 m) gives a better intrusion at the expense of lesser entrainment. The plume trajectories are compared in Figure 6.

Discussion and applications

The present study considers only a few aspects of the entire system. The laboratory study was essentially an initial-value problem with a quasi-stationary period before the basin began to react to the changes due to the outlet. In the real world, the basin is enormous, and the feedback will be much less. A more difficult question to answer is the variability of the stratification in the fjord system due to internal tides and variable water masses in the outer fjord. This type of low-frequency variability will undoubtedly cause a thicker spreading of the mixing water produced by the submerged plume. The value of the present study, however, lies in the control of the boundary conditions to enable a quantification of the alternatives that are available in producing an artificial upwelling of nutrient-rich seawater to the euphotic zone.

Application to the Samnangerfjord

To evaluate the potential value of such a venture, the model, validated to the laboratory simulations, is applied to a summer of field observations of stratification and nutrient distributions in the Samnangerfjord near Bergen, Norway (Aure et al., 2001). From the parametric study, we choose first to run both 1.25 and 1.5 m diameter outlets at 30 m depth and a discharge angle of −35°, and look at the vertical fluxes of N, P and Si through the 15 m depth. The actual density and nutrient profiles are given for 11 dates from March to October 1999. From these calculations, we will show results for a practical application presented by the Bergen Community Energy Works (BKK): a 1.5 m diameter discharge at 35 m depth and −35° angle, and a maximum volume flux of 8 $m^3 s^{-1}$.

Table 5. Nutrient fluxes for a 1.5 m diameter outlet at 30 m depth and an angle of −35°. The fresh water discharge is assumed to have concentrations of 14 mmol m^{-3} of N and Si, but no P

Date (1999)	Q_0 ($m^3\ s^{-1}$)	u_0 ($m\ s^{-1}$)	F	Rho40-15 ($kg\ m^{-3}$)	Z_0 (m) (m)	E (15 m)	N FLUX ($kg\ d^{-1}$)	P FLUX ($kg\ d^{-1}$)	Si FLUX ($kg\ d^{-1}$)
16 March	12	6.8	11.6	1.96	14	14.5	1146	134	1445
15 April	4	2.3	3.7	1.4	7	9.4	267	29	246
29 April	3	1.7	2.9	0.68	6	9.4	222	23	218
28 May	18	10.2	16.9	2.87	25	28.4	2430	311	2652
16 June	8	4.53	7.7	1.63	11	11.6	206	10	392
7 July	18	10.2	16.7	4.14	24	28.8	3157	434	3210
4 August	18	10.2	16.6	1.52	15	28.8	3590	447	3749
18 August	18	10.2	17	2.45	23	28.9	1837	104	1969
1 September	1.5	0.9	1.4	3.24		(Too low *F*)			
22 September	3	1.7	3	1.38	0	9.4	70	4	145
20 October	10	5.7	9.4	2.8	17	12.9	928	83	921

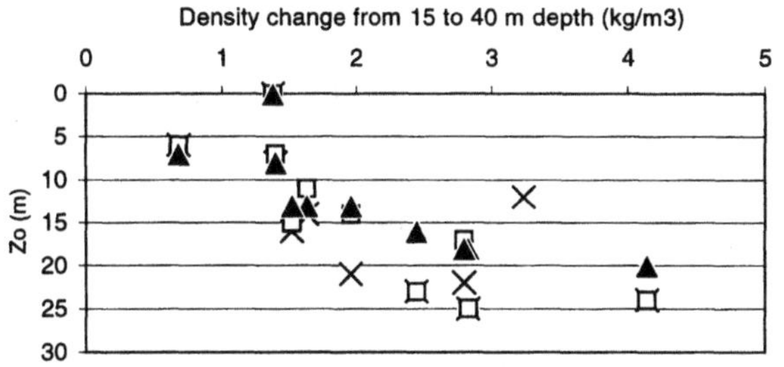

Figure 7. The effect of ambient stratification on the depth of intrusion: crosses – 1.25 m diameter, 30 m depth; squares – 1.5 m diameter, 30 m depth; triangles – 1.5 m diameter, 35 m depth.

The conditions and results of the first computations are given in Tables 4 and 5. It is assumed that the most economical scheme is to place the intake of the discharge pipe behind a dam at 6 m elevation above sea level. The volume fluxes Q_0 are therefore based on the discharge data obtained from BKK, and the total capacity available for a pipe of the chosen diameter with a pressure head of 6 m. Hydraulic head losses of 20% for the 1.25 m pipe and 15% for the 1.5 m pipe are assumed. Thus, the maximum discharge for the 1.25 m pipe is 12 $m^3\ s^{-1}$ and for the 1.5 m pipe is 18 $m^3\ s^{-1}$. The column Rho40-15 in Tables 4 and 5 gives the difference in density between 15 and 40 m depth. It is a measure of the stratification, which is important for the depth of intrusion. Clearly, the deepest intrusions, which are a few meters above Z_0, are associated with the large ambient density gradients (strong pycnoclines).

In the tables, it is seen that the river discharge was low (low *F*) on 1 September. It was too low for the BJET simulation with a 1.5 m diameter outlet. On 22 September, there was also less ambient density and fewer nutrients at the discharge depth due to a coastal downwelling. Even a large discharge will not produce much transport of nutrients to the euphotic zone during such an event. However, it increases the nutrient concentrations in the intermediate layer and increases the vertical fluxes due to wind mixing and other sources of turbulence.

For the cases in which the intrusion is in the euphotic zone ($Z_o < 15$ m), the average fluxes for the computed situations are:

Nutrient	1.25 m diameter ($kg\ d^{-1}$)	1.5 m diameter ($kg\ d^{-1}$)
N	592	917
P	69	108
Si	641	1033

These values include the transports of 14 mmol m^{-3} of N and Si in the river discharge. In contrast to the parametric study in the previous section, the larger diameter gives more vertical flux to the euphotic zone in this case. This is due to the larger discharges during high river flow.

The effect of the ambient stratification is evident in the data of Tables 4 and 5. The equilibrium depth (Z_0) is plotted against the density difference ($\rho_{40} - \rho_{15}$) in Figure 7. The suggested arrangement (1.5 m diameter) is satisfactory for values of $\rho_{40} - \rho_{15} < 2$ kg m^{-3}.

Above this value the intrusion lies below the euphotic zone. For the 1.25 m diameter pipe, the corresponding stability limit is about 1.6 kg m^{-3}. Noting the apparent disparity of the 1 September calculation at $\rho_{40} - \rho_{15}$ = 3.24, it should be pointed out that the actual density gradients are nonlinear, and that much of the spread in Figure 7 is probably due to our simplification of the ambient stratification.

According to List (1979: 342), the rise of the discharge increases with the fourth roots of the vertical momentum flux (discharge velocity), which is negative in our application, and the buoyancy discharge, which is proportional to the volume flux. For the buoyancy dominated plume the rise is inversely proportional to the ambient stratification to the 3/8 power. Since the *downward* discharge velocity for the 1.25 m diameter outlet is 44% larger than for the 1.5 m diameter outlet, there are occasions where the intrusions are deeper. This occurs for high F discharges.

A practical example

From the Bergen Community Energy Works (BKK) we have obtained technical data for an actual discharge arrangement that accounts for all practical installation issues (valve, turns, trash rack, dam in river, etc.). This gives a maximum flux of 8 m^3 s^{-1} through a standard pipe of 1.2 m diameter. To apply the knowledge gained in the parametric study and the other comparisons, the outlet is placed at 35 m depth with an angle of −35° and a diameter of 1.5 m (in an expansion from the main supply pipe).

Computations of equilibrium depths (Z_o) and vertical transports of N, P and Si at 15 m depth are given in Table 6. Comparing these results with those of Table 5 shows that there is a more stable nutrient flux (entrainment) and intrusion depth. This is due to the increased available potential energy with the greater depth of discharge, and that the maximum discharge (8 m^3 s^{-1}) is reached more often.

The results are compared with the 30 m depth discharges in Figure 7. For this arrangement, the statistically averaged vertical flux to the euphotic zone, above 15 m depth, is 467 kg d^{-1} N, 46 kg d^{-1} P and 555 kg d^{-1} Si.

Other applications

The above results are interpreted for the Samnangerfjord. Calculations of the photosynthesis imply that this will add significantly to the primary production in the fjord. There are two major, submerged discharges of fresh water from hydro-power plants in Norway: Vangen and Jostedal (McClimans, 1981; Gjerp & McClimans, 1990), both with low values of F. These outlets are not designed to secure an intrusion deep in the euphotic zone. In both cases, the intrusion depths are less than 5 m. Due to the large diameters of the outlets, the intrusion depth is not as sensitive to the ambient stratification. Here, improvements can be made to optimise the artificial upwelling of nutrients by constructing arrangements that divide the plumes into several smaller plumes (diffusor). This would give more entrainment of the deeper water, and deeper intrusions.

An aspect of the arrangement that has not been treated in detail is its use as a spreader for artificially fertilising the fjord. The source water can be supplied with the desired nutrients. Furthermore, if the effect of the arrangement gives excessive growth, it is possible to throttle the supply of fresh water. A throttle valve must be placed at the intake for maintenance in any event.

In more restricted regions (e.g. land-locked fjords) near large municipalities, the water quality may not be acceptable for, e.g. fish farming. The system studied here can also be applied to improve water quality up to the point of overmixing (Stommel & Farmer, 1953).

Optimisation by active control

The above results are for a passive system. The stochastic nature of the ambient conditions makes this type of system sub-optimal. It must be tailored to the average conditions. By monitoring the variability of the stratification, nutrient distribution and fresh water supply, it should be possible, in principle, to optimise further by using active control of depth, diameter and water discharge. The easiest thing to control is the valve for fresh water flux. The results imply that we will usually want the maximum amount of fresh water, but with strong stability, a lower discharge will insure a higher intrusion at the expense of less entrainment. Regulating the depth of the discharge is more complicated, and will require some sort of telescope along a rail mounted on the bottom slope. This too can be regulated to give a better intrusion depth in the presence of variable density stratification. It is thus necessary to monitor the density stratification to provide the controls with the necessary steering parameters. In the case of a fjord where the stratification is determined primarily by the salinity of the water, this may be

Table 6. Nutrient fluxes for a 1.5 m diameter outlet at 35 m depth and an angle of −35°. The fresh water discharge is assumed to have concentrations of 14 mmol m^{-3} of N and Si, but no P

Date (1999)	Q_0 (m^3 s^{-1})	u_0 (m s^{-1})	F	Rho40-15 (kg m^{-3})	Z_0 (m) (m)	E (15 m)	N FLUX (kg d^{-1})	P FLUX (kg d^{-1})	Si FLUX (kg d^{-1})
16 March	8	4.5	7.7	1.96	13	13.3	696	78	907
15 April	4	2.3	3.7	1.4	8	11.5	327	39	303
29 April	3	1.7	2.9	0.68	7	11.7	278	32	270
28 May	8	4.5	7.4	2.87	18	12.9	668	74	768
16 June	8	4.5	7.7	1.63	13	13.1	292	22	430
7 July	8	4.5	7.4	4.14	20	13	925	113	949
4 August	8	4.5	7.3	1.52	13	13.1	1108	125	1173
18 August	8	4.5	7.5	2.45	16	13.1	496	19	645
1 September	1.5	0.9	1.5	3.24		(Too low F)			
22 September	3	1.7	2.9	1.38	0	11.7	75	6	154
20 October	8	4.5	7.5	2.8	18	13.1	768	71	762

done by simply monitoring the difference in salinity between, say, 15 and 40 m depths.

The third parameter that can be regulated is the diameter of the outlet to control the densimetric Froude number. Such a control would have to be coupled to the valve, since the pressure change with a regulated nozzle will affect the volume flux. This would be a more difficult technical problem than the first two parameters, and the gains for optimisation must be evaluated against the costs of the regulatory systems and their maintenance.

Conclusions

The results of the present study show that significant amounts of nutrient-rich seawater can be raised to the euphotic zone in nutrient-poor regions, through controlled artificial upwelling, using available fresh water sources. Comparisons between BJET and the laboratory simulations show that the bottom topography has a significant effect on the trajectory and entrainment of buoyant plumes that are directed to larger depths along the slope. This can be effectively reduced if the discharge is directed away from the bottom slope. Since both entrainment and intrusion depths have to be considered in artificial upwelling systems of this type, optimisation requires a trade-off of diameter and depth, both of which depend on the discharge and the distributions of density and nutrients in the sea. These latter variables are stochastic by nature, and require a number of measurements to choose the 'most probable' efficient discharge.

Acknowledgements

We thank the Norwegian Research Council for support through the MARICULT program and the Bergen Community Energy Works (BKK) for assistance in performing the present applications to the Samnangerfjord. Much of the background material was obtained from projects that were performed for the power industry in Norway. Access to these reports (see reference list) is possible by making contact with SINTEF Fisheries and Aquaculture, 7465 Trondheim Norway and specifying the report number. We would like to thank anonymous referees for making good suggestions for improving the manuscript.

References

Aure, J., S. R. Erga & L. Asplin, 2001. Fjordcult: Økt biologisk produksjon i fjorder ved kunstig oppstrømning av dypvann (Fjordcult:Increased biological production in fjords by artificial upwelling). Fisken og Havet nr. 11–2000. Havforskningsinstituttet (Institute of Marin Research, Norway). 30 pp.

Gjerp, S. A. & T. A. McClimans, 1990. Jostedalutbygging: Etterprøving av hydrofysiske forhold i nærsonen som følge av utslippet til Gaupnefjorden. SINTEF Report STF60 F90040 (in Norwegian).

Eidnes, G., 1999. Tjeldbergodden. Verifisering av kjølevannseffekter. SINTEF Report STF22 F99226 (in Norwegian).

List, E. J., 1979. Turbulent jets and plumes. In Fischer, H. B., E. J. List, R. C. Y. Koh, J. Imberger & N. H. Brooks (eds), Ch. 9: Mixing in Inland and Coastal Waters. Academic Press, New York: 483 pp.

McClimans, T. A., 1980. Density fronts along finite depth pycnoclines. Stratified flows. Second International Symposium on Stratified Flows, Trondheim, Carstens, T. & T. A. McClimans (eds), Tapir 1: 154–163.

McClimans, T. A., 1981. Dykket utslipp fra Vangen kraftverk. Etterprøving. SINTEF Report STF60 F81113 (in Norwegian).

McClimans, T. A. & G. Eidnes, 2000. FJORDCULT laboratory simulation of controlled artificial upwelling in a fjord using a submerged fresh water discharge. SINTEF Report STF22 A00217. ISBN 82-14-01827-7 [See also: McClimans, T. A. & G. Eidnes, 2000. Forcing nutrients to the upper layer of a fjord by a buoyant plume. In Lawrence, G. A., R. Pieters & N. Yonemitsu (eds), Fifth International Symposium on Stratified Flows. University of British Columbia, Vancouver, Canada, 10–13 July 2000. IAHR 1: 199–204.

Nilsson, J., 1991. Innlagring av oppdriftsstråler. Master's thesis, Institutt for konstruksjonsteknikk, NTH (in Norwegian).

Stommel, H. & H. G. Farmer, 1953. Control of salinity in an estuary by a transition. J. Mar. Res. 12: 13–20.

Sørgård, E., 1992. Validation and suggested improvements for the BJET jet/plume model for subsea releases. Det Norske Veritas Research AS, TR 92–2050.

MIX
Papier aus verantwortungsvollen Quellen
Paper from responsible sources
FSC® C105338
FSC
www.fsc.org

If you have any concerns about our products,
you can contact us on
ProductSafety@springernature.com

In case Publisher is established outside the EU,
the EU authorized representative is:
Springer Nature Customer Service Center GmbH
Europaplatz 3, 69115 Heidelberg, Germany

Printed by Libri Plureos GmbH
in Hamburg, Germany